IET MATERIALS, CIRCUITS AND DEVICES SERIES 72

Tensorial Analysis of Networks (TAN) Modelling for PCB Signal Integrity and EMC Analysis

Other volumes in this series:

Volume 2	**Analogue IC Design: The current-mode approach** C. Toumazou, F.J. Lidgey and D.G. Haigh (Editors)
Volume 3	**Analogue–Digital ASICs: Circuit techniques, design tools and applications** R.S. Soin, F. Maloberti and J. France (Editors)
Volume 4	**Algorithmic and Knowledge-Based CAD for VLSI** G.E. Taylor and G. Russell (Editors)
Volume 5	**Switched Currents: An analogue technique for digital technology** C. Toumazou, J.B.C. Hughes and N.C. Battersby (Editors)
Volume 6	**High-Frequency Circuit Engineering** F. Nibler *et al.*
Volume 8	**Low-Power High-Frequency Microelectronics: A unified approach** G. Machado (Editor)
Volume 9	**VLSI Testing: Digital and mixed analogue/digital techniques** S.L. Hurst
Volume 10	**Distributed Feedback Semiconductor Lasers** J.E. Carroll, J.E.A. Whiteaway and R.G.S. Plumb
Volume 11	**Selected Topics in Advanced Solid State and Fibre Optic Sensors** S.M. Vaezi-Nejad (Editor)
Volume 12	**Strained Silicon Heterostructures: Materials and devices** C.K. Maiti, N.B. Chakrabarti and S.K. Ray
Volume 13	**RFIC and MMIC Design and Technology** I.D. Robertson and S. Lucyzyn (Editors)
Volume 14	**Design of High Frequency Integrated Analogue Filters** Y. Sun (Editor)
Volume 15	**Foundations of Digital Signal Processing: Theory, algorithms and hardware design** P. Gaydecki
Volume 16	**Wireless Communications Circuits and Systems** Y. Sun (Editor)
Volume 17	**The Switching Function: Analysis of power electronic circuits** C. Marouchos
Volume 18	**System on Chip: Next generation electronics** B. Al-Hashimi (Editor)
Volume 19	**Test and Diagnosis of Analogue, Mixed-Signal and RF Integrated Circuits: The system on chip approach** Y. Sun (Editor)
Volume 20	**Low Power and Low Voltage Circuit Design with the FGMOS Transistor** E. Rodriguez-Villegas
Volume 21	**Technology Computer Aided Design for Si, SiGe and GaAs Integrated Circuits** C.K. Maiti and G.A. Armstrong
Volume 22	**Nanotechnologies** M. Wautelet *et al.*
Volume 23	**Understandable Electric Circuits** M. Wang
Volume 24	**Fundamentals of Electromagnetic Levitation: Engineering sustainability through efficiency** A.J. Sangster
Volume 25	**Optical MEMS for Chemical Analysis and Biomedicine** H. Jiang (Editor)
Volume 26	**High Speed Data Converters** A.M.A. Ali
Volume 27	**Nano-Scaled Semiconductor & Devices** E.A. Gutiérrez-D (Editor)
Volume 28	**Security and Privacy for Big Data, Cloud Computing and Applications** L. Wang, W. Ren, K.R. Choo and F. Xhafa (Editors)
Volume 29	**Nano-CMOS and Post-CMOS Electronics: Devices and modelling** Saraju P. Mohanty and Ashok Srivastava
Volume 30	**Nano-CMOS and Post-CMOS Electronics: Circuits and design** Saraju P. Mohanty and Ashok Srivastava
Volume 32	**Oscillator Circuits: Frontiers in design, analysis and applications** Y. Nishio (Editor)
Volume 33	**High Frequency MOSFET Gate Drivers** Z. Zhang and Y. Liu
Volume 34	**RF and Microwave Module Level Design and Integration** M. Almalkawi
Volume 35	**Design of Terahertz CMOS Integrated Circuits for High-Speed Wireless Communication** M. Fujishima and S. Amakawa
Volume 38	**System Design with Memristor Technologies** L. Guckert and E.E. Swartzlander Jr.
Volume 39	**Functionality-Enhanced Devices: An alternative to Moore's law** P.-E. Gaillardon (Editor)
Volume 40	**Digitally Enhanced Mixed Signal Systems** C. Jabbour, P. Desgreys and D. Dallett (Editors)

Volume 43 **Negative Group Delay Devices: From concepts to applications** B. Ravelo (Editor)
Volume 45 **Characterisation and Control of Defects in Semiconductors** F. Tuomisto (Editor)
Volume 47 **Understandable Electric Circuits: Key concepts, 2nd Edition** M. Wang
Volume 48 **Gyrators, Simulated Inductors and Related Immittances: Realizations and applications** R. Senani, D.R. Bhaskar, V.K. Singh, A.K. Singh
Volume 49 **Advanced Technologies for Next Generation Integrated Circuits** A. Srivastava and S. Mohanty (Editors)
Volume 51 **Modelling Methodologies in Analogue Integrated Circuit Design** G. Dundar and M.B. Yelten (Editors)
Volume 53 **VLSI Architectures for Future Video Coding** M. Martina (Editor)
Volume 54 **Advances in High-Power Fiber and Diode Laser Engineering** Ivan Divliansky (Editor)
Volume 55 **Hardware Architectures for Deep Learning** M. Daneshtalab and M. Modarressi
Volume 58 **Magnetorheological Materials and Their Applications** S. Choi and W. Li (Editors)
Volume 59 **Analysis and Design of CMOS Clocking Circuits for Low Phase Noise** W. Bae and D.K. Jeong
Volume 60 **IP Core Protection and Hardware-Assisted Security for Consumer Electronics** A. Sengupta and S. Mohanty
Volume 64 **Phase-Locked Frequency Generation and Clocking: Architectures and circuits for modem wireless and wireline systems** W. Rhee (Editor)
Volume 65 **MEMS Resonator Filters** Rajendra M. Patrikar (Editor)
Volume 67 **Frontiers in Securing IP Cores: Forensic detective control and obfuscation techniques** A Sengupta
Volume 68 **High Quality Liquid Crystal Displays and Smart Devices: Vol. 1 and Vol. 2** S. Ishihara, S. Kobayashi and Y. Ukai (Editors)
Volume 69 **Fibre Bragg Gratings in Harsh and Space Environments: Principles and applications** B. Aïssa, E.I. Haddad, R.V. Kruzelecky, W.R. Jamroz
Volume 70 **Self-Healing Materials: From fundamental concepts to advanced space and electronics applications, 2nd Edition** B. Aïssa, E.I. Haddad, R.V. Kruzelecky, W.R. Jamroz
Volume 71 **Radio Frequency and Microwave Power Amplifiers: Vol. 1 and Vol. 2** A. Grebennikov (Editor)
Volume 73 **VLSI and Post-CMOS Electronics Volume 1: VLSI and Post-CMOS Electronics and Volume 2: Materials, devices and interconnects** R. Dhiman and R. Chandel (Editors)

Tensorial Analysis of Networks (TAN) Modelling for PCB Signal Integrity and EMC Analysis

Edited by
Blaise Ravelo and Zhifei Xu

The Institution of Engineering and Technology

Published by The Institution of Engineering and Technology, London, United Kingdom

The Institution of Engineering and Technology is registered as a Charity in England & Wales (no. 211014) and Scotland (no. SC038698).

First published 2020

The Institution of Engineering and Technology
Michael Faraday House
Six Hills Way, Stevenage
Herts, SG1 2AY, United Kingdom

www.theiet.org

British Library Cataloguing in Publication Data
A catalogue record for this product is available from the British Library

ISBN 978-1-83953-049-4 (hardback)
ISBN 978-1-83953-050-0 (PDF)

Typeset in India by MPS Limited

Contents

About the editors

Blaise Ravelo is a professor at NUIST University, China. His research interests include multiphysics and electronics engineering, and he is a pioneer of the negative group delay (NGD) concept. He was a research director of nine PhD students and regularly involved in EU R&D&I projects. He is a member of IET Electronics Letters editorial board as circuit and system subject editor and has (co)authored more than 250 papers in peer-reviewed journals and conferences.

Zhifei Xu is a postdoctoral researcher at the Missouri S&T EMC Laboratory, Missouri University of Science and Technology, Rolla, USA. His research area covers signal integrity, power integrity and EMC analysis with different models on multilayer PCBs. His current publications are focused on the Kron–Branin model for multilayer PCB modelling applications.

Foreword

The influence of Kron's tensorial analysis of networks (TAN) has considerably increased in pertinence and importance in the field of electronic circuits since the very first papers were published, not surprisingly by Olivier Maurice, at that time senior engineer at EADS® Innovation Works. Very surprisingly, I was a co-author (yes, me) [1] of one paper. The only acceptable reason for this usurpation is that I have provided Olivier with the parasitic emission model, simulations and measurements of a giant integrated circuit freshly released by Infineon®, called the *TriCore*. To be honest, I did not remember I was a co-author, in fact, I should never have been. My punishment is to write the preface of his collaborative book, nearly 15 years later.

Before talking about the book, let me recall two souvenirs: at the time, we were both at EADS Innovation Works, we went to *Nucletudes* for a meeting around microwave effects on integrated circuit. It was rush hour, cars stopped everywhere, stress everywhere, no hope and no solution. Like electromagnetic compatibility (EMC) problem-solving sometimes. Olivier making a clean, straight and efficient drive in the middle of the mess using his motorcycle, me at the back, and we arrived just in time. This could sum up Olivier's long-term strategy regarding the use of most efficient and appropriate method, taking into account the context and its requirements, without violating rules, just taking reasonable risks. *Just think different, and go for it.* The second souvenir is the feedback he gave me when I asked him what was his primary motivation to become R&D director of IRSEEM laboratory in Rouen, after having served as senior engineer at GERAC: he simply told me "have fun". I recently had the pleasure to meet Blaise Ravelo during the memorable EMC Compo 2019 in China (just a few weeks before the CoVid-19 alert sounded), and the thing I can guess is that these two theoreticians (to avoid using the term of "gentle theoretical monsters") had a "lot of fun" to imagine and construct this book. They co-advised the thesis of Zhifei Xu, which was titled *Tensorial analysis of multilayer printed circuit boards: computations and basics for multiphysics analysis*. Surviving to the joint coaching of Olivier and Blaise gives me the conviction that Zhifei is indestructible, and the definitive choice to serve as third co-author of this book.

Olivier published numerous papers in the field of TAN (just have a look at his online CV in HAL open archives [2]) and, within a period of 2 years, published three books in close relationship with TAN and Kron's method, two in French and one in English [3]. It sounds clear that the Olivier's paradigm forms the basic skeleton of this present book, although the idea of publishing this book originated from Blaise and Zhifei. It might be around this period that I realized that I had in the old library of my department a book from Gabriel KRON. I immediately shipped the book (that probably survived certain death) titled *Equivalent Circuits of Electrical Machinery* from John Wiley and Sons, New York, 1951 to Olivier Maurice—Who else?

Contents from the early publications from Olivier Maurice about Kron's method and its application to the field of EMC stood the test of time and continued to constitute unique information to basic understanding of printed circuit boards (PCBs) and electronic circuit modelling in the presence of electromagnetic coupling and interference. In this book, a considerable growth of information, case studies and new applications have required the difficult decisions of Olivier Maurice, Blaise Ravelo and Zhifei Xu about what to develop, what to include while maintaining a reasonable length and overall content coherence. Following the initial chapters are several chapters that provide the theoretical underpinnings, modelling principles and exploitation strategies for handling various integrity and compatibility aspects of PCB: signal integrity, power integrity, coupling with Electromagnetic waves, radiation, etc.

Each of the 15 chapters is dedicated to a specific theoretical or application domain, offering an impressible palette of application examples. Some chapters have benefited from the contribution of researchers and experts from various fields. Yang Liu, ALTRAN®, France, contributed to several chapters, and Christel Cholachue and Amélie Simeons, TENNECO®/Federal Mogul®, France, co-authored the chapter on probe/cable coupling. Fayu Wan, Nanjing University of Information Science & Technology (NUIST), China, analysed analogue-to-digital-converter susceptibility, and Toufic Abboud and Benoit Chaigne, IMACS, Ecole Polytechnique, France, co-authored the chapter on PCB modelling. Dominik Schröder, Sven Lange, Christian Hangmann and Christian Hedayat from Fraunhofer ENAS, Paderborn, Germany, discuss the links between far-field prediction with near-field measurements. It must have been a challenge to include all these contributions and form this otherwise unavailable resource of state-of-the art information on Kron's TAN applications in electronics, from which an EMC modelling and analysis plan may be constructed.

It is a very honourable and meaningful goal of the three main authors, with the help of their co-authors, to provide a comprehensive understanding of the background, theory and applications of the TAN approach for the efficient application to EMC problem modelling and solving.

Etienne Sicard, PhD.
Institut National des Sciences Appliquées (INSA), Toulouse, France
Toulouse (France), March, 2020

References

[1] O. Maurice, E. Sicard (2005). Use of the tensorial analysis of network to compute the emission in TEM CELL from a microcontroller. Study of the mesh connectivity. EMC COMPO 2005 Munich Symposium.

[2] https://cv.archives-ouvertes.fr/olivier-maurice.

[3] O. Maurice (2017). Elements of theory for electromagnetic compatibility and systems. https://hal.archives-ouvertes.fr/hal-01722155/.

Chapter 1
General introduction

Blaise Ravelo[1], *Olivier Maurice*[2], *and Zhifei Xu*[1]

Abstract

This chapter describes the general context of the tensorial analysis of networks (TAN) modelling for printed circuit board (PCB). The summary of the chapter's contents is presented. The fundamental steps to establish the TAN formalism dedicated to the modelling of PCBs for the signal integrity (SI), power integrity (PI) and electromagnetic compatibility (EMC) investigations are described. They synthesize the different contributions in each chapter of the book. The main content concerns the topics of

- the nonlinear conducted EMC characterization of digital component,
- the TAN primitive elements,
- basic elements to practice Kron's method,
- the frequency-domain modelling of PCB with lumped and distributed elements,
- the time-domain analysis of PCB SI and PI,
- the conducted and radiated EMC emission of PCB, and
- the conducted and radiated EMC susceptibility.

Keywords: Methodology, analytical method, numerical method, PCB analysis, SI analysis, PI analysis, EMC analysis, experimental test techniques

1.1 Preliminary introduction

The present chapter is a panoramic overview of the book contents. The summaries describe the technical objects of 14 main chapters. There are few technical interdependences between the chapters. Based on the indicated keywords and abstracts, the readers can directly focus on their chapters of interest in function of summaries.

[1]IRSEEM/ESIGELEC, Rouen, France
[2]ArianeGroup, Paris, France

1.2 Chapter 2: Basic knowledge to practice TAN for PCB SI/PI/EMC investigation

Olivier Maurice, Zhifei Xu, Yang Liu, and Blaise Ravelo

Abstract

The tensorial approach remains one of the less explored circuit theories to solve the modern printed circuit board (PCB) problem. Chapter 2 of the proposed book will open the door allowing us to the non-specialist and even non-expert of circuit design to learn and to practice with their own easy way and rapidly the tensorial analysis of networks (TAN) approach. The fundamental approach to use the tensor algebra will be introduced. The elementary and basic knowledges necessary to elaborate the TAN concept will be defined. An easy concept of physical, classical, graph topological and tensorial object representation of the primitive elements to treat the PCB signal integrity (SI), power integrity (PI) and electromagnetic compatibility (EMC) problems will be described. Several basic illustrative examples of key understanding lumped RC, RL, LC and RLC network-based circuits will be treated in the chapter. The methodologies integrating the topological graphs of the problem will be explained. The different basic steps to practice the TAN approach are indicated. For the concrete comprehension, the different practical cases for solving PCB SI, PI and EMC problems are introduced.

Keywords: TAN approach, Kron's method, Kron–Branin's model, primitive elements, modelling methodology, PCB analysis, SI analysis, PI analysis, EMC analysis, circuit and system theory

1.3 Chapter 3: PCB primitive components analysis with TAN

Zhifei Xu, Blaise Ravelo, Yang Liu, and Olivier Maurice

Abstract

This chapter introduces the basic elements necessary for the printed circuit board analysis with tensorial analysis of networks approach. The elements are first presented based on the classical electrical schemes. Then, the equivalent graphs are presented. The tensor models are established based on the physical law governing each element.

Keywords: Kron's method, TAN, tensor objects, lumped elements, distributed elements

1.4 Chapter 4: Analytical calculation of PCB trace *Z/Y/T/S* matrices with TAN approach

Blaise Ravelo, Zhifei Xu, Yang Liu, and Olivier Maurice

Abstract

Despite the numerous works done about the transmission line modelling of the printed circuit board (PCB) electrical interconnects, huge efforts are still open to the signal integrity, power integrity and electromagnetic compatibility engineers to predict the PCB trace effects during the design phase. The extraction method of the interconnect network as a multiport system will be defined based on a topological algebra. The existing theory and the simulation tools are either not flexible enough or do not allow one to understand the electrical behaviours of the PCB interconnects. The system and circuit theory about the impedance (Z), admittance (Y), transfer (T) and scattering (S) matrices will be provided in the present chapter. The study will be applied to different topologies of interconnect structures as multiport graph topologies. Pedagogical approaches enabling one to practice easily the tensorial analysis of networks (TAN) concept in function of the interconnect structures will be treated in this chapter. The theory representing the interconnect as tensorial objects in branch and mesh spaces will be provided. The problem resolution via the calculations of mesh currents into the $Z/Y/T/S$-matrices will be developed. Illustrative application examples of microstrip, coplanar and multilayer PCB interconnect will be presented. The possibility of using the TAN model for predicting the interconnect behaviour in broadband frequency from DC to several GHz will be demonstrated at the end of the chapter.

Keywords: TAN model, Kron's method, Kron–Branin's model, multiport system, SIMO topology, Z-matrix, Y-matrix, T-matrix, S-matrix, electrical interconnect, analytical calculation, modelling methodology, circuit and system theory

1.5 Chapter 5: Fast *S*-parameter Kron–Branin's modelling of rectangular wave guide (RWG) structure via mesh impedance reduction

Blaise Ravelo and Olivier Maurice

Abstract

The present chapter elaborates an innovative Kron–Branin's model of WG S-parameters. The graph topology is drawn from the equivalent electrical 1D circuit of the wave guide. The branch and mesh space analyses are introduced to determine the main unknowns of the problem represented by the contravariant mesh currents. Then, the mesh impedance reduction method is originally developed. Then, the rigorous tensorial equations enabling one to calculate rapidly the *S*-parameters are presented. Application examples are explored to validate the fast *S*-parameters modelling.

Keywords: TAN approach, Kron–Branin's model, frequency-domain analysis, modelling methodology, wave guide (WG) theory

1.6 Chapter 6: Time domain TAN modelling of PCB lumped system with Kron's method

Zhifei Xu, Yang Liu, Blaise Ravelo, and Olivier Maurice

Abstract

The time domain (TD) modelling of the lumped components-based low and medium speed printed circuit board (PCB) is examined in the present chapter. The TD translation of the physical-classical electric circuit-tensorial analysis of networks (TAN) graph dictionary will be established. Then, the workflow of the methodology describing the routine algorithm of the TD Kron's model will be introduced. By considering pulse input signals, the computation methods of the TAN method resolution based on the time-different iterative discrete solver are established. The feasibility of the TAN TD model will be verified with examples of PCB systems excited by the previously cited test signals. Discussion will be made about the accuracy, advantages and limits of the TAN TD computation methods.

Keywords: TAN approach, Kron's method, Kron–Branin's model, frequency domain analysis, modelling methodology, low and medium speed PCB, SI analysis, PI analysis, EMC analysis

1.7 Chapter 7: Direct time-domain analysis with TAN method for distributed PCB modelling

Zhifei Xu, Blaise Ravelo, Jonathan Gantet, Nicolas Marier, and Olivier Maurice

Abstract

The time domain (TD) tensorial analysis of networks (TAN) modelling of high-speed printed circuit board (PCB) system will be developed in the present chapter. Similarly, to the previous chapter, the dictionary of the TD TAN primitive elements based on distributed elements will be presented. The TD translation of the Kron–Branin method integrating the signal propagator operator will be used in the present case. Then the workflow of the routine algorithm illustrating how to analyse the high-speed PCB system will be described. By considering pulse waveform, discrete mathematical solvers will be developed. The feasibility of the TAN TD model will be verified with examples of SI and PI analyses of high-speed PCB systems excited by the previously cited test-signal waveforms. Discussion will be made about the accuracy, advantages and limits of the TAN TD computation methods.

Keywords: TAN approach, Kron's method, Kron–Branin's model, frequency domain analysis, modelling methodology, high speed PCB system, SI analysis, PI analysis, EMC analysis

1.8 Chapter 8: Coupling between EM field and multilayer PCB with MKME

Zhifei Xu, Yang Liu, Blaise Ravelo, Jonathan Gantet, Nicolas Marier, and Olivier Maurice

Abstract

This chapter is focused on modelling of radiated electromagnetic compatibility (EMC) coupling onto the multilayer printed circuit board (PCB). Kron's method integrates the electromagnetic (EM) emission, Taylor's and field-to-interconnect coupling models. The equivalent graph of the field-to-interconnect coupling is established. The modelling methodology consists in defining the primitive subnetwork elements. These primitive elements are represented by vias, interconnect lines and pads. Kron's graph equivalent to the EMC problem is elaborated. Finally, the coupling voltages are calculated via the tensorial equation translated from the graph. The radiated EMC Kron's model is validated with a four-layer PCB from 0.4 to 1.4 GHz by two scenarios of EM radiation. As proof of concept, a prototype of four-layer PCB was designed, fabricated, tested and simulated in full wave with a commercial three-dimensional EM tool. For the first case, the multilayer PCB was illuminated by plane wave emission propagating in different directions. The numerical computation from Kron's formalism was compared with simulation and measurement. The other case is the field-to-interconnect coupling between a microstrip I line PCB, as an EM field emitter, and the multilayer PCB, as a receiver, in 1-m distance. For both cases, the simulated and calculated voltage couplings onto the multilayer PCB are in good agreement.

Keywords: Kron's method, modelling, methodology, multilayer PCB, EMC coupling

1.9 Chapter 9: Conducted emissions (CEs) EMC TAN modelling

Olivier Maurice, Zhifei Xu, Yang Liu, and Blaise Ravelo

Abstract

The electromagnetic compatibility (EMC) conducted emission (CE) of printed circuit boards (PCBs) constitutes the most attractive research topic of EMC research engineers due to the fascinating challenging aspect related to the electronic component and PCB design complexity. The present chapter proposes an attempted EMC CE theory of the conducted EMC PCB emission. In this case, the PCB is assumed as a hybrid system comprising

- Lumped devices: frequency dependent *R(f)*, *L(f)* and *C(f)* components;
- Active components: as integrated circuits, including the die and packaging parameters that may behave as a nonlinear device;
- Passive elements: vias, pads and anti-pads, and interconnect TLs. The EMC CE model of the PCB system will be developed by considering some different standards perturbation signals.

The originality of the present EMC CE theory will be the elaboration of the transfer impedance matrices relating the contravariables represented by the active component internal activities as current tensor sources and the covariables voltage tensors. The EMC theorization will provide the way to establish the twice covariables transfer-impedance tensors. To highlight the feasibility of the EMC CE tensorial analysis of networks (TAN) model illustrative, examples of PCB used in automotive and PC will be treated. Discussion will be made on the strength and the weakness of the developed model in function of the PCB complexity and also the bandwidth of the EMC noises. Then, the accuracy, advantages and limits of the EMC CE TAN model will be made at the end of the chapter.

Keywords: TAN approach, Kron's method, Kron–Branin's model, frequency domain analysis, modelling methodology, PCB analysis, SI analysis, PI analysis, EMC analysis

1.10 Chapter 10: PCB-conducted susceptibility (CS) EMC TAN modelling

Olivier Maurice, Blaise Ravelo, and Zhifei Xu

Abstract

The complementary aspect of the electromagnetic compatibility (EMC) emission is the susceptibility issues. The present chapter will be focused on the EMC conducted susceptibility (CS) of hybrid printed circuit board (PCB) system with specifications described in the previous chapter. The tensorial description of the PCB susceptible components as mathematical sensitive functions integrating subdomain aspect will be originally introduced. The subdomain functions act as sigmoidal mathematical functions depending on the specification of the EMC perturbations. After the analytical description of the susceptibility functions, the workflow of the tensorial analysis of networks (TAN) modelling methodology indicating the routine algorithm of the EMC analysis will be provided. Then, random risk analyses table, including the objective functions indicating the EMC severity and the damage severity, will be addressed. To validate the EMC CS TAN model, illustration examples of PCB system with the prediction of EMC severity quantification with classes A, B and C will be discussed. Then, the accuracy, advantages and limits of the EMC CS TAN model will be made at the end of the chapter.

Keywords: TAN approach, Kron's method, Kron–Branin's model, frequency-domain analysis, modelling methodology, PCB analysis, SI analysis, PI analysis, EMC analysis

1.11 Chapter 11: PCB-radiated susceptibility (RS) EMC TAN modelling

Zhifei Xu, Blaise Ravelo, Yang Liu, and Olivier Maurice

Abstract

The radiated electromagnetic compatibility (EMC) analyses constitute one of the major challenging issues of the electronic printed circuit board (PCB) designers and EMC engineers. A few methods are currently available for the comprehension of the EMC phenomena between the EM-field interactions with the PCBs. Once again, this chapter will propose some key solutions against this EMC mechanism misunderstanding by means of the tensorial analysis of networks approach. In this chapter, a supplementary mathematical tensorial concept of moment space will be exploited to establish the metric of the EM wave interactions and the PCB hardware components, including the electrical interconnections. Some fundamental cases of scenarios with plane wave radiation interactions and planar PCBs will be developed. The interaction between PCB and other electronic structures as interconnect wires and another PCBs will also be treated. Then, the assessment of the signal-to-noise ratio in function of the radiated emissions will be formulated and validated with numerical and practical results.

Keywords: TAN model, radiated EMC investigation, PCB modelling, radiated emission, radiated susceptibility

1.12 Chapter 12: TAN model of loop probe coupling onto shielded coaxial short cable

Christel Cholachue, Amélie Simoens, Olivier Maurice, and Blaise Ravelo

Abstract

This chapter introduces coaxial cable *S*-matrix modelling with tensorial analysis of networks (TAN). The main objective of the modelling is to determine the shielding effectiveness (SE) of shielded cable coupled with a loop probe. The TAN methodology from the equivalent graph elaboration to the Kron's branch and mesh space analyses and ended by *Z*-matrix is elaborated. The SE is formulated innovatively from the *S*-matrix. A proof-of-concept constituted by a centimetre-length braid shielded cable illuminated by a proximate millimetre-radius circular was designed and simulated with commercial full-wave simulator. The wide band computed and simulated results with parametric analysis in function of the loop probe position are in good agreement.

Keywords: Coaxial cable, circuit theory, near-field coupling, tensorial analysis of networks (TAN), modelling method

1.13 Chapter 13: Nonlinear behaviour conducted EMC model of an ADC-based mixed PCB under radiofrequency interference (RFI)

Fayu Wan

Abstract

This chapter develops a modelling of nonlinear (NL) behaviour of a mixed circuit consisted of an analogue-to-digital converter (ADC). Under normal operation, the test circuit is disturbed by radio-frequency interference (RFI). The NL model of the electromagnetic compatibility (EMC) behaviour is established with the consideration of memory effect and nonlinearity. The developed NL EMC model enables one to predict the direct shift behaviour of ADCs. As proof-of-concept (POC), an ADC demo board is realized and tested. The calculated behavioural model and measurement results are in good agreement. The model helps one to better understand the behaviour of the digital and mixed electronic circuits under RFI. The present chapter is organized in three main sections. The second section of this chapter is focused on the proposed mixed circuit as a microcontroller (μC) NL-conducted EMC with electromagnetic interference signal analysis. For better understanding, the study is focused essentially on the case of RFI higher than the ADC's sampling rate. The relationship between DC offset and sampling rate will be established analytically. The mathematical model enables one to predict the μC behaviour. The analytical approach is described with the NL model represented by the output voltage expressed in polynomial function of input. The third section of this chapter examines the validation results with μC-based mixed circuit POC. Comparison between simulations and measurements by considering the NL RFI model is discussed. Last, the fourth section of this chapter is the conclusion.

Keywords: Mixed PCB, conducted EMC, analogue-to-digital convertor, radio-frequency interference (RFI), nonlinear behaviour modelling, direct shift

1.14 Chapter 14: Far-field prediction combining simulations with near-field measurements for EMI assessment of PCBs

Dominik Schröder, Sven Lange, Christian Hangmann, and Christian Hedayat

Abstract

Using near-field (NF) scan data to predict the far-field (FF) behaviour of radiating electronic systems represents a novel method to accompany the whole RF design process. This approach involves so-called Huygens' box as an efficient radiation model inside an electromagnetic (EM) simulation tool and then transforms the scanned NF measured data into the FF. For this, the basic idea of the Huygens box principle and the NF-to-FF transformation are briefly presented. The NF is measured on the Huygens box around a device under test using an NF scanner, recording the magnitude and phase of the site-related magnetic and electric components. A comparison between a fullwave simulation and the measurement results shows a good similarity in both the NF and the simulated and transformed FF. Thus, this method is applicable to predict the FF behaviour of any electronic system by measuring the NF. With this knowledge, the RF design can be improved due to allowing a significant reduction of EM compatibility failure at the end of the development flow. In addition, the very efficient FF radiation model can be used for detailed investigations in various environments and the impact of such an equivalent radiation source on other electronic systems can be assessed.

Keywords: PCB, near-field-to-far-field transformation, Huygens' box, virtual equivalent emission model for co-simulative EMI characterization, radiated EMC

1.15 Chapter 15: Element of information for numerical modelling on PCB

Toufic Abboud and Benoît Chaigne

Abstract

The present chapter provides the key information about the numerical modelling of printed circuit board (PCB). The constituting key elements are defined and analytically expressed in function of the basic parameters. The purposed model is applied to the calculation of a small PCB with investigation on the meshing effect on the convergence. The main focus of the study is to analyse the inherent and critical points in function of the scale variation. The meshing effects on the PCB 3D modelling will be investigated.

Keywords: PCB, numerical modelling, convergence, 3D modelling, meshing

1.16 Chapter 16: General conclusion

Olivier Maurice, Blaise Ravelo, and Zhifei Xu

Abstract

The contents of these chapters illustrate the investigation on the unfamiliar Tensorial analysis of networks approach proposed in this book. This final chapter summarizes the main potential contributions of all chapters (Chapters 2–15) of this book. In the fields of the electromagnetic compatibility (EMC), signal integrity and power integrity engineering, the book provides new ways to analyse the printed circuit board (PCB) problems. One of the main modellings is developed in the first 11 chapters (Chapters 2–13). Chapter 14 presents a technique of EMC conducted susceptibility analysis of digital components used regularly in the PCBs. Then, investigations on the EM NF radiations from the planar PCBs based on the scanning technique are developed in Chapters 15 and 16. Then, the concluding technical chapter proposes an overview of PCB numerical modelling.

Keywords: Methodology, analytical method, numerical method, PCB analysis, SI analysis, PI analysis, EMC analysis, experimental test techniques

Chapter 2
Basic knowledge to practice TAN for PCB SI/PI/EMC investigation

Olivier Maurice[1], Zhifei Xu[2], Yang Liu[2], and Blaise Ravelo[2]

Abstract

The tensorial approach remains one of the less explored circuit theories to solve the modern printed circuit board (PCB) problem. Chapter 2 of the proposed book will open the door allowing us to the non-specialist and even non-expert of circuit design to learn and to practice with their own easy way and rapidly the tensorial analysis of networks (TAN) approach. The fundamental approach to use the tensor algebra will be introduced. The elementary and basic knowledges necessary to elaborate the TAN concept will be defined. An easy concept of physical, classical, graph topological and tensorial object representation of the primitive elements to treat the PCB signal integrity (SI), power integrity (PI) and electromagnetic compatibility (EMC) problems will be described. Several basic illustrative examples of key understanding lumped RC, RL, LC and RLC network-based circuits will be treated in the chapter. The methodologies integrating the topological graphs of the problem will be explained. The different basic steps to practice the TAN approach are indicated. For the concrete comprehension, the different practical cases for solving PCB SI, PI and EMC problems are introduced.

Keywords: TAN approach, Kron's method, Kron–Branin's model, primitive elements, modelling methodology, PCB analysis, SI analysis, PI analysis, EMC analysis, circuit and system theory

2.1 TAN principles

We present the tensorial analysis of networks (TAN) principles in the simplest form as possible to highlight its advantages without going deeper in its meanings, which

[1]ArianeGroup, Paris, France
[2]IRSEEM/ESIGELEC, Rouen, France

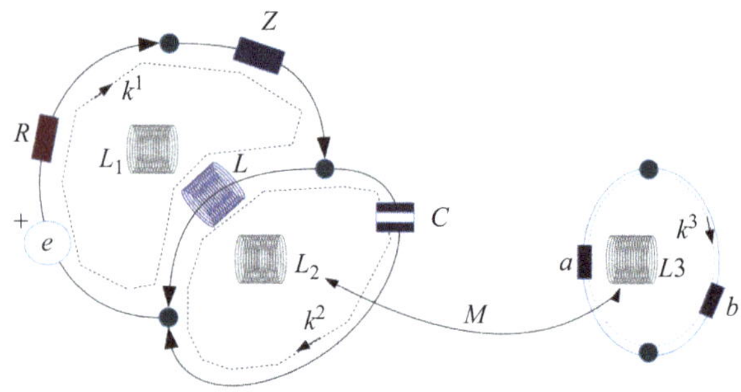

Figure 2.1 Circuit decomposition

is out of the purpose of this book. However, some more detailed descriptions will be given in various chapters in order to introduce specific approaches attached to the chapter subject.

Once an electronic circuit is defined, it can be decomposed in meshes or current loops. An example of circuit and its decomposition in meshes is given Figure 2.1.

In this circuit, we see various kinds of components belonging to branches. These branches are connected in meshes. Sometimes, a branch can be shared between two meshes. When a group of branches are connected to make a mesh, a new inductance created by this closed circulation of branches belongs to the mesh itself. For this reason, it is located at the centre of the mesh.

A component can be itself an inductance represented by the symbol L as shown in Figure 2.1. In this case, it belongs to the branch. All the graphs associated with the circuits are connex. It means that they cannot have open branches with a node connected to only one branch.

In our example, we have two circuits or two graphs (also called networks).

The impedance of a mesh is equal to the summation of all the impedances of the branches that belong to the mesh. In our example, the impedances of the three meshes associated with the mesh currents k^1, k^2, k^3 (we will see later why we put the index up) are (p is Laplace's operator, R, a and b are resistances and L, L_1, L_2, L_3 are inductances):

- mesh 1: $z_{11} = R + Z + (L + L_1)p$;
- mesh 2: $z_{22} = (L + L_2)p + \frac{1}{Cp}$;
- mesh 3: $z_{33} = a + b + L_3 p$.

These three impedances are the three diagonal components of the operator z that will solve the whole circuit. The three components k^1, k^2, k^3 are the three components of the vector $\mathbf{k}$ that describes all the currents in the problem studied. This vector can be identified by its components using the abstract notation k^i. i is called a mute index and is located up right to the symbol k. In classical vectorial notation, we may write

$$\mathbf{k} = k^1\mathbf{u}_1 + k^2\mathbf{u}_2 + k^3\mathbf{u}_3 \text{ [A]} \tag{2.1}$$

where $\mathbf{u}_i$ is the vectorial base of the currents in the mesh space. We can define a dual base $\mathbf{v}^j$ with

$$\mathbf{u}_i \cdot \mathbf{v}^j = \delta_i^j \tag{2.2}$$

This dual base allows to develop a covector of mesh electromotive forces (EM forces) $\mathbf{e}$: $\mathbf{e} = e_1\mathbf{v}^1 + e_2\mathbf{v}^2 + e_3\mathbf{v}^3$ [V]. This covector is identified by its components writing e_q. q is a mute index pointing out the three components of the covector $\mathbf{e}$. We can compute the scalar product:

$$\mathbf{e} \cdot \mathbf{k} = e_i k^j \mathbf{v}^i \cdot \mathbf{u}_j = e_i k^j \delta_j^i \text{ [W]} \tag{2.3}$$

Now we can define arbitrarily a metric operator ζ as

$$e_i = \zeta_{ij} k^j \text{ [V]} \tag{2.4}$$

In the case of orthogonal spaces, z_{ij} is purely diagonal. In general when the covector components depend on all the vector components, we have $z_{ij} \neq 0,\ i \neq j$. With these variables, we can define

$$e_i k^j = \zeta_{ij} k^i k^j = P \text{ [W]} \tag{2.5}$$

where P as an invariant, it means that whatever the choice of ζ, the scalar product (2.5) remains the same. The covector components e_i have their index down. ζ is a tensor twice covariant (with two index down). Note that each time we use the contracted product. When two indices are equal, one up and one down, we realize the product of the two terms for the various values of the index. For example,

- $u_i v^i = \sum_{i=1,2,\ldots,N} u_i v^i$;
- $T_{ij} K^j = \sum_{j=1,2,\ldots,N} T_{ij} K^j$;
- $Q_a^{bc} Q_d^{hc} = \sum_{c=1,2,\ldots,N} Q_a^{bc} Q_d^{hc}$.

The only way to transform a vector in a covector, i.e., a current in voltage is to go through a metric, i.e., an impedance operator ζ. By the same process, the only way to transform a covector in a vector is to go through the inverse of a metric, i.e., an admittance operator Y.

When we inverse a tensor, the superscript becomes subscript and inversely the subscript becomes a superscript. In addition, the index element positioning must be reversed. For example, $(\zeta_{uv})^{-1} = Y^{vu}$ or $(T_i^{jk})^{-1} = L_{kj}^i$.

When a branch is shared between mesh a and mesh b, it creates an extra-diagonal component of ζ at ζ_{ab} and ζ_{ba} with a value representing the impedance of the shared branch.

When a cord constitutes a link between a mesh a and a mesh b, it creates an extra-diagonal component of ζ at ζ_{ab} and ζ_{ba} to show which value is the impedance function of the cord (on our example, it is a mutual inductance with function $\zeta_{23} = \zeta_{32} = -Mp$).

The relation between the number of nodes N, branches B, meshes M and connex networks R is defined by Poincaré's relation: $-N + B - M + R = 0$.

Once a metric or impedance operator ζ is defined in the mesh space, the problem is solved. Solutions of current are obtained by $e_u = \zeta_{uv} k^v \Rightarrow k^v = Y^{vu} e_u$. In the tensorial equation $e_u = \zeta_{uv}k^v$, the mesh currents k^v are the unknown variables and e_u, the known sources.

A current source can be defined making a "virtual" mesh of unknown EM force U and known mesh current J. The mesh J is added to the other meshes by using the same process of construction. In final solving the system, e_u has one component unknown: U and the current vector k^v has one component imposed: J.

A cord γ is fundamentally defined as an operator as $e_i = \gamma_{ij}\left(k^j\right)$. It is in a general way a function applied to a current. For example, $\gamma = \exp\left(-\alpha\bullet\right)$ and

$$\gamma\left(k\right) = \exp(-\alpha k)\ \ [\mathrm{V}] \tag{2.6}$$

In addition, tensor indices must be the same for each member in the equation. When there are contracted products, the index not involved in the products must remain unchanged. For example,

$$e_a + h_a = T^d_{abc}Q^{bc}U_d \tag{2.7}$$

We can change the base using a transform matrix Λ. A matrix has one superscript and one subscript. For example,

$$\Lambda^v_u Q^u = K^v \tag{2.8}$$

and

$$\zeta_{uv}K^v = \zeta_{uv}\Lambda^v_u Q^u \tag{2.9}$$

and if

$$P = \zeta_{uv}k^u k^v \Rightarrow P = \zeta_{uv}\Lambda^u_\alpha Q^\alpha \Lambda^v_\beta Q^\beta = \zeta_{\alpha\beta}Q^\alpha Q^\beta \tag{2.10}$$

What may be the base used for describing the problem, the invariant remains unchanged. That is a fundamental property of tensorial algebra.

For more details, the authors are suggesting to read [1].

2.2 Electronic world and electronic scaling

The electromagnetic laws are applied at the printed circuit board (PCB) level. While time passing, electronic dimensions decrease continuously, but working frequencies increase. Finally, the ratio between the wavelength and the typical dimensions inside the PCB remains quite constant. We study these relations as references to establish fundamental expressions for electromagnetic compatibility (EMC), signal integrity (SI) or power integrity (PI) applied to PCB. Then we look in the next chapters how these fundamental relations are used in the three domains SI, PI and EMC.

SI focuses on signal propagation from point to point.

PI focuses on power supply distribution and decoupling.

EMC focuses on coupling between signals and components disturbances.

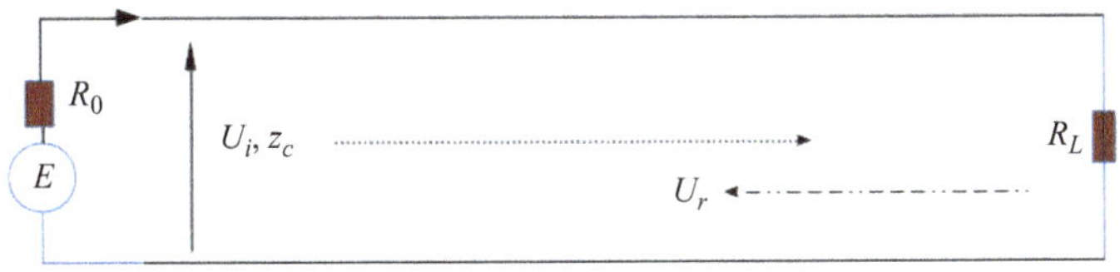

Figure 2.2 Typical line

2.2.1 Propagation

For better understanding, we consider a simple structure of propagation as a transmission line. This line is x-meter long. We connect a pulse generator on the input of this line (see Figure 2.2). If R_0 is the internal impedance of this generator, E its amplitude and if z_c is the characteristic impedance of the line, the signal U_i injected in the line has for amplitude,

$$U_i = \frac{z_c}{z_c + R_0} E \ [\text{V}] \tag{2.11}$$

This signal propagates into the line and reach its extremity. Anywhere in the line of impedance z_c, the wave is associated with a generator of amplitude $2U_i$ and impedance z_c. The transmitted voltage U_p remains

$$U_p = 2U_i \frac{z_c}{2z_c} = U_i \ [\text{V}] \tag{2.12}$$

The voltage wave U_i reaches the end of the line. At this location, the transmitted signal U_t is given by (p is Laplace's operator):

$$U_t = \frac{R_L}{R_L + z_c} 2U_i e^{-(x/c)p} \ [\text{V}] \tag{2.13}$$

where c is the propagation speed in the line. Remember that the exponential function depending on $-x/cp$ translates a propagation delay.

This creates a reflected wave U_r defined by

$$U_t = U_i + U_r \tag{2.14}$$

This leads to

$$U_r = U_t - U_i = \frac{R_L}{R_L + z_c} 2U_i e^{-(x/c)p} - U_i e^{-(x/c)p} = \left(\frac{R_L - z_c}{R_L + z_c} \right) U_i e^{-(x/c)p} \ [\text{V}] \tag{2.15}$$

By denoting,

$$\sigma = \left(\frac{R_L - z_c}{R_L + z_c} \right)$$

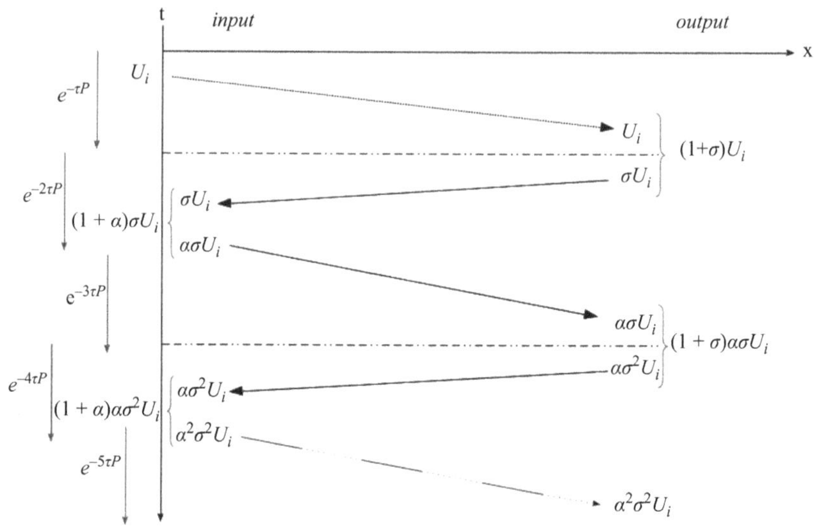

Figure 2.3 Bergeron's diagram

we have

$$U_r = \sigma U_i e^{-(x/c)p} \ [\mathrm{V}] \tag{2.16}$$

This reflected wave travels back to the input of the line, delayed. Seen at the input, the voltage induced by this reflected wave is

$$U_r(0) = \sigma U_i e^{-(2x/c)p} \ [\mathrm{V}] \tag{2.17}$$

Figure 2.3 shows the traveling of voltage waves along the line in progressive (e_p) and backward (e_R) modes. The values in the tab are the moduli of the voltage waves. When we look at the incident voltage at the end of the line, we find this development:

$$\begin{cases} e_p = U_i e^{-\tau p}\left(1 + \alpha\sigma e^{-2\tau p} + \alpha^2\sigma^2 e^{-4\tau p} + \cdots\right) \\ e_R = \sigma U_i e^{-2\tau p}\left(1 + \alpha\sigma e^{-2\tau p} + \alpha^2\sigma^2 e^{-4\tau p} + \cdots\right) \end{cases} \tag{2.18}$$

with $\tau = x/c$, p being Laplace's operator.

We can define a propagation kernel G with

$$G = \sum_{n=0}^{+\infty} \alpha^n \sigma^n e^{-2n\tau p} \tag{2.19}$$

and so

$$\begin{cases} e_p = G U_i e^{-\tau p} \\ e_R = G\sigma U_i e^{-2\tau p} \end{cases} \tag{2.20}$$

It can be mentioned that anywhere the incident voltage may be modelled using a generator $2U_i$ and an impedance z_c. Let us take a look at the end of the line. Using the previous technique, the transmitted voltage U_t on the load should be given by

$$U_t = 2e_p\left(\frac{R_L}{R_L + z_c}\right) \ [\mathrm{V}] \tag{2.21}$$

By limiting G to its first order, we obtain

$$e_p = U_i e^{-\tau p} \ [\mathrm{V}] \tag{2.22}$$

This implies that

$$U_t = \left(\frac{R_L}{R_L + z_c}\right) 2U_i e^{-\tau p} \ [\mathrm{V}] \tag{2.23}$$

Now if we compute $(1+\sigma)U_i$ we find

$$\left[1 + \left(\frac{R_L - z_c}{R_L + z_c}\right)\right] U_i e^{-\tau p} = U_t \ [\mathrm{V}] \tag{2.24}$$

Fortunately, both reasonings give the same result.

If we look at the line input, we can compute the transmitted voltage. As the expression is quite complex, let us consider that $\sigma = -1$ and $G \approx 1$. Then,

$$2e_R = -2U_i e^{-2\tau p} \ [\mathrm{V}] \tag{2.25}$$

If $z_c = R_0$, the current on the input mesh (made of the generator branch and the lien input branch) is

$$k^1 = \frac{E - 2U_i e^{-2\tau p}}{2z_c} \ [\mathrm{A}] \tag{2.26}$$

giving

$$U_t = E - z_c k^1 = \frac{E}{2} - U_i e^{-2\tau p} \ [\mathrm{V}] \tag{2.27}$$

as $U_i = E/2$ we obtain

$$U_t = \frac{E}{2}\left(1 - e^{-2\tau p}\right) \ [\mathrm{V}] \tag{2.28}$$

We can keep in mind this quantity for the next step of our analytical demonstration. When the time is less than twice the travel time of the voltage wave through the line, the input voltage is $E/2$. When the time exceeds this limit, the total voltage becomes zero, the reflected voltage wave suppressing the injected voltage wave (this comes from the fact that we chose $\sigma = -1$).

We understand that all the parameters occurred in the line can be described by observing only its extremities. The two branches represent the input and output of the line. When the line is used, i.e., connected to a generator and a load, the generator branch connected on the input makes a first mesh and the load connected on the output makes a second mesh. Finally, a line is a two-mesh system. But we would like to determine it in an easier way, that is, a more easy expression for the interaction between the two meshes than the one using G, which is a complicated function. That is the purpose of the next section.

2.2.2 Lines and microstrips modelling

Each kind of voltage wave can be considered as a voltage wave coming from a generator located at each extremity of the line. The incident voltage u_i comes from a generator located on the line input, and the reflected voltage u_r comes from a generator located at the end of the line. Voltage and current anywhere in the line are defined by

$$\begin{cases} u(x,t) = u_i(t - \frac{x}{v}) + u_r(t + \frac{x}{v}) \\ z_c i(x,t) = u_i(t - \frac{x}{v}) - u_r(t + \frac{x}{v}) \end{cases} \tag{2.29}$$

This leads to

$$\begin{cases} u(x,t) + z_c i(x,t) = 2u_i(t - \frac{x}{v}) \\ u(x,t) - z_c i(x,t) = 2u_r(t + \frac{x}{v}) \end{cases} \tag{2.30}$$

Using (2.30), we can set the expression of the reported generators in input ($x = 0$) and output ($x = X$) of the line: $2e_p = 2u_i(t - (x/v))$ and $2e_R = 2u_r(t + (x/v))$. They are in series with z_c and use $x = X$ or $x = 0$. We obtain the set of equations called Branin's equations:

$$\begin{cases} w_p = 2e_p = [u(0) + z_c i(0)]\, e^{-\tau p} \\ w_R = 2e_R = [u(X) - z_c i(X)]\, e^{-\tau p} \end{cases} \tag{2.31}$$

Figure 2.4 shows the equivalent schematic associated with Branin's equations. This schematic represents the line alone. The two branches must be added to connect the generator on one input and the load on the other.

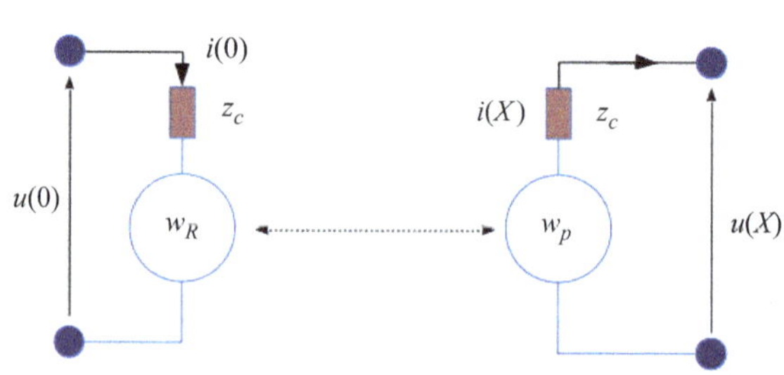

Figure 2.4 Branin's modelling

2.2.3 *Some particular applications*

Some applications of Branin's modelling can be developed connecting a generator and a load to Branin's model. The structure created by these connections is made of two meshes. It is a typical two-dimensional problem.

Let us imagine a generator made of an EM force E and a resistance R_0. We have also a resistance R_L that is a load. We dispose of four branches: one branch for the generator, one for the load and two for a line modelled by a Branin's model. Figure 2.5 shows this collection of *primitive objects*.

Starting from this collection of four branches, we can make connections in order to construct an electrical circuit made of one load connected to a generator through a line. To do that, we connect branch 1 to branch 2, and branch 3 to branch 4. Soldering branch 1 on branch 2 we make a first mesh numbered 1. The second mesh is made of branches 3 and 4. All these terms are synthesized in a matrix called connectivity matrix C defined by

$$C = \begin{bmatrix} 1 & 0 \\ 1 & 0 \\ 0 & 1 \\ 0 & 1 \end{bmatrix} \tag{2.32}$$

Between the current on the branches i^x and the current on the meshes k^y, we have the matrix relation:

$$i^x = C^x_y k^y \tag{2.33}$$

The currents $\{i^1, i^2, \ldots, i^4\}$ are the components of the current vector i^x. The currents $\{k^1, k^2\}$ are the components of the current vector k^y. Going from a representation in the branch space to a representation in the meshes space, we apply just a change of space where the matrix C is the change of space matrix.

Before, to connect all the branches between each other, we can associate all of them with a generic model. This model includes an operator z that gives the relation between the flux i^v in the branch and an external excitation e_u. Kirchhoff's law describes these relations, including the potential across the branch V_u by noting

$$e_u = V_u + z_{uv} i^v \ \text{[V]} \tag{2.34}$$

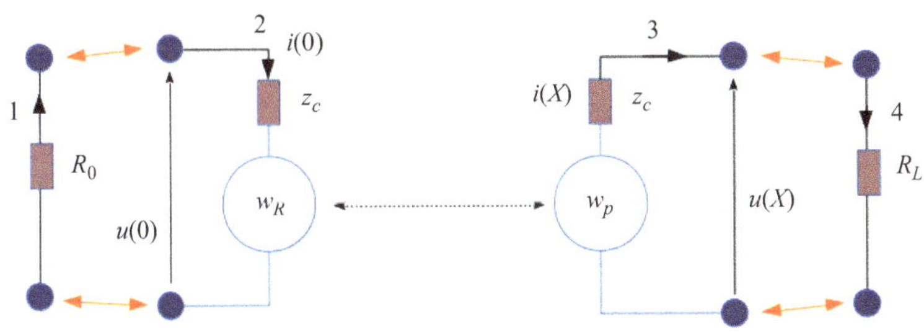

Figure 2.5 Primitive objects

where e_u is the external excitation. It is an EM force coming from an electromagnetic field. The transformed circuit is equivalent to apply the connectivity to the previous equation. We write

$$e_u = V_u + z_{uv} C^v_\alpha k^\alpha \ [\mathrm{V}] \tag{2.35}$$

We can multiply all members by the inverse of the connectivity:

$$C^u_\beta e_u = C^u_\beta V_u + C^u_\beta z_{uv} C^v_\alpha k^\alpha \ [\mathrm{V}] \tag{2.36}$$

$C^u_\beta e_u = e_\beta$, which is the excitation in the meshes space.

Because the integration of voltages along a mesh leads to zero, we have $C^u_\beta V_u = 0$. As

$$\zeta_{\beta\alpha} = C^u_\beta z_{uv} C^v_\alpha \ [\Omega] \tag{2.37}$$

We finally obtain

$$e_\beta = \zeta_{\beta\alpha} k^\alpha \ [\mathrm{V}] \tag{2.38}$$

How this equation is declined in the case of Branin's model?

Looking at the circuit in Figure 2.5, we can understand that the first mesh has for impedance $R_0 + z_c$. It models the line input. A second mesh of impedance $z_c + R_L$ models the line output. In a 2D-space based on the mesh currents k^1 and k^2, the impedance operator ζ of this space associated with the circuit described in Figure 2.5 is given for the two meshes of the line extremities by

$$\zeta = \begin{bmatrix} R_0 + z_c & 0 \\ 0 & z_c + R_L \end{bmatrix} \ [\Omega] \tag{2.39}$$

It remains to add the coupling between these two meshes, i.e., the kernel G that propagates the signal across the line. How can we determine the components of G? In any case, the components of G can be determined computing the EM force on one target depending on the flux that creates this EM force, flux which is the source.

If we consider the second equation of (2.31), we can easily obtain the expression of w_R. The voltage $u(X)$ is given by

$$u(X) = R_L k^2 \ [\mathrm{V}] \tag{2.40}$$

And the current $i(X) = k^2$. Finally,

$$w_R = \left[R_L k^2 - z_c k^2\right] e^{-\tau p} \ [\mathrm{V}] \tag{2.41}$$

which leads to (remember that the reported generator on the input of the line is the opposite to the input current)

$$G_{12} = -\frac{w_R}{k^2} = [z_c - R_L] e^{-\tau p} \ [\Omega] \tag{2.42}$$

By the same process we also find

$$G_{21} = -[z_c - R_0]\,e^{-\tau p}\ [\Omega] \tag{2.43}$$

The kernel is defined by

$$G = \begin{bmatrix} 0 & -[z_c - R_L]\,e^{-\tau p} \\ -[z_c - R_0]\,e^{-\tau p} & 0 \end{bmatrix}\ [\Omega] \tag{2.44}$$

(*Note:* When we find the expression of a reported generator coming from some coupling process, the coupling impedance is obtained with a minus sign. Effectively, if $w = zi$ this implies that in the global equation $w' + w = z'i' \Rightarrow w' = z'i' - zi$.)

Note that if a source power supply is applied to mesh 1, it appears on mesh two after the propagation delay. Remark that if $z_c = R_0 = R_L \Rightarrow G_{ij} = 0$. But the nominal transmission function of the line transmits an input signal to the output after a propagation delay. This is mathematically demonstrated if you replace $u(0)$ by a signal $V(0)$ in the first equation of (2.31)

$$V(X) = V(0)e^{-\tau p}\ [\mathrm{V}] \tag{2.45}$$

So, each time we have a propagation structure, the source vector must be completed by the propagated sources obtained on matched loads.

The reader that may want to test Branin's model can find hereafter (Annexe 2.A) an example of computation realized in python to model a line of load conditions R_L and R_c.

2.2.4 Lossy propagation model

In difference to the previous case, the present paragraph is importantly elaborated by taking into account to the propagation losses. It is a chance because taking into account rigorously losses in the lines is difficult.

The amplitude decreasing comes from two different effects:

- the losses through Joule's dissipations in the line,
- the effect of propagation speed dispersion depending on frequencies.

Let us first take a look at the effect of dispersion. It means that depending on the frequency, the speed of propagation changes. We can imagine a signal s made of three components:

$$s(t) = 1,5 + \sin\left(2\pi \frac{t}{400}\right) + \sin\left(2\pi \frac{3t}{400}\right)\ [\mathrm{V}] \tag{2.46}$$

If this signal is injected in a line for which the dispersion k is defined by

$$k = \frac{\omega}{c}\ [\mathrm{A}] \tag{2.47}$$

For a line of length x with a matched source $s(t)$ and load, the signal V_s sent to the line output is (using the Laplace's transform):

$$V_s(p) = s(p)e^{-jkx} \ [\mathrm{V}] \tag{2.48}$$

where V_s is completely identical to $s(p)$, delayed. Let us denote the propagation speed, v_1, for frequency, ω_1, and propagation speed, v_2, for frequency, ω_2, which is assumed to be higher than $\omega_1 (\omega_2 > \omega_1)$. Knowing these parameters, the output signal can be expressed as:

$$V_s = 1,5 + \sin\left(2\pi \frac{t}{400}\right) \delta_{\frac{\omega_1}{v_1}x} + \sin\left(2\pi \frac{3t}{400}\right) \delta_{\frac{\omega_2}{v_2}x} \neq s(t) \ [\mathrm{V}] \tag{2.49}$$

The nominal signal is shown in Figure 2.6.

If the difference of frequencies implies a delay of 20 s, the signal reaching the output of the line, corresponding to the nominal signal shown in Figure 2.6, should behave as the curve shown in Figure 2.7.

It can be seen that some peaks are higher than before, whereas others are lower. That is the effect of dispersion.

Now we can estimate the influence of Joule losses.

Joule losses can be generated from the wires or from the dielectrics. For the wires, we must estimate the losses in high frequencies beyond to the skin cut-off frequency. The losses in continuous R_{DC} are given by (the wire is of radius r, conductivity σ and permeability μ)

$$R_{DC} = \frac{1}{\sigma \pi r^2} \ [\Omega][\mathrm{m}]^{-1} \tag{2.50}$$

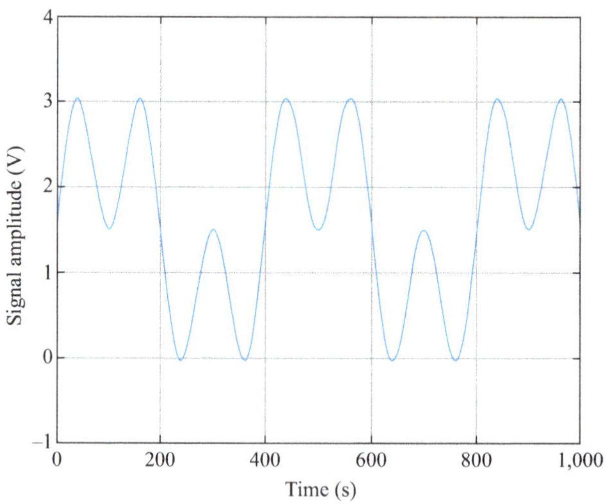

Figure 2.6 Nominal signal s(t)

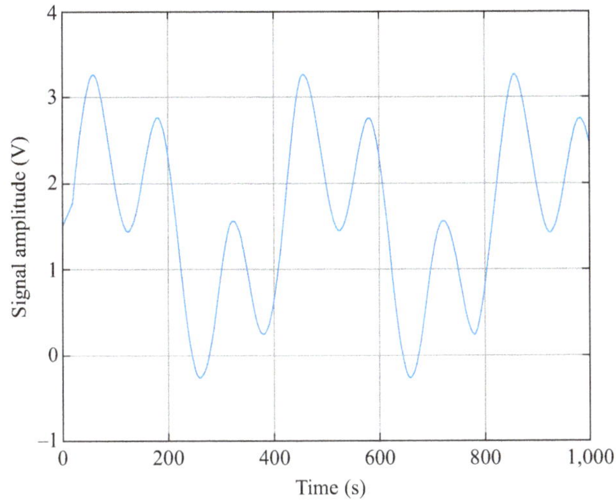

Figure 2.7 Output signal s(t)

While beyond the skin cut-off frequency, we have

$$R_{HF} = \frac{1}{\sigma 2\pi r}\sqrt{\pi f \mu \sigma} \ \ [\Omega][\mathrm{m}]^{-1} \tag{2.51}$$

A simple and efficient model consists in adding half this resistance to each extremity of the line. In that case, the diagonal components of ζ become

$$\zeta = \begin{bmatrix} R_0 + \frac{1}{2}(R_{DC} + R_{HF}) + z_c & 0 \\ 0 & R_L + \frac{1}{2}(R_{DC} + R_{HF}) + z_c \end{bmatrix} [\Omega] \tag{2.52}$$

But the kernel G remains the same. The effect of losses is to decrease the equivalent level transmitted into the line and to the load.

The complex part of the permittivity creates losses in the dielectric. A part of the current is dissipated in this conductance. Therefore, we can write

$$\varepsilon = \varepsilon' - j\varepsilon'' \ [\mathrm{F}][\mathrm{m}]^{-1} \tag{2.53}$$

The capacitance is then given by (α is a geometrical function and j the imaginary number)

$$C = \alpha\varepsilon = \alpha\left(\varepsilon' - j\varepsilon''\right) \ [\mathrm{F}] \Rightarrow z(C) = \frac{1}{j\alpha(\varepsilon' - j\varepsilon'')\omega} \ [\Omega] \tag{2.54}$$

The impedance is equivalent to a capacitance C_0 in parallel with a conductance g defined by

$$C_0 = \alpha\varepsilon' \, [\mathrm{F}] \quad g^{-1} = \alpha\varepsilon''\omega \, [\Omega]^{-1} \tag{2.55}$$

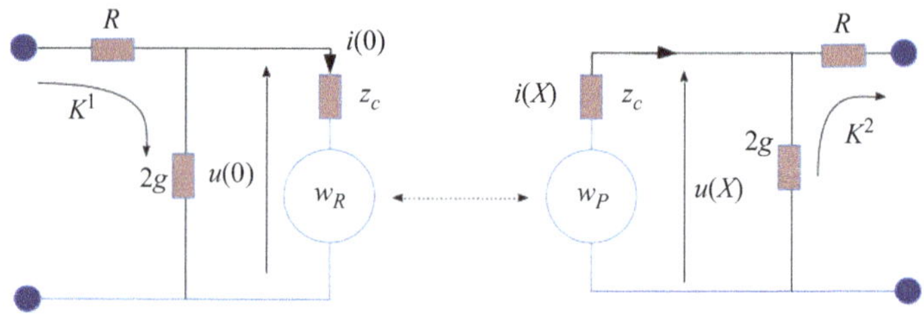

Figure 2.8 Lossy Branin's model

In that case, the dielectric losses are reported by two conductances of value $(2g)^{-1}$ located at each extremity of the line. In Branin's model, they are in parallel with z_c. G remains unchanged. The impedance operator for Branin's model becomes

$$\zeta = \begin{bmatrix} 2g + z_c & 0 \\ 0 & 2g + z_c \end{bmatrix} \; [\Omega] \tag{2.56}$$

Under this topology, the current in the line i^0 becomes cross talked with the lossless one.

All this is synthesized in lossy Branin's model shown in Figure 2.8.

Branin's equation (2.31) remains unchanged.

The system can be reduced to become similar to lossless Branin's model. We have to use Thèvenin's model on the line extremities. Using this modelling, the impedance operator (which includes G) becomes for the lossy line:

$$\zeta = \begin{bmatrix} \frac{2g(R+R_0)}{2g+R+R_0} + z_c & -[z_c - R_L]\,e^{-\tau p} \\ -[z_c - R_0]\,e^{-\tau p} & \frac{2g(R+R_L)}{2g+R+R_L} + z_c \end{bmatrix} \; [\Omega] \tag{2.57}$$

With $R = R_{DC} + R_{HF}$.

2.2.5 Asymptotic behaviour without propagation

Let us denote the delay for any propagation structure τ. If $p = \alpha + j\omega$, we consider an excitation $e(p)$ propagating into this structure. For a matched line, the output signal is given by

$$s(p) = e(p)e^{-\alpha\tau}e^{-j\omega\tau} \; [\mathrm{V}] \tag{2.58}$$

Noting: $\bar{s}(p) = s(p)e^{-\alpha\tau}$ and $\tau = 1/\omega_\tau$, then

$$s(p) = \bar{s}(p)e^{-j(\omega/\omega_\tau)} \; [\mathrm{V}] \tag{2.59}$$

Using Fourier's series for $\bar{s}(p)$ we obtain

$$s(p) = \sum_0^N [A_n \cos(n\omega_0 t) + B_n \sin(n\omega_0 t)]\, e^{-j(n\omega_0/\omega_\tau)} \; [\mathrm{V}] \tag{2.60}$$

Then if $\forall n,\ \omega_\tau \gg n\omega_0 \Rightarrow e^{-j(n\omega_0/\omega_\tau)} \to 0$. There are no propagation effects and $s(p) = \bar{s}(p)$. Note that the fact to reduce the propagation delay to zero does not change anything for causality. And causality is the fundamental property of physical space with the conservation law for energy.

From the moment we do not consider propagation, all our arguments are no longer valid. No wave means no propagation impedance and z_c does not exist anymore. The propagation speed is infinite and we must look at the line with another regard. $\exp(-\alpha\tau) = \exp(-\alpha x/v)$ but the speed being $+\infty$, $\exp(-\alpha\tau) = 1$.

Calling a "short line" a line for which our assumption ($\forall n,\ \omega_\tau \gg n\omega_0$) remains true, we can characterize the impedance of this short line in short circuit or open circuit. Let us take a look at the first case.

We excite the short line in open circuit using a generator of $e_0 = 1$ volt and self-impedance equal to R_0. Measuring the output signal $s(p)$, we obtain a Laplace form given by

$$s(p) = e_0 \left[\frac{1}{p} - \frac{1}{p + \omega_\tau} \right] \ [\mathrm{V}] \tag{2.61}$$

if $R_0 \gg R$, R being the wire resistances, we can approximate $\omega_\tau \approx 1/(R_0 C)$.

In a more practical view, if we imagine that we measure the short line in short circuit, exciting the line with a short-circuit current i^0 we obtain for the output current $i(p)$:

$$i(p) = i^0 \left[\frac{1}{p} - \frac{1}{p + \omega_a} \right] \ [\mathrm{A}] \tag{2.62}$$

with $\omega_a \approx R_0/L$, where L is the inductance of the short line. For a wire of diameter d and conductivity σ, the DC resistance is known. Finally, the equivalent schematic of the short line is shown in Figure 2.9 and leads to the impedance operator ζ_{sl}. (Have you noticed the inductance located at the centre of the mesh? This inductance belongs to the circuit loop and not to any branch. It is not a lumped element or any inductive component but the inductance created by the closed circulation of the current k^1. So it is more than legitimus to locate it at the centre of the mesh, as we can work on the mesh space. It is noteworthy that this situation illustrates a particularly important

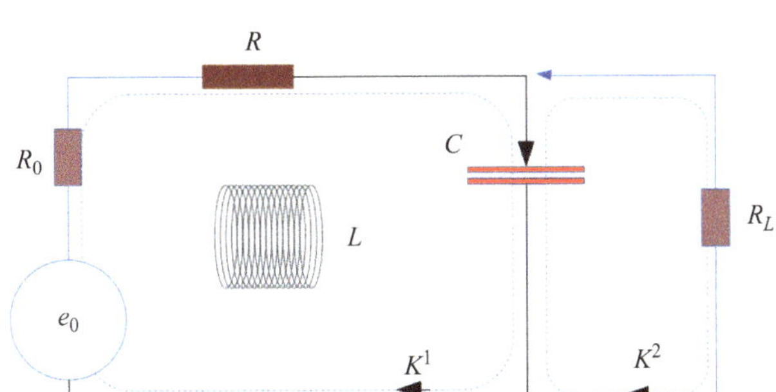

Figure 2.9 Short line model

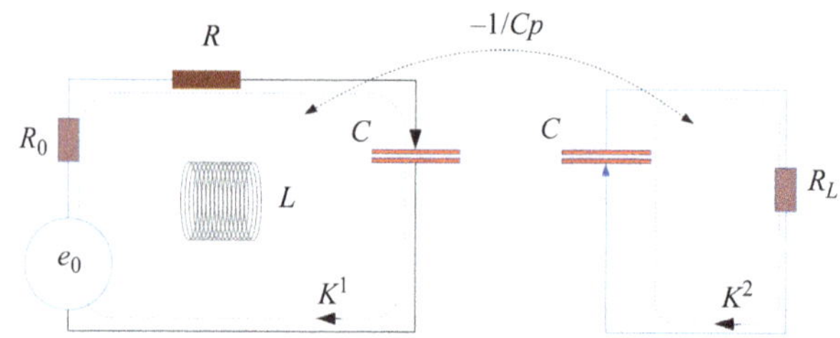

Figure 2.10 Short line model

advantage of TAN compared to the classical nodal methods. This configuration of middle and disconnected inductance cannot be implemented with nodal method.)

$$\zeta_{sl} = \begin{bmatrix} R_0 + R + Lp + \frac{1}{Cp} & -\frac{1}{Cp} \\ -\frac{1}{Cp} & \frac{1}{Cp} + R_L \end{bmatrix} \; [\Omega] \tag{2.63}$$

Looking at this operator, we understand that the structure presented in Figure 2.10 has the same operator.

It means, looking at this figure, that the kernel G for the short line becomes

$$G = \begin{bmatrix} 0 & -\frac{1}{Cp} \\ -\frac{1}{Cp} & 0 \end{bmatrix} \; [\Omega] \tag{2.64}$$

The criterion often used to employ one model rather than another is based on the comparison between the shortest signal wavelength λ_c and the line length x. If $\lambda_c > 10x$, the short line model is used. If we consider typical numerical data represented by signals made of pulses during at least a time τ, we have

$$\lambda_c = v\tau \; [\mathrm{m}] \tag{2.65}$$

where v is defined by

$$v = \frac{c}{\sqrt{\varepsilon_r}} \; [\mathrm{m}][\mathrm{s}]^{-1} \tag{2.66}$$

where c is the light speed and ε_r is the relative permittivity.

Another fundamental concept comes from the signal decomposition. The signal $s(p)$ can always be written as

$$s(p) = s_0 + s_\omega(p) \; [\mathrm{V}] \tag{2.67}$$

s_0 is the signal part when $\omega = 0$. It is defined by

$$s_0 = \int_0^{+\infty} dt\, s(t)e^{-\alpha t} \; [\mathrm{V}] \tag{2.68}$$

Consecutive to what was said, there is no propagation applied to s_0. A computation for the "DC" part s_0 of the signal $s(p)$ can be made independently from the computation made for $s_\omega(p)$. Then the results can be added to obtain the complete solution. There are no contradiction with the causality theorem because in practice, the DC part cannot be observed independently from the dynamic part s_ω.

s_0 can also be called the polarization signal of $s(p)$. This polarization signal is fundamental for nonlinear and small signals analysis. Component models can depend on it, and it is often set by the power supplies.

2.2.6 Field coupling modelling

We have already studied interactions between two meshes disconnected, i.e., without any branch going from one mesh to the other. These two meshes constitute two connex networks. In their graphs, it does not exist any nodes not connected with two branches. It means that all the branches can be enclosed in a mesh. This connexity is a fundamental property of the graph that we manipulate in physics.

Now, let us consider a simple mesh of current, k, constituted by an RC-network with one resistance, R, and one capacitor, C. With two close networks consisted of circuits with these two elements, we can subsequently create by the loop characterized, for example, by an inductance, L. Now we excite this mesh with an EM force e.

The electrokinetic energy T associated with this structure is defined by

$$T = \frac{1}{2}Lk^2 \text{ [J]} \tag{2.69}$$

while the potential energy U developed in the branch of current i is defined by

$$U = \frac{1}{2C}q^2 \text{ [J]} \tag{2.70}$$

and the dissipation F by

$$F = \frac{1}{2}R\int_t dtk^2 \text{ [J]} \tag{2.71}$$

We can use these relations to solve Lagrange's equation:

$$\frac{d}{dt}\frac{\partial T}{\partial k} + p\frac{\partial U}{\partial q} + \frac{d}{dt}\frac{\partial F}{\partial k} = e \text{ [V]} \tag{2.72}$$

As for $1/pk = q$ (p being Laplace's operator), by replacement we obtain

$$p(Lk) + \frac{1}{Cp}k + Rk = e \text{ [V]} \tag{2.73}$$

which is nothing else than $e = \zeta k$. It means that the equations established using the TAN are the solutions of Lagrange's equations. In the terms of energy, T and U give the relation between the field of matter k and the stored energy of the interaction fields while F gives the energy dissipated by radiation. L represents the inertia of the circuit. It is the force that tends to the opposite to the creation of currents by an external impulse. U is the potential energy acquires by the circuit through the field. F is the energy radiated by the circuit through heating.

But the major function saying how an external force acts on the circuit is the induced EM force e.

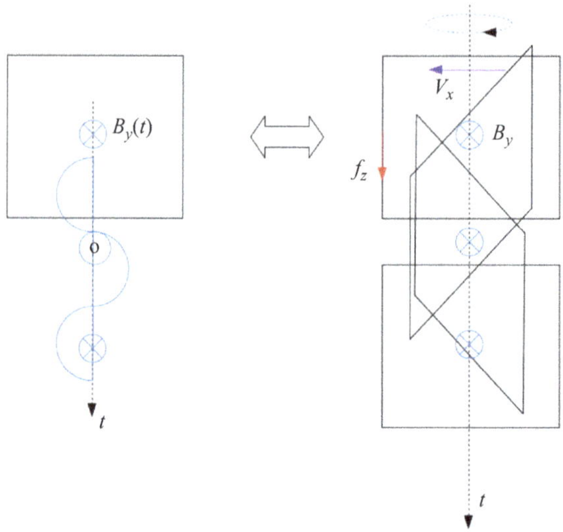

Figure 2.11 Lorentz's law

If we look on a loop illuminated by a variable magnetic field, we understand looking at Figure 2.11 that it is similar to a loop moving in a constant magnetic field. The force f created by a moving load in fields is given by Lorentz's law:

$$\mathbf{f} = q\mathbf{E} + q\mathbf{v} \times \mathbf{B} \text{ [N]} \tag{2.74}$$

The integration of the force f_z over a distance z gives an energy—a work, W. For the electric part we can find immediately the corresponding power P by

$$W = \int_z dzqE_z = qu_z \Rightarrow P = \frac{d}{dt}W = -u_z i \text{ [W]} \tag{2.75}$$

For the magnetic part we make

$$W = \int_z dzf_z = \int_z dzqv_xB_y \Rightarrow P = -i\frac{d}{dt}XZB_y \text{ [W]} \tag{2.76}$$

If we use these relations in our simple mesh for e, it means that $i = k$. But to calculate the dissipated power from the relation (2.73), we have to multiply each member par k, obtaining

$$p\,(Lk)\,k + \frac{1}{Cp}kk + Rkk = ek \text{ [V]} \tag{2.77}$$

We can underline that as $i = k$ then e is given by

$$e = -u_z - p\left(XZB_y\right) \text{ [V]} \tag{2.78}$$

It is induced from external fields illuminating the circuit. If we know the current at the origin of this field, we can write the ratio between the EM force and the current, giving the definition of the electromagnetic interaction between two elements of any circuits, and between a circuit and the external world. This interaction is called a cord and can be any function of the form e/k.

An element of current s having the dimension of a moment at a point O creates a potential vector A (in weber per meter $[\mathrm{Wb\,m}]^{-1}$) on location B as

$$A_\alpha(B) = \mu_{\alpha\beta}\frac{s^\beta(O)}{4\pi R_{OB}}\exp\left(-\frac{R_{OB}}{c}p\right)\ [\mathrm{Wb}][\mathrm{m}]^{-1} \tag{2.79}$$

while the time integration of this moment creates a scalar potential ψ defined by

$$\psi_\sigma(B) = \frac{1}{p}\left(\frac{s^\beta(O)}{4\pi\varepsilon^{\beta\sigma}R_{OB}}\right)\ [\mathrm{V}] \tag{2.80}$$

The scalar ψ does not depend on any delay. This interaction is instantaneous, and we have seen that this is not in contradiction with causality. ε and μ are the tensors of dielectric permittivity and magnetic permeability. The EM force is obtained through

$$u_z = \int_z dz\,[\nabla\psi]_z \quad B_y = [\nabla\times\mathbf{A}]_y\ [\mathrm{V}] \tag{2.81}$$

Four kinds of interactions are often useful: the capacitive interaction, the mutual inductance coupling, common ground coupling and the crosstalk. Let us take a look at these four fundamental interactions for PCB. Each time it concerns the computations of components of G. For the capacitive, mutual inductance or common ground couplings, the wavelength of the interaction field is large compared to the length of the receiver (which can be a line, an electronic component, etc.). If not, the coupling is a crosstalk coupling involving both kinds of fields.

2.2.6.1 Capacitive coupling

Having a geometrical element existing on a PCB, we can distribute arbitrary loads on it in order to compute the scalar potential. We must remember fundamental Gauss's law:

$$\int_s ds_\alpha\varepsilon^{\alpha\beta}E_\beta = \int_v dv\rho_v\ [\mathrm{C}] \tag{2.82}$$

It means that knowing the total load enclosed in a volume v and knowing the medium permittivity $\varepsilon^{\alpha\beta}$, the scalar product of the electric field by the normal on all points to the surface s enclosing the loads is equal to the net number of loads.

On the other hand we know that

$$\psi = -\int_x \mathbf{E}\cdot d\mathbf{x}\ [\mathrm{V}] \tag{2.83}$$

By definition, the capacitive coupling C is defined by

$$C = \frac{q}{\psi}\ [\mathrm{F}] \tag{2.84}$$

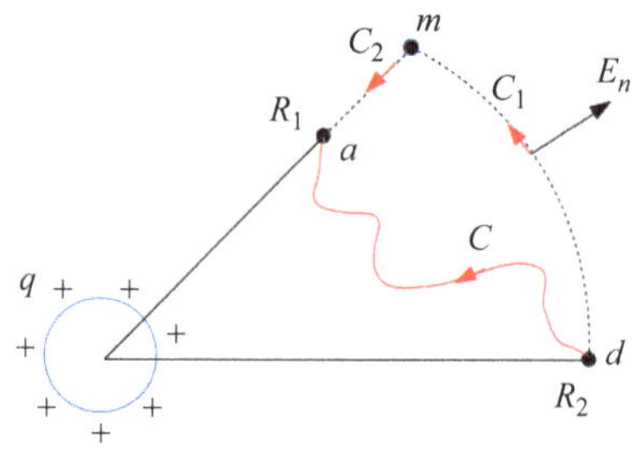

Figure 2.12 Gauss's law

For example [2], if we consider a cylindrical wire of radius r, length 1 m and homogeneous medium ε_0, under the assumption that the field is homogeneous also, based on Gauss's theorem, it can be demonstrated that if the wire wears a load q then

$$2\pi\varepsilon_0 r E_n = q \text{ [C]} \tag{2.85}$$

where E_n is the normal component of the electric field in reference to the surface of the wire. Now if we want to compute the voltage between two points near this wire, how may we proceed? Consider Figure 2.12 where we represent a curvilinear trajectory between two points d and a. We want to compute relation (2.83) for this circulation of the electric field.

We know that the work W for a conservative force does not depend on the trajectory. So the result must be the same using trajectories C_1 and C_2 rather than C. But on C_1, the field E_n is everywhere perpendicular to the trajectory vector dC_1. So

$$-\int_{C_1} d\mathbf{C}_1 \cdot \mathbf{E}_n = 0 \text{ [V]} \tag{2.86}$$

Finally,

$$\psi = -\int_{C2} d\mathbf{C}_2 \cdot \mathbf{E}_n = -\int_{R_2}^{R_1} dR \frac{q}{2\pi\varepsilon_0 R} = \frac{q}{2\pi\varepsilon_0} \ln\left(\frac{R_2}{R_1}\right) \text{ [V]} \tag{2.87}$$

Golden rule: to determine the direction of the electric field, you just have to consider a positive load then looking in what direction it goes. As $\mathbf{f} = q\mathbf{E}$, the force for this positive load is in the same direction as the field.

If the induced voltage is ψ_2 coming from load q^1 it creates the coupling capacitance:

$$C^{12} = \frac{q^1}{\psi_2} = \left[\frac{1}{2\pi\varepsilon_0} \ln\left(\frac{R_2}{R_1}\right)\right]^{-1} \text{ [F]} \tag{2.88}$$

If we have an electrostatic problem with three bodies, two wires over a reference wire as shown Figure 2.13, the capacitance matrix associated with this structure is

$$C^{ij} = \begin{bmatrix} C^{11} & C^{12} \\ C^{21} & C^{22} \end{bmatrix} \text{ [F]} \tag{2.89}$$

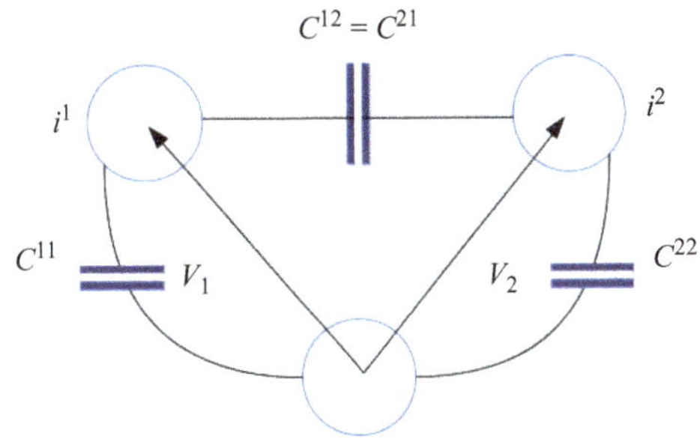

Figure 2.13 Three-body interaction

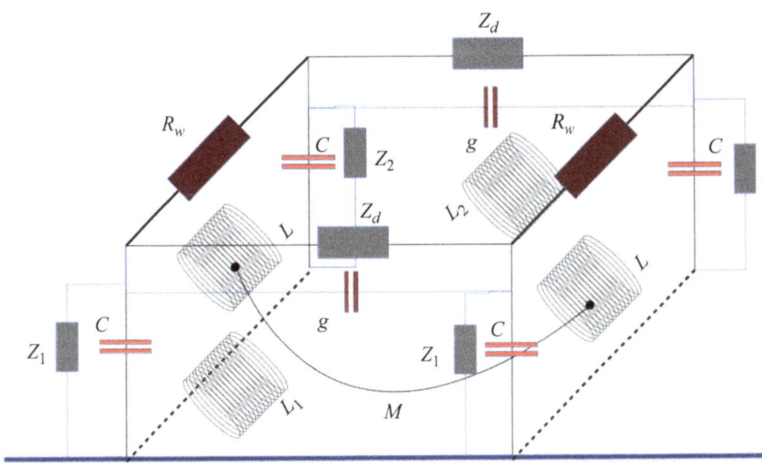

Figure 2.14 Model of two lines cross talked

leading to the system of equations:

$$\begin{cases} i^1 = pC^{11}\psi_1 + pC^{12}\psi_2 \\ i^2 = pC^{21}\psi_1 + pC^{22}\psi_2 \end{cases} \tag{2.90}$$

But while the electrostatic equation $q = C\psi$ gives us the way to compute the capacitance, it does not give the interaction functions. These interactions are defined in the mesh space and remain to be determined. We take the example of two identical short lines cross talked as presented Figure 2.14.

The structure is made of four meshes. Two meshes associated with the short lines, including the impedances z_1, z_2, R_w, L, C, one mesh associated with the differential loads g, Z_d, R_w, M and two meshes made with the limit conditions of the lines

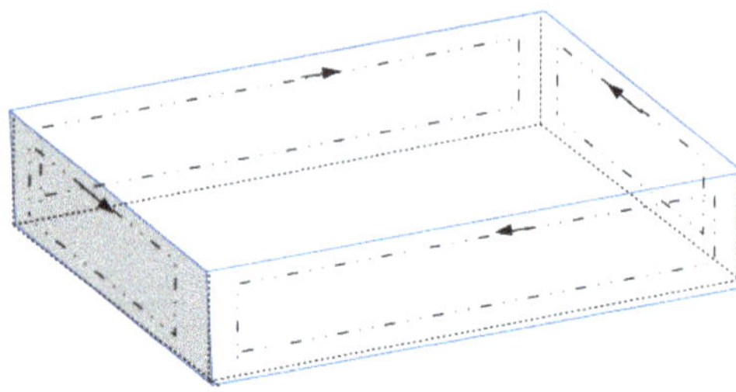

Figure 2.15 Two lines cross talked: mesh currents

z_1, z_2, Z_d, g, c (we include the ground impedance in R_w). At the beginning, the two lines are separated and defined by the operator:

$$\zeta = \begin{bmatrix} \frac{z_1}{z_1Cp+1} + R_w + \frac{z_2}{z_2Cp+1} + Lp & 0 \\ 0 & \frac{z_1}{z_1Cp+1} + R_w + \frac{z_2}{z_2Cp+1} + Lp \end{bmatrix} \; [\Omega] \tag{2.91}$$

In this operator, the interaction kernel G is equal to zero, which is coherent with the fact that the two lines are supposed not coupled.

Now we locate these two lines in the same harness, using them for a differential link. It means to add the two meshes in limit conditions to construct this new structure. We have to choose directions for the mesh currents, as shown in Figure 2.15.

For this structure, we obtain the impedance operator:

$$\zeta = \begin{bmatrix} \frac{z_1}{z_1Cp+1} + R_W + \frac{z_2}{z_2Cp+1} + Lp & \frac{z_1}{z_1Cp+1} & \frac{z_2}{z_2Cp+1} & -Mp \\ \frac{z_1}{z_1Cp+1} & \frac{2z_1}{z_1Cp+1} + \frac{Zd}{Zdgp+1} + L_1p & 0 & \frac{z_1}{z_1Cp+1} \\ \frac{z_2}{z_2Cp+1} & 0 & \frac{2z_2}{z_2Cp+1} + \frac{Zd}{Zdgp+1} + L_2p & \frac{z_2}{z_2Cp+1} \\ -Mp & \frac{z_1}{z_1Cp+1} & \frac{z_2}{z_2Cp+1} & \frac{z_1}{z_1Cp+1} + R_W + \frac{z_2}{z_2Cp+1} + Lp \end{bmatrix} [\Omega] \tag{2.92}$$

We clearly see through the organization of ζ that the capacitive interaction previously studied is not finally the interaction field. In fact, the capacitive coupling is associated with a branch and not a cord. This effect is caused by from the fact that it is intrinsically linked with the longitudinal electric field. Another way to understand this is observing that the central structure, including g and M, can be considered as another line—a differential line—and so is not a specific cord. The components of ζ demonstrate that the cords come from the common mode impedances shared between the limit conditions and the cross-talked lines.

But our operator ζ is not complete. It lacks the common ground couplings.

2.2.6.2 Common ground couplings

Looking at Figure 2.14, we understand that both lines use the same ground. Due to this sharing, a common branch is also shared between their two meshes. If R_G is the resistance of this ground, it must be added to the interaction existing between the two meshes, in components ζ_{14} and ζ_{41}. Finally,

$$\zeta = \begin{bmatrix} \frac{z_1}{z_1Cp+1} + R_w + \frac{z_2}{z_2Cp+1} + Lp & \frac{z_1}{z_1Cp+1} & \frac{z_2}{z_2Cp+1} & -Mp - R_G \\ \frac{z_1}{z_1Cp+1} & \frac{2z_1}{z_1Cp+1} + \frac{Zd}{Zdgp+1} + L_1p & 0 & \frac{z_1}{z_1Cp+1} \\ \frac{z_2}{z_2Cp+1} & 0 & \frac{2z_2}{z_2Cp+1} + \frac{Zd}{Zdgp+1} + L_2p & \frac{z_2}{z_2Cp+1} \\ -Mp - R_G & \frac{z_1}{z_1Cp+1} & \frac{z_2}{z_2Cp+1} & \frac{z_1}{z_1Cp+1} + R_w + \frac{z_2}{z_2Cp+1} + Lp \end{bmatrix} [\Omega] \quad (2.93)$$

Often, people are encouraged to add an inductance to this ground impedance. But inductance does not belong to a branch. To define an inductance, we need to be able to compute a magnetic flux ϕ and know the current J making this flux; we can compute the inductance $L = \phi/J$. We may wonder at how is this flux with an open graph computed. In fact the expression of inductance for a single branch comes from the asymptotic behaviour of a two-branch system where we make far one of the two branches mathematically. For the two lines, inductances and capacitances are often computed knowing their characteristic impedance z_c and propagation speed v (in the next equation, μ_0 is the magnetic permeability).

$$z_c = \frac{\mu_0 v}{2\pi} \ln\left(\frac{d}{r}\right) \ [\Omega] \tag{2.94}$$

After what

$$L = \frac{z_c}{v} \ [\mathrm{H}][\mathrm{m}]^{-1} \ C = \frac{1}{v z_c} \ [\mathrm{F}][\mathrm{m}]^{-1} \tag{2.95}$$

If we consider the case of a two wires line of radii r and distance between centres d, its characteristic impedance is given by

$$z_c = \frac{\mu_0 v}{2\pi} \ln\left(\frac{d}{r}\right) \ [\Omega] \tag{2.96}$$

and the inductance is

$$L = \frac{\mu_0}{2\pi} \ln\left(\frac{d}{r}\right) \ [\mathrm{H}]\,[\mathrm{m}]^{-1} \tag{2.97}$$

We may think that as a consequence, the inductance of a single wire can be estimated by $L' = L/2$, saying that making two wires in series creates the line. But fundamentally, the inductance is the electromagnetic inertia. It translates the tendency of the charged particle to create forces that resist the force which makes them move. It comes from Lenz's law. But if we have a single branch under the assumption of short line, we cannot have current on this single branch. And so the inductance cannot exist. In the case of a wavelength very short compared to the wire length l, the wire becomes an antenna. We can compute its resistance R and capacitance C. Its inductance is determined knowing the first mode of resonance $\lambda = 2l = c/f_r$, and $L = (4\pi^2 f_r^2 C)^{-1}$.

The ground impedance R_G is not only a resistance, and it includes an imaginary part. This imaginary part can be seen somewhere as a self-inductance of the branch. When the magnetic field propagates in a metal region, the attenuation is different than that in free space. The exact computation depends on each geometry. But we can estimate its influence quite simply with a first-order approximation. Before that, let us see an example [3] consisted in skin effect in a cylindrical conductor.

We have a cylindrical conductor of radius x. It wears a current density σ. The current intensity flowing through a portion of depth dx in the cylinder is given by

$$dI_x = 2\pi\sigma x\, dx \ [\mathrm{A}] \tag{2.98}$$

The magnetic field created by this current at a distance r from the axe ($r > x$) is

$$dH = \frac{dI_x}{2\pi r} = \frac{\sigma x\, dx}{r}\ [\mathrm{A}]\,[\mathrm{m}]^{-1} \tag{2.99}$$

And so, the total magnetic field created by all the currents for which $x < r$ is given by

$$H = \frac{1}{r}\int_0^r dx\,\sigma x\ [\mathrm{A}]\,[\mathrm{m}]^{-1} \tag{2.100}$$

By derivating, we obtain

$$\frac{dH}{dr} + \frac{1}{r}H - \sigma = 0\ [\mathrm{A}]\,[\mathrm{m}]^{-2} \tag{2.101}$$

The EM force on a length l of the conductor for a resistivity ρ is

$$e = \rho\sigma l\ [\mathrm{V}] \tag{2.102}$$

At a distance dr far from the previous one, we obtain

$$e + de = (\rho\sigma + \rho\, d\sigma)\,l\ [\mathrm{V}] \tag{2.103}$$

The closed-loop circuit between these two locations in r and $r + dr$ gives a total EM force equal to de. This EM force must be equal to the EM force coming from the magnetic field variation. So

$$de = \rho l\, d\sigma = \mu\frac{dH}{dt} l\, dr\ [\mathrm{V}] \tag{2.104}$$

i.e.:

$$\frac{dH}{dt} = \frac{\rho}{\mu}\frac{d\sigma}{dr}\ [\mathrm{A}]\,[\mathrm{m}]^{-1}\,[\mathrm{s}]^{-1} \tag{2.105}$$

By derivating (2.101) depending on t, we obtain

$$\frac{\partial^2 H}{\partial r\partial t} + \frac{1}{r}\frac{\partial H}{\partial t} - \frac{\partial\sigma}{\partial t} = 0\ [\mathrm{A}]^2\,[\mathrm{m}]^{-3}\,[\mathrm{s}]^{-1} \tag{2.106}$$

or

$$\frac{\rho}{\mu}\frac{\partial^2\sigma}{\partial r^2} + \frac{\rho}{\mu r}\frac{\partial\sigma}{\partial r} - \frac{\partial\sigma}{\partial t} - 0\ [\mathrm{A}]\,[\mathrm{m}]^{-2}\,[\mathrm{s}]^{-1} \tag{2.107}$$

Using Laplace's transform, this becomes

$$\frac{\rho}{\mu}\frac{\partial^2\sigma}{\partial r^2} + \frac{\rho}{\mu r}\frac{\partial\sigma}{\partial r} - p\sigma = 0\ [\mathrm{A}]\,[\mathrm{m}]^{-2}\,[\mathrm{rd}]\,[\mathrm{s}]^{-1} \tag{2.108}$$

This differential equation can be solved using Bessel's functions $\mathbf{J}_0$. Noting $k^2 = \mu\omega/\rho$, the solutions are

$$\sigma = A\mathbf{J}_0(krj^{3/2})\ [\mathrm{A}]\,[\mathrm{m}]^{-2} \tag{2.109}$$

This result can be used in numerical implementation but remains difficult to memorize and use. Let us consider now a rectangular conductor (like our ground) of height h, depth w and length l. Its DC resistance is

$$R_{DC} = \rho\frac{l}{wh}\ [\Omega] \tag{2.110}$$

while the magnetic flux inside the conductor leads to an inductance L_G approximated by

$$L_G = \mu\frac{hl}{w}\ [\mathrm{H}] \tag{2.111}$$

As we know that the frequency is increasing, the current flows in a limited part of the conductor must be given by the skin depth (we have seen in the cylinder case how the current density decreases inside the conductor). The skin depth δ is defined by

$$\delta = (\pi f\mu\sigma)^{-1}\ [\mathrm{m}] \tag{2.112}$$

where σ is the conductor conductivity, f is the frequency and μ, as before, the magnetic permeability. With the increase in frequency, the resistance becomes

$$R_{HF} = \frac{l}{\sigma 2\pi w\delta}\ [\Omega] \tag{2.113}$$

and the inductance becomes

$$L_G = \mu\frac{\delta l}{w}\ [\mathrm{H}] \tag{2.114}$$

As we have seen previously, the resistance model is easy to use. But the inductance part is not so easy to use, because when $f \to 0$ it is not defined. In fact the inductance influence can be quite well simulated taking

$$L_G \approx \mu\frac{l}{w}he^{-\sqrt{\pi f\mu\sigma}h}\ [\mathrm{H}] \tag{2.115}$$

In general, this self-inductance is neglected because it is very small compared to the free-energy inductance associated with the line over the ground plane (in our case). But it explains the light increase of inductance at low frequencies. For the resistance, there are no difficulties and we obtain, for the ground plane,

$$R_G = R_{DC} + R_{HF}\ [\Omega] \tag{2.116}$$

Finally, if z_G is the share ground plane impedance, we have

$$z_G = R_G + L_G p\ [\Omega] \tag{2.117}$$

That is, this impedance must be taken into account as interaction impedance in ζ.

$$\zeta = \begin{bmatrix} \frac{z_1}{z_1Cp+1} + R_w + \frac{z_2}{z_2Cp+1} + Lp & \frac{z_1}{z_1Cp+1} & \frac{z_2}{z_2Cp+1} & -Mp - z_G \\ \frac{z_1}{z_1Cp+1} & \frac{2z_1}{z_1Cp+1} + \frac{Zd}{Zdgp+1} + L_1p & 0 & \frac{z_1}{z_1Cp+1} \\ \frac{z_2}{z_2Cp+1} & 0 & \frac{2z_2}{z_2Cp+1} + \frac{Zd}{Zdgp+1} + L_2p & \frac{z_2}{z_2Cp+1} \\ -Mp - z_G & \frac{z_1}{z_1Cp+1} & \frac{z_2}{z_2Cp+1} & \frac{z_1}{z_1Cp+1} + R_w + \frac{z_2}{z_2Cp+1} + Lp \end{bmatrix} [\Omega] \quad (2.118)$$

2.2.6.3 Mutual inductance coupling

The mutual inductance between two separate loops is perfectly defined by Neumann's formula:

$$M = \frac{\mu_0}{4\pi} \oint_{C1} \oint_{C2} \frac{d\mathbf{C}_1 \cdot d\mathbf{C}_2}{R_{12}} \text{ [H]} \tag{2.119}$$

where C_1 and C_2 are circulations along each loop as presented in Figure 2.16.

A problem may appear when the two loops share a branch. In this case, Neumann's formula diverges and cannot be used. In that case, the problem is quite difficult. We can estimate the inductance of each loop separately. The inductance can be computed using tables for specific geometries, measurements or computing the ratio ϕ/i where ϕ is the magnetic flux inside the loop and i, the courant on the loop. In general, best formula linking the current and the magnetic field to compute the flux is Biot and Savard's one:

$$B(i) = \mu_0 \frac{i\,dx}{4\pi R^2} \sin\theta \text{ [T]} \tag{2.120}$$

where $i\,dx$ is the moment of current source, R the distance to the point where the field is computed and θ the angle between the moment of current and the vector joining the moment and the observation point. For example, for a circular loop of radius R, we have (we suppose the field homogeneous on the total surface of flux):

$$B(i) = \mu_0 \int_{\theta=0}^{2\pi} d\theta \frac{iR}{4\pi R^2} = \mu_0 \frac{i}{2R} \text{ [T]} \tag{2.121}$$

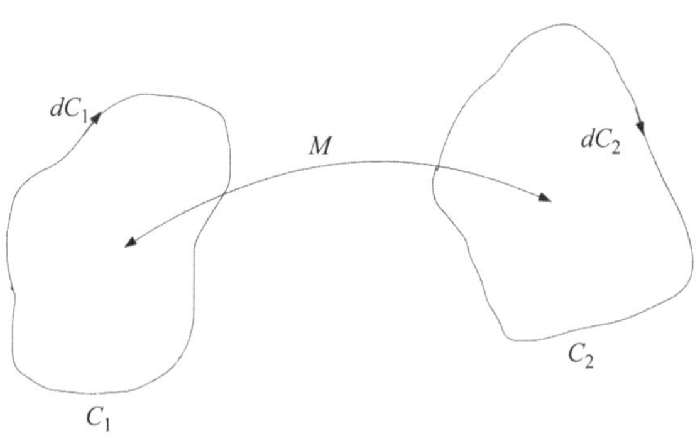

Figure 2.16 Mutual inductance

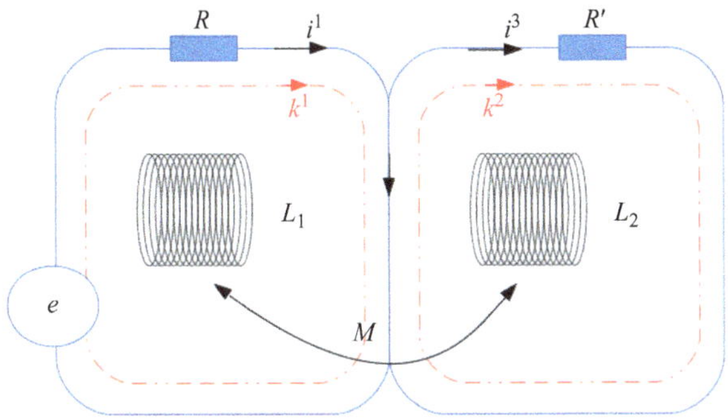

Figure 2.17 Mutual inductance with shared branch

and

$$\phi = \frac{\pi R^2 B(i)}{i} = \mu\pi R\frac{i}{2} \text{ [W]} \Rightarrow L = \frac{\phi}{i} = \mu\pi\frac{R}{2} \text{ [H]} \tag{2.122}$$

Now how can we proceed for two loops sharing one branch?

Figure 2.17 shows two connected loops having each their self-inductance and coupled through a mutual inductance M.

This simple circuit can be solved classically using the derivative current theorem. We know that without M,

$$i^3 = i^1 \frac{R + L_1 p}{R + R' + L_1 p + L_2 p} \text{ [A]} \tag{2.123}$$

The currents induced by the source e radiate a potential vector to the three receiving branches where i^3 runs. So the double integration of Neumann's relation must be reduced to the closed circulation C_1 for one loop and an open circulation C_2 for the receiving loop, excluding the shared branch. If we consider squared loops, as represented in Figure 2.18, we have three geometrical situations:

- parts of the circulations C_1 and C_2 are in parallel;
- parts of the circulations C_1 and C_2 are in the opposite directions;
- parts of the circulations C_1 and C_2 follow the same direction.

This involves three kinds of scalar products in Neumann's integral. For example, if we compute the mutual inductance between the two squared loops in Figure 2.18, we have to integrate the ways $c_{11}c_{22} - c_{11}c_{23} + c_{13}c_{23} - c_{13}c_{21} + c_{12}c_{22} - c_{14}c_{22}$. Each time the distance d between each point of each segment must be computed.

As we will see in various cases in this book, the computation of mutual inductances is not an easy task. But the engineer searches for the order of values. There is a

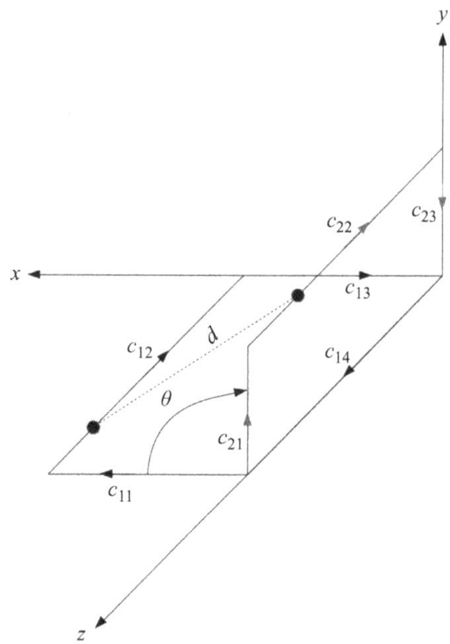

Figure 2.18 Mutual inductance with shared branch: integration

second golden rule: the mutual inductance cannot be higher than the self-inductance. The relation

$$M = k\sqrt{L_1 L_2} \text{ [H]} \tag{2.124}$$

remains true in any cases, often with a coefficient k very low, except with magnetic materials. Typically the ratio M/L is near to the coefficient k, it means around $1/10$, $2/10, \ldots$

2.2.6.4 Crosstalk coupling

Crosstalk coupling means that the dimensions of the receiver are wide compared to the wavelength. When the emitter is larger than the wavelength, a first job consists in computing the field emitted by this object on one point of a short receiver. Knowing the vector potential formula, this job is supposed to be known. So the fact to have an emitter larger than the wavelength is not a particular case if the receivers remain short compared to the wavelength. We retrieve the previous situations but written in dependence with the fields. In any case, we can use the two next relations giving the EM force coming from the magnetic field and the MM (magnetomotive) force coming from the electric field.

EM force induced by the magnetic field
When an incident magnetic field illuminates a loop of section S, self-inductance L and impedance Z, it creates an EM force e defined by

$$e = -p\mu(\mathbf{S} \cdot \mathbf{H}) \text{ [V]} \tag{2.125}$$

The equivalent schematic of the coupled circuit leads to the equation linking the EM force e and Foucault's current induced in the loop k:

$$e - Lpk = Zk \text{ [V]} \tag{2.126}$$

MM force coming from the electric field
When an electric field illuminates a dielectric of a capacitance in a circuit made of one capacitance C, one inductance L and one impedance Z, it creates an MM force f defined by

$$f = -p\varepsilon(\mathbf{S} \cdot \mathbf{E}) \text{ [A]} \tag{2.127}$$

The equivalent schematic of the coupled circuit leads to the equation linking the MM force f and the current induced in the circuit k:

$$\frac{1}{Cp}f - Lpk = \left(\frac{1}{Cp} + z\right)k \text{ [V]} \tag{2.128}$$

These two fundamental relations are always used in particular for short receivers (receivers constituted by small dimensions compare to the wavelength). The field can come from anywhere and any number of sources. As in these models, the field is supposed to be homogeneous over the surface **S**, we do not care of its origin.

And now what about large receivers? In that case, the result always comes from an integration of the field over directions of space. There are some particular cases like crosstalk between lines, antennas, where objects have a particular geometry, but in the general case, the approach is to integrate the field with its phase changing all along the direction of integration (by principle, as the object is long compared to the wavelength). Looking at (2.125) and (2.127), it means to write $S = h\,dx$, where h is a constant height, for example, and dx, the direction of integration.

Let us take an example. We consider two matched lines of length X in parallel following an axis x. The distance between the two lines is d. The height of the two lines over the ground plane is h. The second line wears a current i^2 coming from a step function excitation defined by

$$i^2(p) = i_0\frac{1}{p} \text{ [A]} \tag{2.129}$$

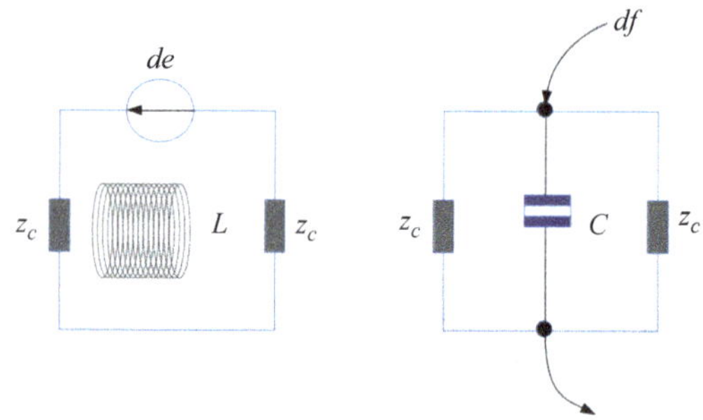

Figure 2.19 Crosstalk circuits

This current is associated with a magnetic field. We estimate the perpendicular component to the first line section of this magnetic field by

$$H_T = \frac{i^2(p)}{\pi} \frac{h/2}{d^2 + (h/2)^2} \text{ [A] [m]}^{-1} \tag{2.130}$$

The EM force de induced on a small part of the first line is given by

$$de \approx -p\mu h\, dx \frac{i_0}{p\pi} \left(\frac{h/2}{d^2 + (h/2)^2} \right) e^{-(x/v)p} \text{ [V]} \tag{2.131}$$

Looking at Figure 2.19, we can calculate the voltage reported on the input of the first line e_1 (i.e., "near-end" crosstalk. v is the propagation speed on the two lines):

$$e_1(0,p) = -\left(\frac{z_c}{2z_c + Lp\, dx} \right) \int_0^X dxp\mu h \frac{i_0}{p\pi} \left(\frac{h/2}{d^2 + (h/2)^2} \right) e^{-(x/v)p} e^{-(x/v)p} \text{ [V]} \tag{2.132}$$

We obtain with $Lp\, dx \rightarrow 0$:

$$e_1(p) = -\frac{\mu h i_0}{4\pi} \left[\frac{h/2}{d^2 + (h/2)^2} \right] \frac{v}{p} \left(1 - e^{-2(X/v)p} \right) \text{ [V]} \tag{2.133}$$

Noting,

$$\beta = \left[\frac{h/2}{d^2 + (h/2)^2} \right]$$

This gives us a first crosstalk impedance operator for the near-end magnetic coupling $\mathscr{H}_{12}^{NE} = 2e_1(p)/i^2(p)$:

$$\mathscr{H}_{12}^{NE} = -\frac{\mu h v}{2\pi} \beta \left(1 - e^{-2(X/v)p} \right) \text{ [}\Omega\text{]} \tag{2.134}$$

For the far-end crosstalk, we make

$$s_1(p) = \frac{\mu h i_0}{\pi} \beta \int_0^X dx e^{-\frac{x}{v}p} e^{-((X-x)/v)p} \text{ [V]} \tag{2.135}$$

or

$$s_1(p) = \frac{\mu h i_0}{\pi} \beta X e^{-(X/v)p} \text{ [V]} \tag{2.136}$$

which leads this time to the far-end magnetic coupling impedance operator $2s_1(p)/i^2(p)$:

$$\mathscr{H}_{12}^{FE} = 2p\frac{\mu h}{\pi} \beta X e^{-(X/v)p} \text{ [}\Omega\text{]} \tag{2.137}$$

Looking at the electric field coupling, we have to compute the work of the electric field coming from the charges on the second line, on the height of the first line. Figure 2.20 shows the operation realized.

On average, the potential u_L developed under the line can be approximated by

$$u_L(x) = \beta\frac{q}{2\pi\varepsilon} = -\beta\frac{i^2(p)}{2\pi\varepsilon p} e^{-(x/v)p} \text{ [V]} \tag{2.138}$$

The induced current di is obtained by

$$di = \frac{1}{2}\frac{Cdxpu_L(x)}{1 + z_c/2Cdxp} \approx Cdxpu_L(x) \text{ [A]} \tag{2.139}$$

which is nothing but the elementary MM force as $df = -p\varepsilon SE = -pCu_L$.

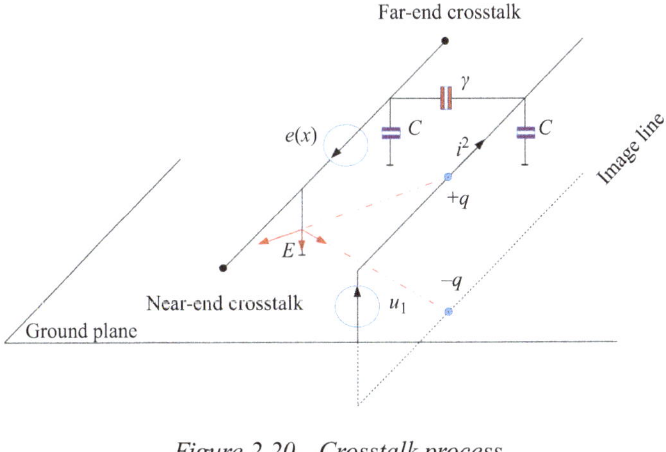

Figure 2.20 Crosstalk process

By integrating on the left, we obtain the near-end MM crosstalk force:

$$f_1 = \int_0^X dfe^{-(x/v)p} = \frac{\beta C}{8\pi\varepsilon}\frac{i_0 v}{p^2}\left(1 - e^{-2(X/v)p}\right) \text{ [A]} \tag{2.140}$$

Integrating on the right gives the far-end MM crosstalk force:

$$g_1 = \int_0^X dfe^{-((X-x)/v)p} = -\frac{\beta C}{4\pi\varepsilon}\left(\frac{i_0}{p}\right)Xe^{-(X/v)p} \text{ [A]} \tag{2.141}$$

These relations give the transfer functions for the electrical coupling:

$$\mathcal{Q}_{12}^{NE} = \frac{\beta C}{8\pi\varepsilon}\frac{v}{p}\left(1 - e^{-2(X/v)p}\right) \quad \mathcal{Q}_{12}^{FE} = -\frac{\beta C}{4\pi\varepsilon}Xe^{-(X/v)p} \tag{2.142}$$

The last operation to do is to incorporate these new sources in Branin's model of the first line (the process will be completely symmetric and the same coupling exists from the line 1 to 2). Vabre [4] has synthesized the coupling relations between two neighbouring lines with the structure defined in Figure 2.21.

Vabre's equations that use the $RLC\gamma M$ lumped elements are

$$\begin{cases} e_2(p) = \alpha\left(\frac{K+1}{4}\right)\left[u_i(p) - u_i(p)e^{-2\tau p}\right] \\ s_2(p) = -\alpha\left(\frac{K-1}{2}\right)\tau p u_i(p)e^{-\tau p} \end{cases} \tag{2.143}$$

The two lines are matched with a characteristic impedance R_c. $\tau = X/v$. α and K are coefficients defined by

$$\alpha = \frac{\gamma}{C+\gamma} \quad K = \frac{M}{\alpha L} \tag{2.144}$$

We have to incorporate these results in Branin's model. It means that we must add sources in Branin's model. To reach the level e_2, for example, an equivalent schematic of a generator $2e_2$ with an internal impedance R_c. u_i is the incident wave on the source line. We have $u_i = R_c k^i$, where k^i is the incident current. Finally, the cord operators are defined by

Figure 2.21 Coupled lines structure

$$G = \begin{bmatrix} 0 & 0 & \alpha\left(\frac{K+1}{2}\right) R_c\left[1 - e^{-2\tau p}\right] & -\alpha\,(K-1)\,\tau p R_c e^{-\tau p} \\ 0 & 0 & -\alpha\,(K-1)\,\tau p R_c e^{-\tau p} & \alpha\left(\frac{K+1}{2}\right) R_c\left[1 - e^{-2\tau p}\right] \\ \alpha\left(\frac{K+1}{2}\right) R_c\left[1 - e^{-2\tau p}\right] & -\alpha\,(K-1)\,\tau p R_c e^{-\tau p} & 0 & 0 \\ -\alpha\,(K-1)\,\tau p R_c e^{-\tau p} & \alpha\left(\frac{K+1}{2}\right) R_c\left[1 - e^{-2\tau p}\right] & 0 & 0 \end{bmatrix} [\Omega] \tag{2.145}$$

The two 2×2 matrices in diagonal are the two Branin's model of the two lines previously studied. The complete impedance tensor is

$$\zeta = \begin{bmatrix} R_0 + R_c & -\left(R_c - R_L\right) e^{-\tau p} & \alpha\left(\frac{K+1}{2}\right) R_c\left[1 - e^{-2\tau p}\right] & -\alpha\,(K-1)\,\tau p R_c e^{-\tau p} \\ -\left(R_c - R_0\right) e^{-\tau p} & R_L + R_c & -\alpha\,(K-1)\,\tau p R_c e^{-\tau p} & \alpha\left(\frac{K+1}{2}\right) R_c\left[1 - e^{-2\tau p}\right] \\ \alpha\left(\frac{K+1}{2}\right) R_c\left[1 - e^{-2\tau p}\right] & -\alpha\,(K-1)\,\tau p R_c e^{-\tau p} & R_0 + R_c & -\left(R_c - R_L\right) e^{-\tau p} \\ -\alpha\,(K-1)\,\tau p R_c e^{-\tau p} & \alpha\left(\frac{K+1}{2}\right) R_c\left[1 - e^{-2\tau p}\right] & -\left(R_c - R_0\right) e^{-\tau p} & R_L + R_c \end{bmatrix} [\Omega] \tag{2.146}$$

This operator solves completely the problem of two cross-talked lines. Note that changing the parameter, the two lines can have different parameters, not being identical like in the example.

We have seen through the example of two lines the general method to compute the interaction between any parts of two electronic systems. We can know investigate the electronic components themselves then the interfaces between the electronic components and the transmission media. Having knowledge of this material, we are ready to investigate on the cases of SI, PI and EMC problems for PCBs, the case that gives interest to us: SI, PI and EMC for PCB.

2.2.7 Components modelling

The object of this paragraph is not to study all the electronic component models. The objective in SI, PI or EMC is not exactly the same as for any hardware analysis. We want to reproduce correctly the signal exchanged in the electronics. We do not care of how they was manufactured. It is important to accept this affirmation because the computation time can be drastically reduced using macromodels for the electronics rather than electronically exact models.

We are interested in three particular models:

- model for nonlinear behaviours in general;
- ICEM model;
- IBIS model.

2.2.7.1 Model for nonlinear behaviour in general

One technique is useful to reach this objective of macromodelling by exploiting the approach of "subdomain."

We can define a domain as an operator $\overset{q}{\mathscr{D}}_{[a,b]}$. It is defined by

$$\overset{q}{\mathscr{D}}_{[a,b]} = 1 \ if \ q \in [a+d, b-d] \tag{2.147}$$

where d is the distance and $\overset{q}{\mathscr{D}}_{[a,b]}$ is the infinitely derivable function. Three kinds of function are often used: Gaussian functions, sigmoïde functions and arctangent functions. Here we principally use the sigmoïde function analytically defined by

$$\overset{q}{\mathscr{D}}_{[a,b]} = \left[1 + ce^{-a(q-q_1)}\right]^{-1} - \left[1 + ce^{-a(q-q_2)}\right]^{-1} \tag{2.148}$$

- The coefficient a defines the value of $\left[1 + ce^{-a(q-q_1)}\right]^{-1}$ at the boundary when $q = q_1$ or $q = q_2$.
- The coefficient c defines how the operator $\left[1 + ce^{-a(q-q_1)}\right]^{-1}$ goes from 0 to 1 or 1 to 0; it is similar to rise or fall time.
- The shorter the rise time, the shorter the distance d. But d cannot be reduced to zero. In this case, the function $\left[1 + ce^{-a(q-q_1)}\right]^{-1}$ may be a pulse with infinite rise time, not derivable.

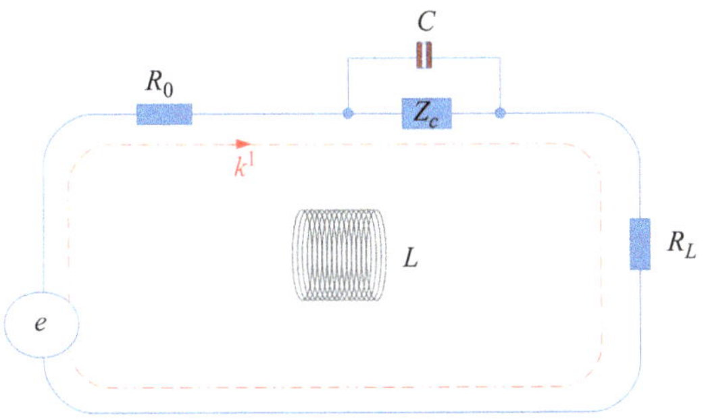

Figure 2.22 Commutator circuit

The operator $\left[1 + ce^{-a(q-q_1)}\right]^{-1}$ is able to define intermediate states where the component belongs to two states at the same time. It is a physically possible situation when, for example, a component leaves a state partially to join another.

As an example, we can apply this macromodel to a commutator. The commutator command is a voltage u. If $u \in [0, 1]$ [V], the commutator behaves like a resistance of 1 Ω. If $u \in [-1, 0]$ [V], the commutator behaves like a resistance of 100 kΩ. The best way to model this kind of commutator is to fix its high value and to use a domain function to reduce this value to 1 Ω. Making this way, the commutator cannot take an unknown value. So in this case,

$$\overset{u}{\mathscr{D}}_{[0,1]} = \left[1 + ce^{-a(u-0,2)}\right]^{-1} - \left[1 + ce^{-a(u-1,5)}\right]^{-1} \tag{2.149}$$

for a we can take 10 and for c: 0,5. The impedance operator of the commutator z_c is so defined by

$$z_c = 10^5 - \overset{u}{\mathscr{D}}_{[0,1]}\left(10^5 - 1\right) \ [\Omega] \tag{2.150}$$

For illustrating this mechanism, we take the example of a commutator used to control some high-frequency signal. The circuit considered is shown in Figure 2.22.

The capacitor is represented by the parasitic elements of the commutator (1 nF). The inductance is the magnetic coupling over the circuit (10 nH). The resistance is matched with the high-frequency generator (50 Ω) that works between 1 MHz and 1 GHz.

We want to compute the circuit transfer function depending on the commutator state. The circuit has three states: one "on" ($[ON\rangle$), one "off" ($[OFF\rangle$) and a transition state. In a double loop, we compute for each level of the commutator command between 0.1 and 0.9 V the circuit transfer function. We obtain the set of curves presented in Figure 2.23. We see that before the frequency of ≈20 MHz, the transfer function depends directly of the commutator command. While due to its parasitic capacitance and the circuit inductance, the transfer function does not depend on all of the commutator up to this frequency. The curves were traced thanks to a python program using the domain functions previously described. The program listing is given in Annexe 2.B.

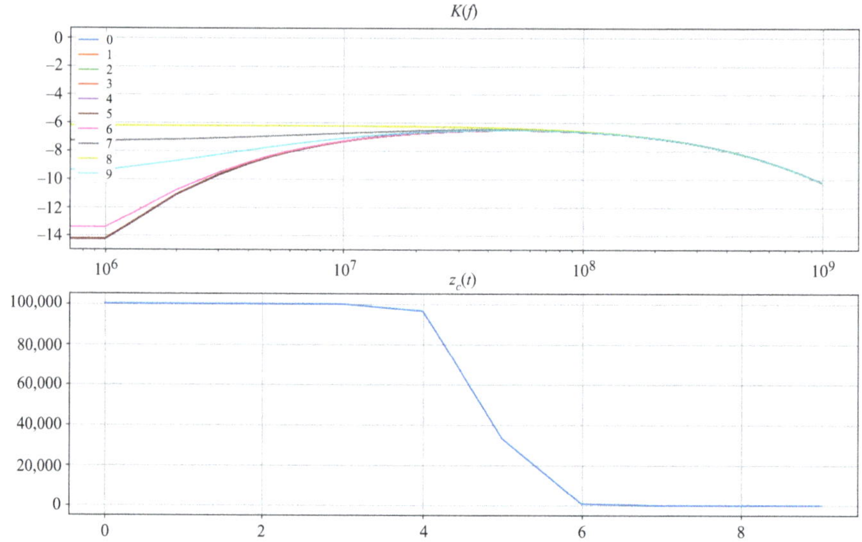

Figure 2.23 Commutator circuit responses

The curve downside shows the commutator impedance depending on time, while the curve upside shows the transfer function for each time step.

We must remember that Laplace's transform as Carson's one can be applied under the assumption that no element in the time domain depends on time or changes during the time. So in our approach we consider that all components remain constant during a time step, set by some parameters, including polarization signals (DC power supply, DC command of transistors, etc.). Under this condition, the circuit can be studied in Laplace's expressions for various values of polarization or other parameters playing the same role.

2.2.7.2 ICEM model

An important model for EMC is the ICEM model. ICEM means *integrated circuit (IC) emission model*. It was elaborated by the UTE WG47 group in France in order to be able to model noises coming from large IC-like microprocessors. This model is used to predict the conducted emissions of an IC. Many publications on this model are already available (see, for example, [5]).

ICEM basics consist in a noise source, an internal impedance and an interface circuit between its kernel and the PCB. The noise source called *IA* (for internal activity) is a source of current linked with the internal IC activity. This means that the model may depend on the IC activity. That is why in the standard, the model development is supposed to make the IC working in typical activities, in order to obtain a pertinent noise source $IA(f)$. That is a real difficulty today with ICs. In fact the EMC of these IC's cannot be decorrelated from their software. In the conception phase, the engineers are obliged to use generic models like ICEM, where it can be some significant differences

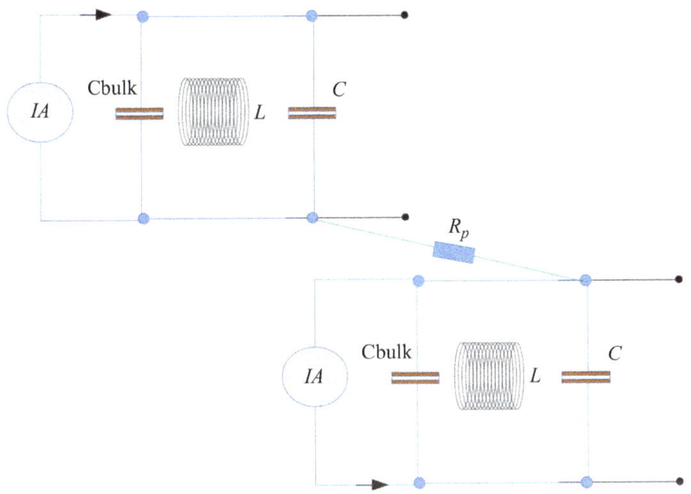

Figure 2.24 Typical ICEM model

with the measurements depending on the real activity of the IC in its equipment. Anyway, we have no other solutions today and in practice, using ICEM prediction gives sufficiently accurate results to continue trusting in this approach.

The internal impedance is associated with all the transistor grid capacitors used in the IC. It is reduced to this capacitor named "bulk capacitor *C*."

The last element of the ICEM model is the package equivalent circuit or "PDN" (for *passive description network*). It is an "RLC" network representing the interface between the chip (IA and bulk capacitor) and the external world, i.e., the PCB. Some other components can be involved to take into account the bridge between various parts of the IC. In particular, each power supply access has its own ICEM model. Between these models, a resistance R_p can translate exchanges of currents between ports. Finally, an example of ICEM model for two ports of an IC is shown in Figure 2.24.

2.2.7.3 IBIS model

Historically, IC-based SI analysis can be performed via IBIS model. As ICs is becoming more and more large with a great number of access and fast signals, it was necessary to have models in order to conduct the SI conception of electronic boards, including these ICs.

IBIS means input/output *(I/O) buffer information specification*. Version 7 is available since March 15, 2019 on **IBIS.org**. IBIS is probably the most used model for SI and I/O specifications. It was created by INTEL®. The I/O specifications are described in an IBIS file with extension .ibs, following rules detailed in the IBIS standard. It is an ASCII file containing the information on the IC interfaces described in a textual form. Keywords are defined between brackets and followed by measurements. Often, details are given in a preamble, on the conditions used to test the component. Today, specialized firms work on IBIS model developments. The success of this standard makes sometimes difficult the control of its model quality. Some careful attention must be paid, depending on their origins.

Well-known manufacturers give the IBIS model of their devices on their website. But if you read carefully the information, they often not guarantee these models. The major reason is that often they were not constructed by themselves, but by foreign companies in India, for example. Anyway, even if they ask for some modifications sometimes, they remain a base to start some SI study using ICs.

Between many keywords, some of them have a particular interest for us. The first one is the keyword [package]. Under this keyword we have a description of what was called the *PDN* for ICEM. Figure 2.25 shows a part from an IBIS file giving a package description.

This is a first-order description, available for SI, in the frequency working band of the component, but it can be too weak for an outoff band behaviour as in EMC. That is why, often, ICEM description is of second order when IBIS one is of first order.

IBIS gives also $V(I)$ curves for the different I/O pins of the component. They are tables giving the correspondences V, I, sometimes for typical, minimum and maximum values. The keyword [pin] lists the pins concerned by the V/I curves given after. Figure 2.26 gives another extract from an IBIS file showing this kind of description.

The input impedance of the corresponding pin is also given for typical, minimum and maximum values (C_{comp}).

```
[Package]
| variable        typ                   min                   max
R_pkg     2.0000m             1.0000m              4.0000m
L_pkg     0.2000nH            0.1000nH             0.4000nH
C_pkg     2.0000pF            1.0000pF             4.0000pF
|
```

Figure 2.25 Typical IBIS package description

```
[Model]  out
Model_type I/O_open_drain
Polarity        Inverting
Vinl =   0.7500V
Vinh =   3.0000V
C_comp    6.0000pF          6.0000pF            6.0000pF
|
|
[Temperature Range]  25.0000          85.0000            -40.0000
[Voltage Range]       3.0000V          2.5000V             5.5000V
[Pulldown]
|Voltage   I(typ)              I(min)              I(max)
|
  -1.50    -68.3300mA        -43.8900mA          -0.1344A
  -1.40    -62.6100mA        -40.2600mA          -0.1212A
  -1.30    -57.8300mA        -37.0200mA          -0.1125A
  -1.20    -53.1400mA        -33.8500mA          -0.1041A
```

Figure 2.26 Typical IBIS V/I function

The equivalent circuit that can be made starting from an IBIS file is very similar to the one used in ICEM. It includes a package, which can be considered as an independent part of the device (as being the link between the chip and the PCB). An equivalent impedance of the studied pin, and the I–V characteristic of the signal under given measurement conditions. So in that case, the package model is associated with an RLC mesh with the series resistance, the capacitor in parallel and the inductance belonging to the loop. But IBIS standard specifies that matrices can be defined. It can be band, sparse or full matrices and describe the self-impedance of each I/O or the coupled impedances between pins. The TAN model coming from these descriptions can be elaborated in two steps. In a first step, one mesh is associated with each pin, made of the self-impedances described in the IBIS file. Then each time an inductance L_{ij} exists between two pins, a cord of values $L_{ij}p$ is added between the two corresponding meshes. Then in a second step we make as many meshes as there are coupling capacitors between pins. These meshes have for impedance

$$\frac{1}{C_i p} + \frac{1}{C_j p} + \frac{1}{\gamma_{ij} p}$$

where C_i is the capacitor of a first pin, C_j is the capacitor of a second pin and γ is the coupling capacitor between these two pins. Between these added meshes and the first ones, a cord of value $1/(C_i p)$ exists between one added mesh and the mesh, including C_i. Another cord exists with value $1/(C_i p)$ between the added mesh and the mesh, including C_j. So two cords are associated with each added mesh. The whole network is the equivalent schematic in the mesh space of the ICs, constructed with the information coming from the IBIS file of the ICs.

IBIS model, ICEM model, etc. are examples of a wide number of available models for components. They have all their advantages and weaknesses, but they are all easy to implement under the TAN formalism. And we can have advantages depending on the domain technique for macromodelling nonlinear behaviour, playing eventually on both temporal and frequency domains.

```
import numpy as np
import pylab as plt

c=0.5
a=40.
N=10
M=1000
fo=1E6
Ro=50.
RL=50.
L=10E-9
C=1E-9
#
k=np.zeros(M,dtype=float)
zc=np.zeros(N,dtype=float)
ax=np.zeros(N,dtype=float)
axf=np.zeros(M,dtype=float)
u=0.
```

```
for t in range(N):
    ax[t]=t
    #if t>4:
    u=(t+1.)/10.
    D=1./(1.+c*np.exp(-a*(u-0.6)))-1./(1.+c*np.exp(-a*(u-1.2)))
    zc[t]=1E5-D*(1E5-1.)
    for f in range(1,M):
        p=np.pi*2.*f*fo
        axf[f]=f*fo
        Zcom=zc[t]/(zc[t]*C*p+1.)
        zeta=Ro+RL+Zcom+L*p
        e=5.*(1./p-1./p*np.exp(-10E-9*p))
        J=e/zeta
        k[f]=20.*np.log10(abs(RL*J/e))
    plt.subplot(2, 1, 1)
    plt.plot(axf, k,label=t)

plt.title(u'k(f)')
plt.xscale('log')
plt.grid(True)
plt.legend()

plt.subplot(2,1,2)
plt.plot(ax,zc)
plt.title(u'zc(t)')
plt.grid(True)

plt.show()
```

Annexe 2.A

```
import numpy as np
import pylab as plt

Rc = 50.
RL = 10.
Ro = 10.
N = 3000
dt = 1E-9


def tg(t, tautt):
    tt = t
    taut = tautt
    if tt > taut:
        return int(tt - taut)
    if tt <= taut:
        return 0


a = 0.5
```

```
c = 3E8
ax = np.zeros(N, dtype=float)
e = np.zeros(N, dtype=float)
k = np.zeros((N, 2), dtype=float)
meme = np.zeros(N, dtype=float)

for t in range(N):
    s = t * dt
    ax[t] = s * 1E9
    e[t] = 1. / (1. + a * np.exp(-c * (s - 100E-9)))
    - 1. / (1. + a * np.exp(-c * (s - 300E-9)))
    meme[t] = e[tg(t, 500)]

for t in range(N):
    zeta = [[Rc + Ro, 0], [0, RL + Rc]]
    T = [[e[t] + (Rc - RL) * k[tg(t, 500), 1]], [e[tg(t, 500)]
     + (Rc - Ro) * k[tg(t, 500), 0]]]
    k[t, :] = np.transpose(np.dot(np.linalg.inv(zeta), T))

plt.subplot(2, 1, 1)
plt.plot(ax, k[:, 1])
plt.title(u'Output current')
plt.grid(True)

plt.subplot(2, 1, 2)
plt.plot(ax, k[:, 0])
plt.title(u'Input current')
plt.grid(True)
plt.show()
```

Annexe 2.B

```
import numpy as np
import pylab as plt

c=0.5
a=40.
N=10
M=1000
fo=1E6
Ro=50.
RL=50.
L=10E-9
C=1E-9
#
k=np.zeros(M,dtype=float)
zc=np.zeros(N,dtype=float)
ax=np.zeros(N,dtype=float)
axf=np.zeros(M,dtype=float)
```

```
u=0.
for t in range(N):
    ax[t]=t
    #if t>4:
    u=(t+1.)/10.
    D=1./(1.+c*np.exp(-a*(u-0.6)))-1./(1.+c*np.exp(-a*(u-1.2)))
    zc[t]=1E5-D*(1E5-1.)
    for f in range(1,M):
        p=np.pi*2.*f*fo
        axf[f]=f*fo
        Zcom=zc[t]/(zc[t]*C*p+1.)
        zeta=Ro+RL+Zcom+L*p
        e=5.*(1./p-1./p*np.exp(-10E-9*p))
        J=e/zeta
        k[f]=20.*np.log10(abs(RL*J/e))
    plt.subplot(2, 1, 1)
    plt.plot(axf, k,label=t)

plt.title(u'k(f)')
plt.xscale('log')
plt.grid(True)
plt.legend()

plt.subplot(2,1,2)
plt.plot(ax,zc)
plt.title(u'zc(t)')
plt.grid(True)

plt.show()
```

References

[1] Maurice, O. *Theoretical application of the tensorial analysis of network for EMC at the system level*. Bookelis, 2017, Aix-en-Provence, France.

[2] Paul, C. R., *Electromagnetic compatibility*. Wiley, second edition, 2006, Hoboken, New Jersey, USA.

[3] Angot, A., *Compléments de mathématiques*. Masson, 1982, Paris, France.

[4] Vabre, J. P. (1975). Monographie sur les lignes couplées. Annales Des Télécommunications, 30(11–12), 421–453.

[5] Levant, J. L., Ramdani, M., and Perdriau, R. (2004). ICEM modelling of microcontroller current activity. Microelectronics Journal, 35(6), 501–507.

Chapter 3

PCB primitive components analysis with TAN

Zhifei Xu[1], Blaise Ravelo[1], Yang Liu[1], and Olivier Maurice[2]

Abstract

This chapter introduces the basic elements necessary for the printed circuit board analysis with tensorial analysis of networks approach. The elements are first presented based on the classical electrical schemes. Then, the equivalent graphs are presented. The tensor models are established based on the physical law governing each element.

Keywords: Kron's method, TAN, tensor objects, lumped elements, distributed elements

3.1 TAN operators for electrical application

An electrical object to analyse can be a tensor, a vector, a matrix, or simply a table of data according to its formatting properties. Based on the tensorial analysis of networks (TAN) approach, there are two types of objects differentiated by the physical nature: the force objects and the flux. Figure 3.1 can be assumed as an illustrative example of these TAN operators.

The interactions can be characterized by the currents in Figure 3.1(a) or by the voltage in Figure 3.1(b). In the first case, the network is described in the spaces of branch currents and mesh currents. In the second case, it is characterized in the spaces of branch voltages and pairs of nodes voltages. The two identical graphs in Figure 3.1 illustrate five dimensions in branch spaces. The space of the network presents a dimension $R = 1$, the node space will be in dimension $N = 3$, and the branch current space has dimension $B = 5$. The size of the mesh current space is $M = B + R - N = 3$ and for the space of pair of node $P = N - R = 2$.

3.1.1 Covariant parameters: voltage tensors

Between each node or branch, there is a source type object. These source objects can be the sets of time-dependent values and related to potential energies. Let u be the

[1]IRSEEM/ESIGELEC, Rouen, France
[2]ArianeGroup, Paris, France

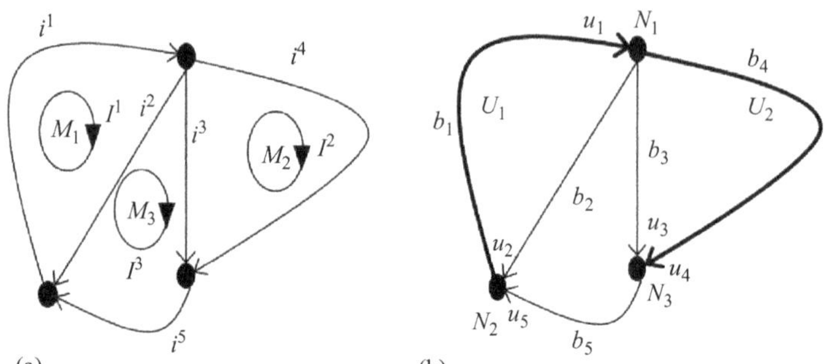

Figure 3.1 Topological expression of different spaces: (a) mesh space and (b) branch and nodal space

co-vector representing these values. This co-vector is represented with a base b^k in the branch space B. We denote as u_k, the co-vector source object with the subscripts situated at the bottom in the space of the branches. The components of the co-vector are also called covariate, which is detailed in Appendix A.

$$u = \sum_{k=1}^{B} u_k b^k \tag{3.1}$$

This relation can be transformed with the Einstein expression:

$$u = u_k b^k \tag{3.2}$$

In the context of electrical network study, the source objects in the branch space are the electromotive forces (EMF) presented at the terminals of each branch b^k of the graph. The voltages u_2 in Figure 3.1 represent the voltage potential difference between the two nodes at the ends of the branch b_2.

3.1.2 Contravariant parameters: current tensors

In every branch of the network, the flux is defined as a contravariant component with a superscript, we take current as, for example i^k, attached to a contravariant in branch space denoted as b_k.

$$i = \sum_{k=1}^{B} i^k b_k \tag{3.3}$$

with the Einstein expression:

$$i = i^k b_k \tag{3.4}$$

We assume this flux as the current flowing in a branch of the graph. In Figure 3.1(a), five currents in the branch space and three currents in the mesh space are represented. For example, the flux traversing the branch b_1 is identified as current i^1. In the mesh space, the current noted I^3 represents the flowing current of the mesh

3. The branch and mesh spaces are represented by lowercase letters for the first case and capital letters for the second case.

There are five branches in the topology of Figure 3.1; in this branch space, we can define the current vector representing the five branch currents as below. The index of each branch is located on the top because the branch currents are contravariant.

$$i^k = \begin{bmatrix} i^1 \\ i^2 \\ i^3 \\ i^4 \\ i^5 \end{bmatrix} \tag{3.5}$$

Similarly, in the same branch space, we define the vector of voltage potential differences u with five components representing the voltages of the five corresponding branches. It is noteworthy that, in the considered branch space, the components of u_c are denoted in subscript because the voltage vector represents a covariant 1-D variable.

$$u_c = \begin{bmatrix} u_1 & u_2 & u_3 & u_4 & u_5 \end{bmatrix} \tag{3.6}$$

We introduce a force object, also called source. This vector is composed of five voltage sources e_a, present on each branch.

$$e_a = \begin{bmatrix} e_1 & e_2 & e_3 & e_4 & e_5 \end{bmatrix} \tag{3.7}$$

Finally, the covariant and contravariant components attached to a branch or a pair of nodes, generating a power instantaneous at each moment. In any network electrical, this power must be an invariant, that is to say, whatever the basis of reference the vector has defined, the power will remain unchanged.

3.1.3 Twice covariant parameters: impedance tensors

These source type objects and fluxes are connected by a fundamental metric operator, transforming a flux into a source type object. It is called impedance operator. Its inverse is the admittance operator. Equation below presents the relation between flux and source in the branch space. e_k is the voltage source present on the k branch and is^p is the current source flowing on p branch.

$$u_k = e_k - z_{kp} i^p \tag{3.8}$$

$$i^p = is^p - y^{pk} u_k \tag{3.9}$$

The impedance operator z_{kp} is, as shown in the position of the indices, a twice covariant tensor represented by a square matrix related to the number of branches considered in the network of Figure 3.1 ($B = 5$). In the same way, the admittance operator y^{kp} contains components twice the contravariant and will have the same size as the impedance operator. The goal here is simply to introduce the objects with the tensor or matrix notation that we will be used in the remaining parts of this manuscript.

$$z_{kp} = \begin{bmatrix} z_{11} & z_{12} & z_{13} & z_{14} & z_{15} \\ z_{21} & z_{22} & z_{23} & z_{24} & z_{25} \\ z_{31} & z_{32} & z_{33} & z_{34} & z_{35} \\ z_{41} & z_{42} & z_{43} & z_{44} & z_{45} \\ z_{51} & z_{52} & z_{53} & z_{54} & z_{55} \end{bmatrix} \tag{3.10}$$

$$y^{kp} = \begin{bmatrix} y^{11} & y^{12} & y^{13} & y^{14} & y^{15} \\ y^{21} & y^{22} & y^{23} & y^{24} & y^{25} \\ y^{31} & y^{32} & y^{33} & y^{34} & y^{35} \\ y^{41} & y^{42} & y^{43} & y^{44} & y^{45} \\ y^{51} & y^{52} & y^{53} & y^{54} & y^{55} \end{bmatrix} \tag{3.11}$$

The elements presented on the diagonal of the matrices represent the impedance or admittance of each branch. Extra-diagonal terms represent the mutual couplings between the branches of the network. For example, z_{24} is the mutual impedance between branches b_4 and b_2. Now we can express the four previous quantities U_k, E_k, I^p, and Z_{kp} in the space of the mesh currents with capital letters notations in Figure 3.1. Higher order tensors can be established through the operators detailed in Appendix A.

3.1.4 Electrical problem metric elaboration

In mesh space, we apply the law of Kirchhoff: the sum of the voltage difference force in a mesh is zero. Finally, $U_k = 0$ reduces the number of matrix size to handle. In addition to this important property, the number of branches is always greater than the number of meshes. For these two reasons, solving the reduced equations in the mesh space will allow to establish fast and efficiently the solutions of the system metric. With the example of Figure 3.1, we obtain the vector in the equation below which is contravariant with three dimensions in the mesh space current:

$$I^p = \begin{bmatrix} I^1 \\ I^2 \\ I^3 \end{bmatrix} \tag{3.12}$$

The vector of the branch sources e_k with $(k = 1, 2, \ldots, 5)$ can be transformed into mesh space as the source vector E_k (with $k = 1, 2, 3$) representing the three voltage sources in the meshes as m_1, m_2, and m_3:

$$E_k = \begin{bmatrix} E_1 & E_2 & E_3 \end{bmatrix} \tag{3.13}$$

Finally, the impedance operator Z_{kp}, a tensor with two rank, with three dimensions is presented. The representation with high rank tensors are explored in Appendix A of this manuscript. The tensor of admittance can be formed in the same way.

$$Z_{kp} = \begin{bmatrix} Z_{11} & Z_{12} & Z_{13} \\ Z_{21} & Z_{22} & Z_{23} \\ Z_{31} & Z_{32} & Z_{33} \end{bmatrix} \tag{3.14}$$

These three matrices are finally linked by the following expression in mesh space represented in tensorial form:

$$E_k = Z_{kp} I^p \tag{3.15}$$

where I^p is the vector of current sources.

$$I^p = \begin{bmatrix} I^1 \\ I^2 \end{bmatrix} \tag{3.16}$$

The matrix of admittance is denoted by Y^{pk}:

$$Y^{pk} = \begin{bmatrix} Y^{11} & Y^{12} \\ Y^{21} & Y^{22} \end{bmatrix} \tag{3.17}$$

We have just presented six different topological spaces: the source type objects in branch and mesh space, the space of branch and mesh current. Once these objects and operators described in the general form of matrices and in different spaces, the assignment of these matrices to tensors is necessary. Before recalling some notations of the tensor algebra applied to our study, it is important to set the definitions of notation used for the object with physical meanings.

3.1.5 Branch space to mesh space conversion

To detail this powerful mathematical support for the study of the big systems, we will begin by presenting the connection matrices allowing the transformation between the branch and mesh spaces. The source and impedance tensors of a network will be integrated simply in the space of branches. And then, by using the connection matrix, these two tensors will be directly transformed into the mesh space. The computations of the solution can be performed in the mesh space. We are focusing on branch and mesh space in this chapter. Through these space transformations, various interactions can be put into the best suited space.

All the ingenuity of TAN lies in the way of transforming a topological space to the representation of a physical phenomenon in the electrical networks. The branch space allows you to simply characterize any complex network since it divided your network into branches. On the other hand, we prefer to solve the equations in the space of mesh offering advantageous properties for the calculation cause the mesh space connecting all the branches as a network. The following very simple electrical network in Figure 3.2 is used to interpret the TAN principle whose topological characters are $B = 3$, $N = 2$, $R = 1$ and $M = B + R - N = 2$.

Let us denote e_a, i^b, and z_{ab}, with $a, b = 1, 2, \ldots, B$, the tensors representing the source, current, and impedance in branch space. And E_m, I^n, and Z_{mn}, with $m, n = 1, 2, \ldots, M$, are the source, current, and impedance in mesh space, respectively. There is a connection matrix denoted as C allowing the transformation between branch and mesh spaces for the electrical parameters implied in (3.18). We suppose:

$$i^a = C^a_n I^n \tag{3.18}$$

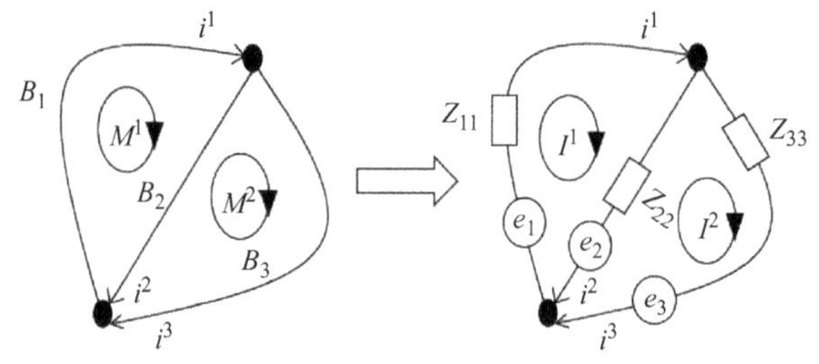

Figure 3.2 Branch and mesh space representation

with the same expression, E_m and Z_{mn} can be presented as

$$E_m = C_m^a e_a \tag{3.19}$$

$$Z_{mn} = C_m^a z_{ab} C_n^b \tag{3.20}$$

where $a = 1, 2, \ldots, B$ and $m, n = 1, 2, \ldots, M$, the expressions can be written in matrix form as

$$i^a = C_n^a I^n \tag{3.21}$$

$$E_m = C_m^a e_a \tag{3.22}$$

$$Z_{mn} = C_m^a z_{ab} C_n^b \tag{3.23}$$

Illustrative example: based on the graph shown in Figure 3.2, we have

$$\begin{aligned} i^1 &= 1I^1 + 0I^2 \\ i^2 &= -I^1 + 1I^2 \\ i^3 &= 0I^1 + 1I^2 \end{aligned} \tag{3.24}$$

Then

$$\begin{bmatrix} i^1 \\ i^2 \\ i^3 \end{bmatrix} = \begin{bmatrix} 1 & 0 \\ -1 & 1 \\ 0 & 1 \end{bmatrix} \begin{bmatrix} I^1 \\ I^2 \end{bmatrix} \tag{3.25}$$

We can identify the connection matrix C with two columns and three rows. Therefore, the source vector in mesh space can be obtained with the equation in (3.19)

$$\begin{bmatrix} E_1 \\ E_2 \end{bmatrix} = \begin{bmatrix} 1 & -1 & 0 \\ 0 & 1 & 1 \end{bmatrix} \begin{bmatrix} e_1 \\ e_2 \\ e_3 \end{bmatrix} \tag{3.26}$$

With the same method, the mesh impedance matrix obtained from branch space can be expressed as

$$Z_{mn} = \begin{bmatrix} 1 & -1 & 0 \\ 0 & 1 & 1 \end{bmatrix} \begin{bmatrix} z_{11} & z_{21} & z_{31} \\ z_{21} & z_{22} & z_{23} \\ z_{31} & z_{32} & z_{33} \end{bmatrix} \begin{bmatrix} 1 & 0 \\ -1 & 1 \\ 0 & 1 \end{bmatrix} = \begin{bmatrix} Z_{11} & Z_{12} \\ Z_{21} & Z_{22} \end{bmatrix} \tag{3.27}$$

where the diagonal coefficients of z_{ab} and Z_{mn} represent the impedance of each considered branch. z_{11} is the proper impedance of branch 1. Z_{11} is the proper impedance of mesh 1. The extra-diagonal coefficients are the coupling impedance between two branches or between two meshes. For example, $z_{23} = 0$ is the coupling impedance between branches 3 and 2. Z_{21} is the coupling impedance between meshes 1 and 2. In the space of the meshes, the law of Kirchhoff makes it possible to write the equation of the network in (3.15). Equation (3.15) finally gives the evolution of the mesh currents in all the network. All the described spaces and their transformations are presented in Figure 3.3.

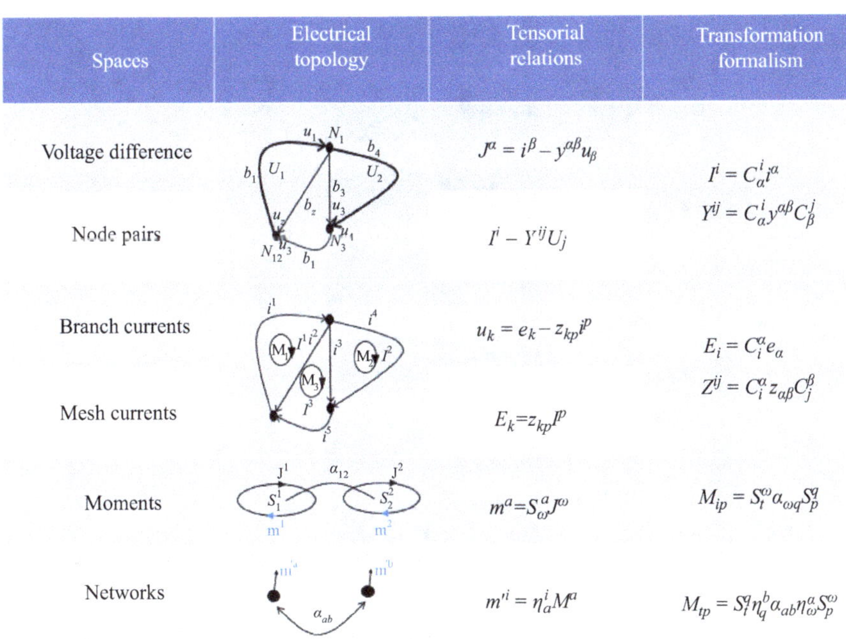

Figure 3.3 Expression of different spaces [1]

3.2 TAN modelling methodology

In brief, the TAN approach is an analysis concept initially dedicated to electrical networks. The constituting components of the system are assumed to be the initial elements. By using the graph topology representation and tensor calculus, the constituting components can be analysed separately and then connected. TAN topology is essentially built with the combination of nodal, branch, mesh, moment, and network spaces [2]. The TAN formalism can be merely implemented in different steps as shown in Figure 3.4. The application of TAN to model analytically the printed circuit board (PCB) interconnect effects can be performed with the following methodology:

1. Step 1: Separate the complex system assumed as the problem to be solved into several parts (segments analysis) with the constituting elements.
2. Step 2: The equivalent electrical model of each composed element must be established.
3. Step 3: This step is focused on the combination of the established models into the equivalent graph topology.
4. Step 4: This step is mainly focused on the analytical investigation of the abstracted graph topology. It should define all the lowest level primitive elements of the equivalent topology such as resistors, capacitors, diodes, and transistors.

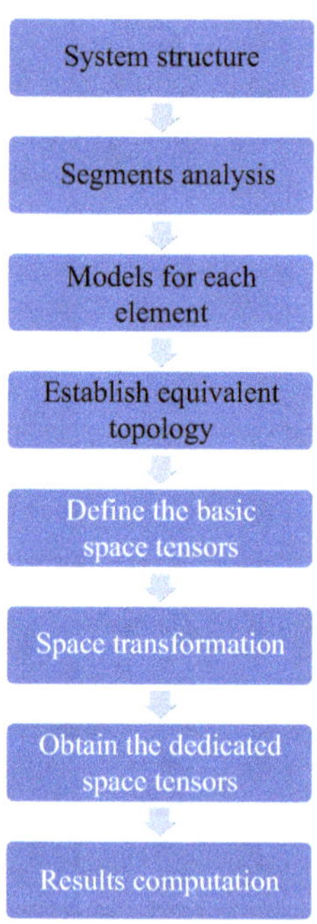

Figure 3.4 Kron's working methodology

5. Step 5: An appropriated mathematical theory, the interconnection equation, can be used to combine all different parts to obtain the mesh space expression of the topology.
6. Step 6: Perform various analysis based on the mesh space expression of the system (Figure 3.5).

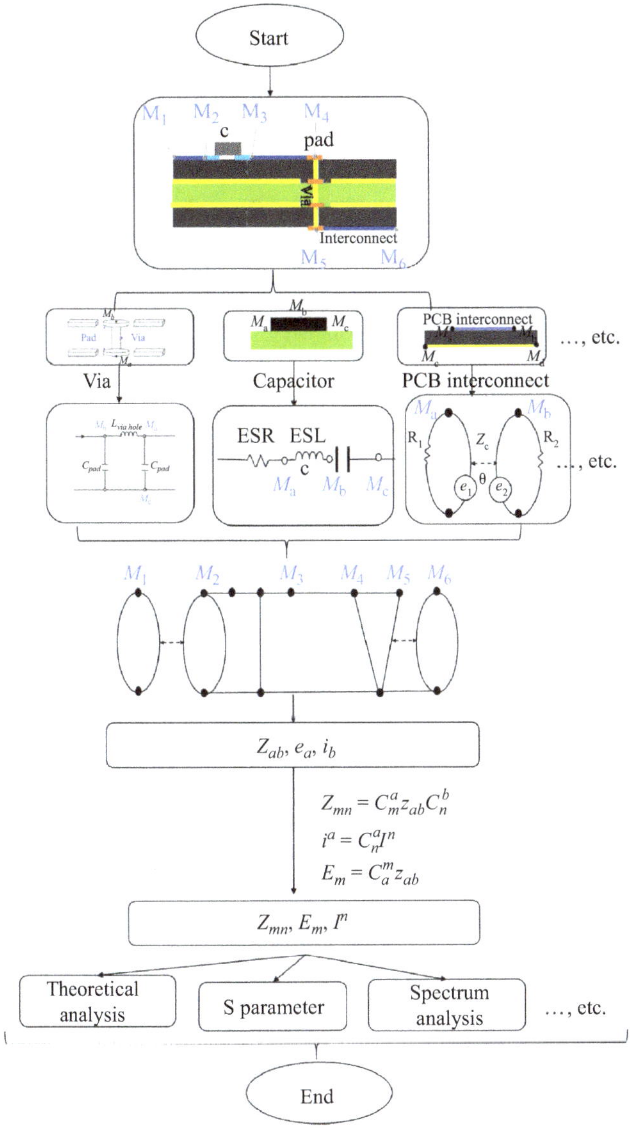

Figure 3.5 Multilayer PCB modelling methodology

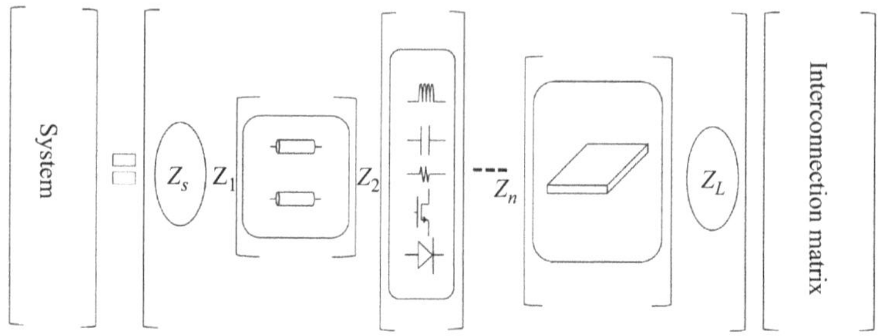

Figure 3.6 Overview of the PCB system diagram

By principle, the PCB electrical structures are constituted by basic lumped (passive or/and active) and distributed interconnections. An example of the representation of these structures can be found in Figure 3.6, including the passive, active, complex structures as power ground planes, and interconnects vias. The overall system can be considered several subsystem blocks consisted of electrical current/voltage sources, interconnects, and loads (lumped linear/nonlinear devices). Based on TAN modelling steps, the multilayer PCB can be modelled in two main steps:

- Step 1 divides the structure into several subnetworks constituted by line interconnects elements, via elements, and so on.
- Step 2 is to combine the subnetworks by defining an interconnection matrix for the sub-elements and then the computational equations are developed with TAN.

3.3 PCB elements modelling

Based on diakoptics analysis, the whole PCB can be a direct summation of various models attached with its parts. Then it remains to add the couplings between these parts. If some circuits appear regularly, it is possible to study theoretically the whole structure by only studying two parts and then benefit of the periodic structure, to know what happened after by reproduction of the previous study. The overall system can be connected by the association of the solution of each elementary block. The system-obtained results can be calculated from the association of elementary solution. The calculation can be mathematically proceeded via the interconnection matrices. Here we presented five different elements in multilayer PCB with their equivalent circuits, topology, and the branch space tensors expression in Figure 3.7; the details of these components can be found in the following sections.

3.3.1 Interconnects

Behind the technological race, the modern PCBs integration density is continuously increasing in addition to the operation frequency. The design methodology for the interconnect is more and more challenging [3]. In addition, undesirable parasitic

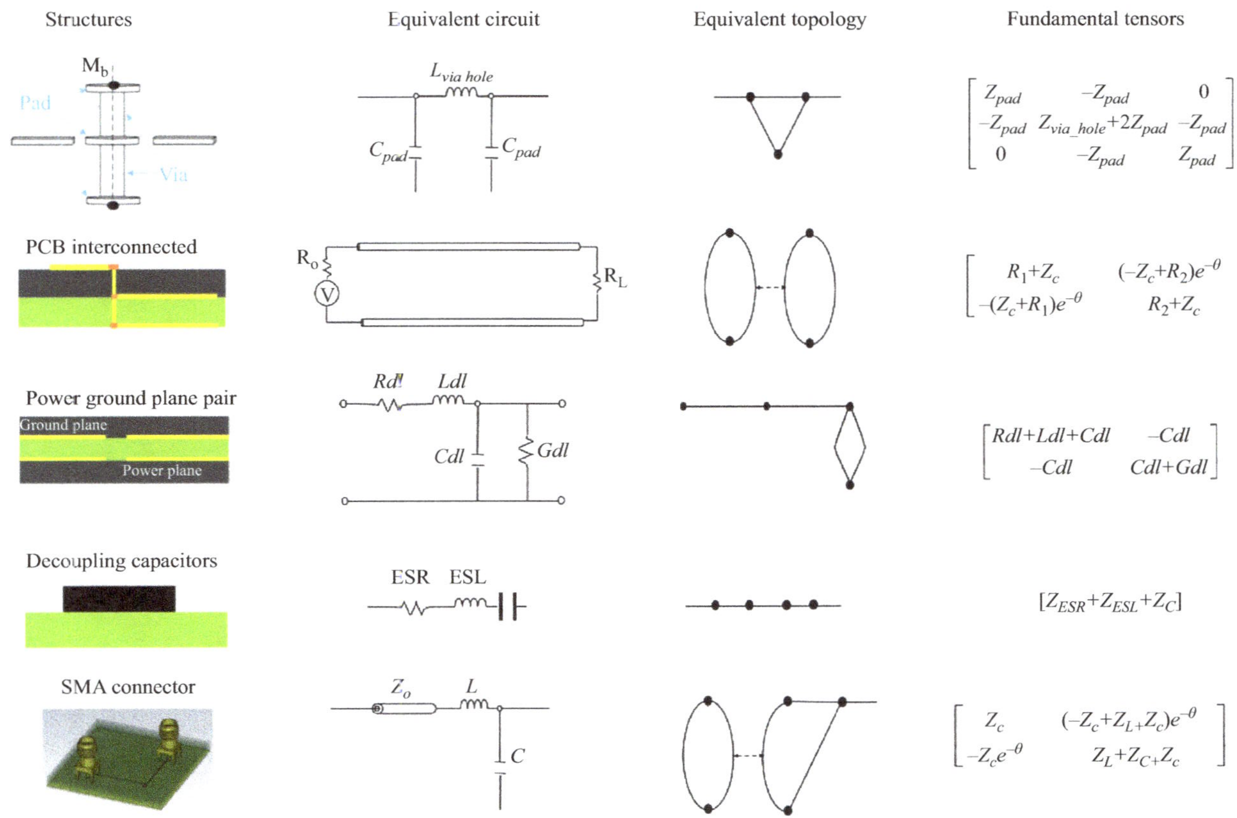

Figure 3.7 Primitive components in a multilayer PCB

and noise effects from the multilayer interconnects can be sources of EMC and SI issues [4,5]. Careful characterization of the traces is required. To characterize the PCB traces, lots of closed-form equations have been developed to calculate the characteristic impedance. In 1977, Bahl and Trivedi proposed a closed-form equation in (3.28) to estimate the characteristic impedance of microstrip lines. Besides this, the Wheeler's equations are also widespread in the world to characterize the microstrips [6]. However, the impedance of the traces, the ε_{eff}, and even the physical parameters change also with the frequencies. The characteristic impedance established by Bahl and Trivedi can be formulated as

$$Z_c = \frac{60}{\sqrt{\varepsilon_{eff}}} \ln\left(\frac{5.98h}{0.8w + t}\right) \tag{3.28}$$

where ε_{eff} is the effective dielectric constant that can be calculated in (3.29), w and t are the width and thickness of microstrip line, respectively, and h is the height of substrate.

$$\varepsilon_{eff} = \frac{\varepsilon_r + 1}{2} + \frac{\varepsilon_r - 1}{2}\left[\left(1 + 12\left(\frac{h}{w}\right)\right)^{-1/2} + 0.04\left(1 - \left(\frac{w}{h}\right)\right)^2\right] \tag{3.29}$$

Besides this, Wheeler's (3.30)–(3.33) are also widespread in the world to characterize the microstrips [6].

$$Z_0 = \frac{\eta_0}{2\pi\sqrt{2E_r + 1}} \ln\left(1 + 4\left(\frac{h}{W_{eff}}(X_1 + X_2)\right)\right) \tag{3.30}$$

$$X_1 = 4\frac{(14E_r + 8)}{11E_r}\left(\frac{h}{W_{eff}}\right) \tag{3.31}$$

$$X_2 = \sqrt{16\left(\frac{h}{W_{eff}}\right)^2\left(\frac{14E_r + 8}{11E_r}\right)^2 + \left(\frac{E_r + 1}{2E_r}\right)\pi^2} \tag{3.32}$$

$$W_{eff} = w + \frac{t}{\pi}\ln\left(\frac{4e}{\sqrt{(t/h)^2 + (t/(w\pi + 1.1t\pi))^2}}\right)\frac{E_r + 1}{2E_r} \tag{3.33}$$

3.3.1.1 Telegrapher's model

The Telegrapher's structure is normally used to analyse the TLs. According to the TL theory, we can consider cascaded elementary RLCG cells as shown next (Figure 3.8).

Resistance

The frequency behaviour of the resistance is known as skin effects, the resistance will stay in constant until the skin effect emergence [7], the skin effect emergence with relation to the frequency is given by

$$f_{skin-effect} = \frac{4\rho}{\pi\mu_0 t^2} \tag{3.34}$$

where $f_{skin-effect}$ is the appearance frequency (Hz) of skin effect, ρ is the resistivity of the trace material, μ_0 is the air magnetic permeability which is equal to $4\pi\,10^{-7}$ H/m,

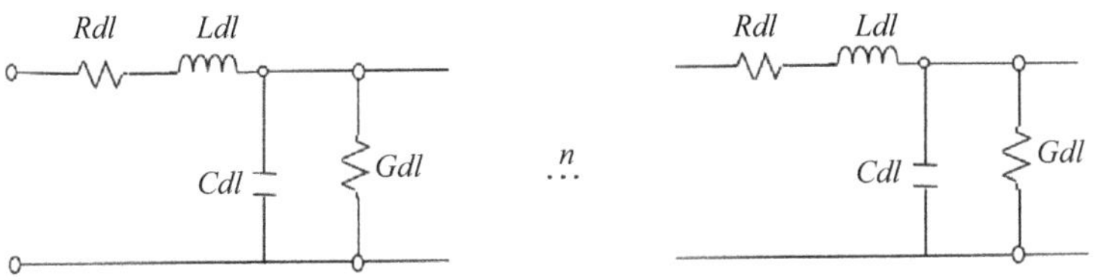

Figure 3.8 RLCG cell

and t is the thickness (m). This equation validates only when $t \ll w$, w is the width of the trace.

The well-known skin depth δ calculation for PCB trace is given as follows:

$$\delta = \frac{1}{\sqrt{\pi \mu \sigma f}} \tag{3.35}$$

The resistance in high frequencies can be written as in equation if we neglect the return resistance by ground plane [7]:

$$R_{RF} = \frac{\rho}{\delta} \tag{3.36}$$

The total resistance can be assumed as the sum of RF resistance and DC resistance. The DC resistance can be given by

$$R_{DC} = \frac{\rho l}{S} \tag{3.37}$$

where l is the length of the conductor and S is the area of the conductor. The typical PCB copper traces has a thickness between 18 and 40 μm; the width is between 0.1 and 0.6 mm. We need to note that the temperature has the influence on the resistance and skin effects of a track.

Inductance

The inductance of PCB trace contains the internal inductance and external inductance. Generally, the internal inductance is neglected. Considering the skin effect of the PCB trace, we can consider that the internal inductance is changing with frequency [7]. The same idea as resistance, the internal inductance will be expressed as the difference of total inductance in high and low frequencies. The following equation is the expression of internal inductance of a wire which is valid only in the case that the conductor is isolated.

$$L_{in}, DC = \frac{\mu_0}{8\pi}\ \mathrm{H/m} \tag{3.38}$$

This is a constant value that does not dependent on the wire diameter; as soon as the wire radius becomes higher than the skin depth, the internal inductance is given by

$$L_{in}, HF = \frac{1}{4\pi r_w}\sqrt{\frac{\mu_0}{\pi \sigma f}}\ \mathrm{H/m} \tag{3.39}$$

with r_w being the conductor radius (m).

Capacitance

One of the most important elements that affects the capacitance is the relative dielectric constant ε_r. We have been always focused on the substrate material FR-4 that has the characteristics shown in [8].

We should note that the FR4 substrate parameter variation will also impact the loss tangent. The FR4 relative permittivity is varying from 4.6 to 4.25 between 1 MHz and 1 GHz. An equation [9] is generally used to reproduce the impact:

$$\varepsilon_r = 4.97 - 0.257\log\left(\frac{f}{10^7}\right) \tag{3.40}$$

Knowing the inductance, the capacitance can be obtained:

$$LC = \mu\varepsilon \tag{3.41}$$

Conductance

The dielectric between the plates of the capacitor allows a leakage current conduction [7]. Generally, a capacitor with a high quality will have a low conduction current. An important parameter to characterize this conductance is called "loss tangent." The conductance is related to the capacitance by

$$G = 2\pi f\tan(\delta) \tag{3.42}$$

This relation is only validated in a homogeneous environment. The way used to characterize the conductance is the introduction in the permittivity of an imaginary part:

$$\varepsilon_r = \varepsilon_r^{'} - j\varepsilon_r^{''} \tag{3.43}$$

The loss tangent is related to the imaginary and real part of the ε:

$$\tan(\delta) = \frac{\varepsilon_r^{''}}{\varepsilon_r^{'}} \tag{3.44}$$

We need to note that the conductance varies also with the temperature as resistance.

3.3.1.2 PCB trace modelling

Now let us consider the resistively ended interconnect structure introduced in Figure 3.9. Based on the microwave circuit theory, this TL can be assumed as n sub-elements in cascade as depicted in Figure 3.9. By principle, for the present interconnect structure, the Kron's method consists of representing the resistively ended TL network composed as the short-circuit ended subnetworks with dotted lines added in the rectangular boxes. The impedance matrix model and the voltage excitation vector associated to this system can be written as

$$Z_{ab} = \begin{bmatrix} R_0 & 0 & 0 & 0 \\ 0 & [Z_1] & 0 & 0 \\ 0 & 0 & [Z_2] & 0 \\ 0 & 0 & 0 & R_L \end{bmatrix} \tag{3.45}$$

The associated subnetworks impedance matrix expressed in (3.45) can be established from Figure 3.9.

$[Z_1]$ and $[Z_2]$ are the matrices represented by the impedance z and admittance y of the TL. In the TL sub-element shown in Figure 3.10, i^1 and i^2 are the mesh fictive currents, z and y are the TL impedance and admittance per unit length for each split cell. Then it yields the following expressions:

$$\begin{cases} zi^1 + (i^1 - i^2)/y = 0 \\ (i^2 - i^1)/y = 0 \end{cases} \Rightarrow e(1)_a = Z(1)_{ab} i(1)^b \tag{3.46}$$

$$Z(k)_{ab} = \begin{bmatrix} z + 1/y & -1/y \\ -1/y & 1/y \end{bmatrix} \tag{3.47}$$

where $k = 1, \ldots, n$.

So far, the subnetworks impedance matrix introduced in (3.45) of Figure 3.9(b) is established by considering $[Z_1]$ and $[Z_2]$. However, it is necessary to connect these sub-elements in order to get the equivalent graph topology. The transformation of Figure 3.9 into Figure 3.11 is a way to proceed with the TAN approach. Figure 3.11 is generated by applying the concept of interconnection of subnetworks. To determine the mathematical abstract equation corresponding to the graph topology drawn in

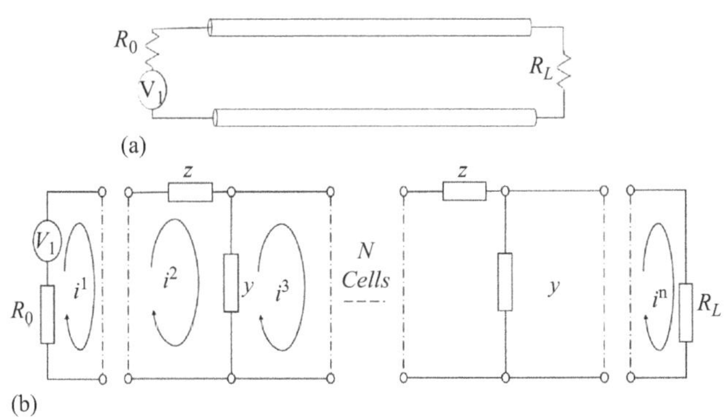

Figure 3.9 (a) Resistively ended network and (b) the equivalent subnetworks constituted graph topology

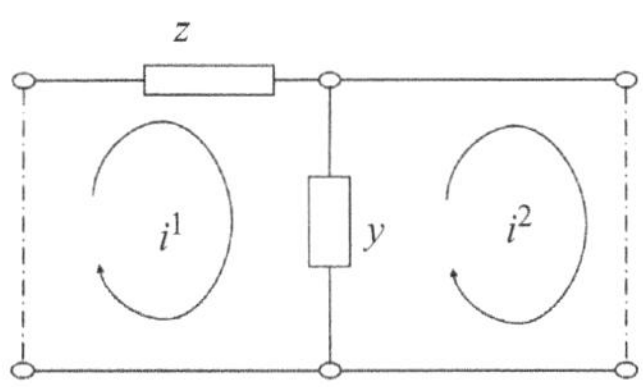

Figure 3.10 TL sub-element graph topology

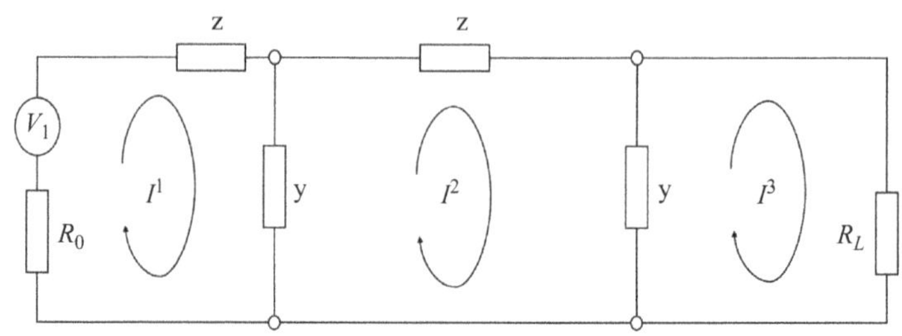

Figure 3.11 TL TAN equivalent graph topology

Figure 3.11, we need to define an interconnection matrix that is denoted as C in the present chapter. With the mesh fictive currents denoted i^b in Figure 3.10, Figure 3.9 must be defined in function of the mesh fictive currents I^n:

$$\begin{cases} i^1 = 1I^1 + 0I^2 + 0I^3 \\ i^2 = 1I^1 + 0I^2 + 0I^3 \\ i^3 = 0I^1 + 1I^2 + 0I^3 \\ i^4 = 1I^1 + 1I^2 + 0I^3 \\ i^5 = 0I^1 + 0I^2 + 1I^3 \\ i^6 = 0I^1 + 0I^2 + 1I^3 \end{cases} \Rightarrow i_b = C_n^b I^n \to C = \begin{bmatrix} 1 & 0 & 0 \\ 1 & 0 & 0 \\ 0 & 1 & 0 \\ 0 & 1 & 0 \\ 0 & 0 & 1 \\ 0 & 0 & 1 \end{bmatrix} \tag{3.48}$$

The impedance matrix Z_{mn} and the voltage vector E_m of Figure 3.11 can be determined systematically.

$$Z_{mn} = \begin{bmatrix} R_0 + Z_1 + 1/Y_1 & -1/Y_1 & 0 \\ -1/Y_1 & 1/Y_1 + Z_2 + 1/Y_2 & -1/Y_2 \\ 0 & -1/Y_2 & 1/Y_2 + R_L \end{bmatrix} \tag{3.49}$$

$$E_m = \begin{bmatrix} V_1 & 0 & 0 \end{bmatrix} \tag{3.50}$$

The mesh fictive currents through R_0 and R_L can be obtained from

$$I^n = Y^{nm} E_m \tag{3.51}$$

3.3.1.3 Kron–Branin model

The Branin's branch space tensor contains Branin's EMF coupling sources between the interconnects. And the Branin's model represents the interconnect representation as a mesh network coupled by EMFs associated to the propagating signals. Figure 3.12 shows a Branin's model for a TL with characteristic impedance Z_c and propagation constant θ.

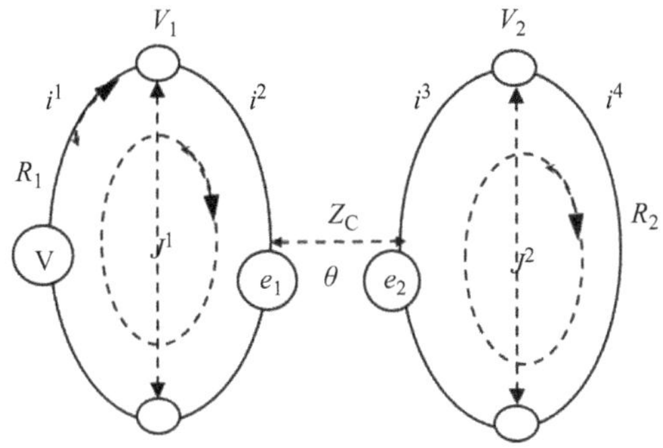

Figure 3.12 Branin's TL model

The TL effect is represented by the EMFs ($e_{1,2}$). Based on the TAN approach, the Branin's branch space expression can be developed:

$$z_{ab} = \begin{bmatrix} R_1 & 0 & 0 & 0 \\ 0 & Z_c & -Z_c i^3 e^{-\theta} & R_2 i^4 e^{-\theta} \\ R_1 e^{-\theta} & -Z_c i^1 e^{-\theta} & Z_c & 0 \\ 0 & 0 & 0 & R_2 \end{bmatrix} \tag{3.52}$$

where

$$\begin{cases} e_1 = R_2 i^4 e^{-\theta} - Z_c i^3 e^{-\theta} \\ e_2 = V - R_1 i^1 e^{-\theta} + Z_c i^2 e^{-\theta} \end{cases} \tag{3.53}$$

And also the connection matrix between branch and mesh can be developed as

$$C = \begin{bmatrix} 1 & 0 \\ 1 & 0 \\ 0 & 1 \\ 0 & 1 \end{bmatrix} \tag{3.54}$$

the mesh impedance and source matrix can be established as

$$Z_{\mu\upsilon} = C_{\mu}^{a} z_{ab} C_{\upsilon}^{b} \tag{3.55}$$

Finally, the system of this Branin's model can be expressed as in the following equation:

$$\begin{bmatrix} V \\ Ve^{-\theta} \end{bmatrix} = \begin{bmatrix} R_1 + Z_c & (-Z_c + R_2)e^{-\theta} \\ -(Z_c - R_1)e^{-\theta} & R_2 + Z_c \end{bmatrix} \begin{bmatrix} I^1 \\ I^2 \end{bmatrix} \tag{3.56}$$

3.3.2 Vias

The via model constituted by the lumped LC-network as shown in Figure 3.13, the basic geometry, and parasitic parameters as the capacitors and inductors can be computed through the closed-form equations [10,11] in (3.59) based on Figure 3.14.

The capacitors in the via network represent the pads of the via, the inductors represent the via holes. The 3D demonstration of the via is shown in Figure 3.15, C_{P1}, C_{anti1}, C_{anti2}, C_{P2}, and $L_{1,2,3}$ are the top layer pad, anti-pad (ground), anti-pad (power), bottom layer pad, and via holes, respectively, in this multilayer structure. The values of the via parameters can be calculated through the following equations.

$$L_{via} = \frac{\mu_0}{2\pi}\left[h \ln\left(\frac{h+\sqrt{r_{via}^2+h^2}}{r_{via}}\right) + \frac{3}{2}\left(r_{via} - \sqrt{r_{via}^2+h^2}\right)\right] \tag{3.57}$$

$$C_{pad} = \frac{1.41\varepsilon_r T D_{pad}}{D_{antipad} - D_{pad}} \tag{3.58}$$

$$\begin{aligned} C_{anti} = {} & \frac{4\varepsilon_r}{\pi}\ln\left[2(r_1-r_0)+\pi(h-r_1)\right]^{2(r_0-r_1)} + \frac{4\varepsilon_r}{\pi}\ln\left[2(r_1-r_0)\right. \\ & \left. +\pi(h-r_1)\right]^{\pi r_1} - \frac{4\varepsilon_r}{\pi}\ln\left[2(r_1-r_0)\right]^{\pi r_1} - \frac{4\varepsilon_r}{\pi}\ln\left[2(r_1-r_0)\right]^{2(r_0-r_1)} \end{aligned} \tag{3.59}$$

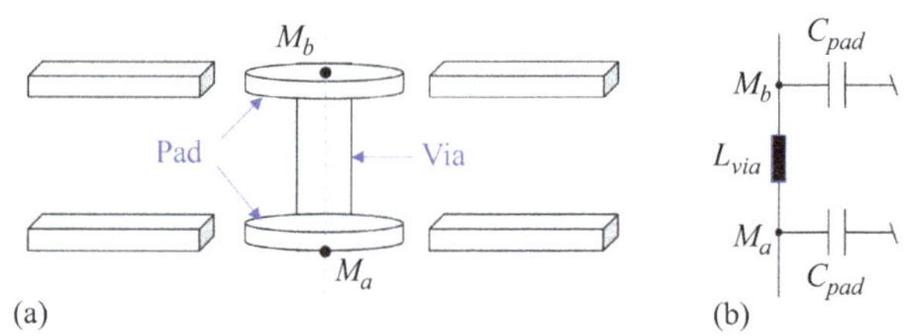

Figure 3.13 (a) Via structure and (b) via model

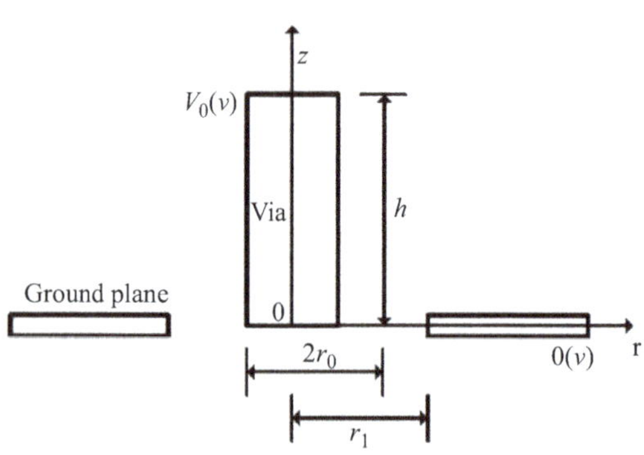

Figure 3.14 Extraction for equivalent capacitor of via structure [10]

where h is the via hole height, r_{via} is the radius of the via hole, D_{pad} is the diameter of via pad, $D_{antipad}$ is the diameter of anti-pad, T is the thickness of dielectric, and ε_r was defined earlier.

3.3.3 Power-ground plane

The power ground plane pair is presented in Figure 3.15 and are implemented in layers 2 and 3. The modelling method is the same as the TL modelling in Section 3.3.1.2. The *RLC* parameters can be calculated from equations in (3.61). The TAN non-connected matrix of the power-ground plane can be easily established from the equation below. In this matrix, the element parameters are $z = R_s + j\omega L_s$, $y = j\omega C_s$, with R_s, L_s, and C_s are per unit length resistance, inductance, and capacitance, respectively.

$$Z(k)_{ab} = \begin{bmatrix} z + 1/y & -1/y \\ -1/y & 1/y \end{bmatrix} \tag{3.60}$$

with $k = 1, \ldots, n$.

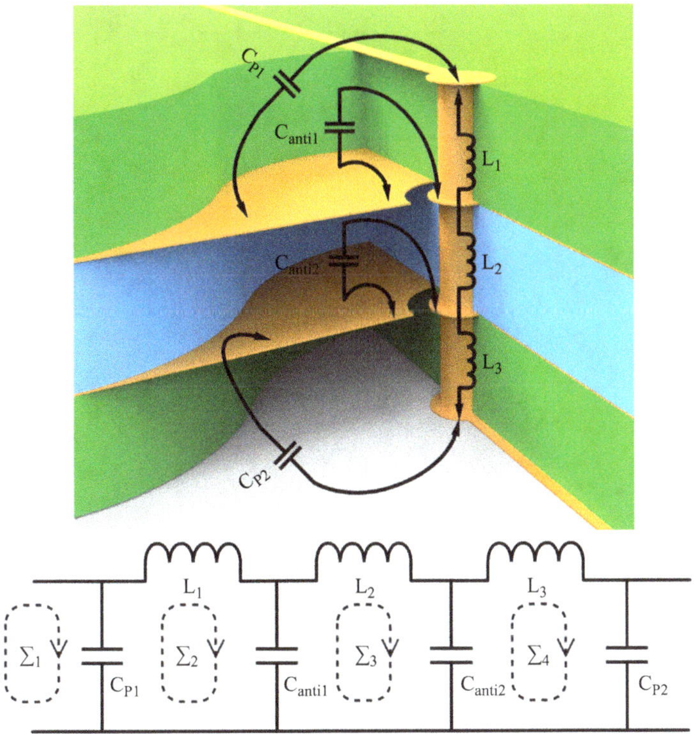

Figure 3.15 Equivalent 3D demonstration of via

$$\begin{cases} R = (\rho/\pi t)\ln(s/r_{via}) \\ L = \mu_0\mu_r h/W \\ C = \varepsilon_0\varepsilon_r lW/h \end{cases} \quad (3.61)$$

where ρ is the conductor resistivity, t is the thickness of conductor, r_{via} is via radius, s is the distance between power pin's via at voltage regulator module (VRM) and integrated circuit (IC), l is the length of the plane in meters, and h is the distance between power and ground plane.

3.3.4 SMA connectors

The SubMiniature version A (SMA) connector can also affect the SI in high speed signal transmission [12]. In this section, the SMA models [14] to multilayer PCB with SMA (Type: SMA edge mount jack receptacle) connector is elaborated. An example of the SMA connector located on the multilayer structure is shown in Figure 3.16.

The part of SMA on the top layer can be considered as a coaxial cable with length to 9.52 mm, the inner dielectric substrate is Teflon with ε_r=2.1 (Figure 3.17). The

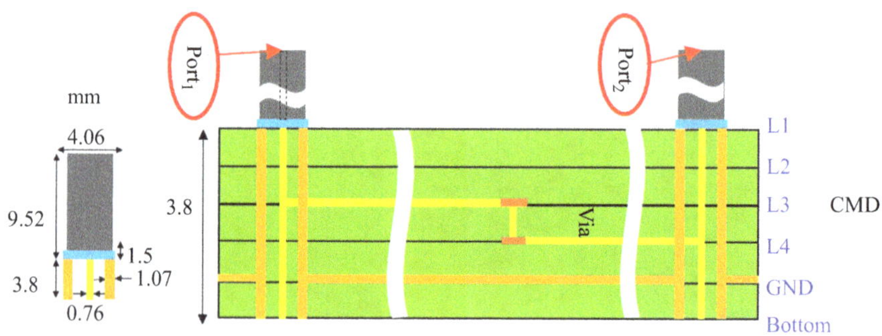

Figure 3.16 Geometry of SMA connector located on the multilayer structure

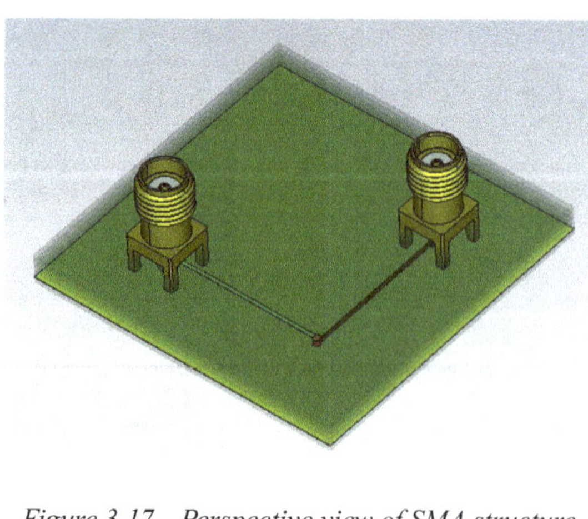

Figure 3.17 Perspective view of SMA structure

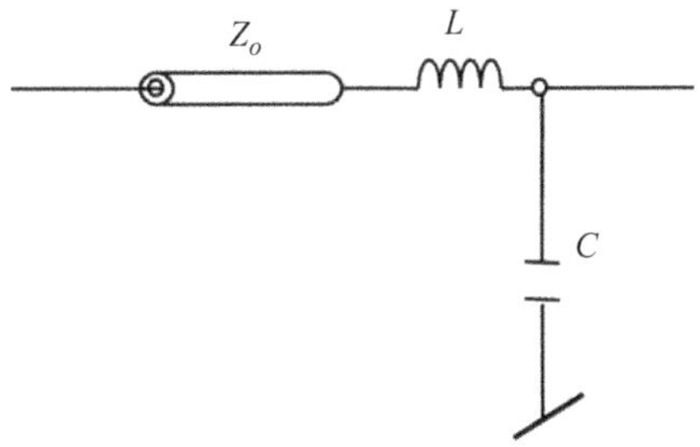

Figure 3.18 Equivalent SMA model

central pin can be represented as the inductance effect. From the top to bottom layer except layer 5, the central pin will generate capacitive effects as shown in Figure 3.18. The via is modelled by LC lumped network representing the pads and via hole.

Where the first lossless TL Z_o represents the upper external (with respect to the multilayer PCB) part of the SMA, Z_w represents the remaining part of the middle pin that can be taken as a via hole, because there is no clear reference conductor for this TL. The characteristic impedance is taken equal to the wave impedance [13]. L and C here represent the middle pin's inductance and capacitance inside the PCB. L and C can be calculated by (3.59). $Z_0 = Z_{SMA} = 50\ \Omega$ and its propagation time delay τ is written as

$$\tau = \sqrt{\varepsilon_r}\frac{h}{c_0} \tag{3.62}$$

where ε_r is the permittivity of the dielectric substrate, h is the height of the SMA's external part, and c_0 is the speed of light. The characteristic impedance Z_w is calculated through [15]

$$Z_w = \sqrt{\frac{\mu_0\mu_r}{\varepsilon_0\varepsilon_r}} \tag{3.63}$$

where ε_r and μ_r are the relative effective permittivity and permeability, respectively, of the substrate.

References

[1] Leman S. Contribution à la résolution de problèmes de compatibilité électromagnétique par le formalisme des circuits électriques de KRON. University of Lille; 2009.

[2] Maurice O, Durand P, Reineix A, and Dubois F. Kron's method and cell complexes for magnetomotive and electromotive forces. International Journal of Applied Mathematics. 2014;44(4):183–191.

[3] Averill RM, Barkley KG, Bowen MA, *et al.* Chip integration methodology for the IBM S/390 G5 and G6 custom microprocessors. IBM Journal of Research and Development. 1999;43(5.6):681–706.

[4] Schuster C, and Fichtner W. Parasitic modes on printed circuit boards and their effects on EMC and signal integrity. IEEE Transactions on Electromagnetic Compatibility. 2001;43(4):416–425.
[5] Kim J, and Li E. Special issue on PCB level signal integrity, power integrity, and EMC. IEEE Transactions on Electromagnetic Compatibility. 2010;52(2): 246–247.
[6] Wheeler HA. Transmission-line properties of a strip on a dielectric sheet on a plane. IEEE Transactions on Microwave Theory and Techniques. 1977;25: 631–647.
[7] Lafon F. Techniques and methodologies development to take into account EMC constraints in Automotive equipment design. Immunity analysis from component until the equipment. INSA Renne, ESEO Group; 2011.
[8] Thierauf SC. High speed circuit board signal integrity. Norwood, MA: Artech House; 2004.
[9] IPC. Controlled impedance circuit boards and high speed logic design; 1996. IPC 2141.
[10] Shang Y, Li C, and Xiong H. One method for via equivalent circuit extraction based on structural segmentation. IJCSI International Journal of Computer Science Issues. 2013;10.
[11] Goldfarb ME, and Pucel RA. Modeling via hole grounds in microstrip. IEEE Microwave and Guided Wave Letters. 1991;1(6):135–137.
[12] Fan J, Ye X, Kim J, *et al.* Signal integrity design for high-speed digital circuits: progress and directions. IEEE Transactions on Electromagnetic Compatibility. 2010;52(2):392–400.
[13] Antonini G, Scogna AC, and Orlandi A. Equivalent circuit extraction for an SMA connector. In: Progress in Electromagnetic Research Symposium. Pisa, Italy; 2004.
[14] Xu Z, Liu Y, Ravelo B, *et al.* Multilayer power delivery network modeling with modified Kron's method (MKM). In: 2017 International Symposium on Electromagnetic Compatibility – EMC EUROPE. Angers: IEEE; 2017.
[15] Brodwin ME. Engineering electromagnetic fields and waves. Journal of the Franklin Institute. 1977;304(2–3):145–146.

Chapter 4
Analytical calculation of PCB trace *Z/Y/T/S* matrices with TAN approach

Blaise Ravelo[1], Zhifei Xu[1], Yang Liu[1], and Olivier Maurice[2]

Abstract

Despite the numerous works done about the transmission line modelling of the printed circuit board (PCB) electrical interconnects, huge efforts are still open to the signal integrity, power integrity and electromagnetic compatibility engineers to predict the PCB trace effects during the design phase. The extraction method of the interconnect network as a multiport system will be defined based on a topological algebra. The existing theory and the simulation tools are either not flexible enough or do not allow one to understand the electrical behaviours of the PCB interconnects. The system and circuit theory about the impedance (Z), admittance (Y), transfer (T) and scattering (S) matrices will be provided in the present chapter. The study will be applied to different topologies of interconnect structures as multiport graph topologies. Pedagogical approaches enabling one to practice easily the tensorial analysis of networks (TAN) concept in function of the interconnect structures will be treated in this chapter. The theory representing the interconnect as tensorial objects in branch and mesh spaces will be provided. The problem resolution via the calculations of mesh currents into the $Z/Y/T/S$-matrices will be developed. Illustrative application examples of microstrip, coplanar and multilayer PCB interconnect will be presented. The possibility of using the TAN model for predicting the interconnect behaviour in broadband frequency from DC to several GHz will be demonstrated at the end of the chapter.

Keywords: TAN model, Kron's method, Kron–Branin's model, multiport system, SIMO topology, Z-matrix, Y-matrix, T-matrix, S-matrix, electrical interconnect, analytical calculation, modelling methodology, circuit and system theory

4.1 Introduction

To meet the public and industrial demands, the modern printed circuit boards (PCBs) integration density and electrical interconnect complexity increases constantly. These modern PCB analyses require deep understanding of signal integrity

[1]IRSEEM/ESIGELEC, Rouen, France
[2]ArianeGroup, Paris, France

(SI) and electromagnetic compatibility (EMC) phenomena. Despite the simulation tool development [1], the modelling of PCB electrical interconnects remains one of the hot research topics of SI and EMC research engineers [2,3].

Because of undesirable effects as electromagnetic interference and coupling [4], the SI and EMC prediction is a key point for analysing the PCB interconnects issues [5]. Modelling methodologies must be developed for predicting the SI and EMC of emerging analogue and mixed PCBs [6,7]. This predictive approach requires innovative insights on interconnects modelling [8].

To face up this technical challenge, EMC engineers use familiar computational methods [9–13]. Therefore, more and more computer-aided design tools have been deployed for the PCB interconnect analyses. Various solvers-based numerical computations as transmission line (TL) matrix [9], method of moments [10], finite element method [11], finite difference time domain, partial element equivalent circuit, Simulation Program with Integrated Circuit Emphasis (SPICE) and also, emerging algorithms as artificial neural network [12] are developed to solve PCB and EM problems. However, most of these familiar computational solvers [1,9–12] are not enough for the interconnect PCB phenomenological basic understandings and process time-consuming. For example, because of the full wave meshing computations, the main causes of interconnect propagation delays are not obviously highlighted. For this reason, analytical approaches based on the TL theory are still necessary.

The most popular interconnect delay estimation is based on the Elmore [13] and Wyatt [14] models. This TL model is based on the first-order lumped RC-network. The RC-model was widely exploited to estimate fast the tree interconnect propagation delay [15,16]. The RC-model was also used to optimize the interconnect wire size [17] and the dominant delay of interconnect trees [18]. However, the TL-lumped RC-model presents up to 30% of propagation delay relative errors. Therefore, more accurate TL-lumped RLC-model, including inductive effects, was proposed for the PCB SI analysis [19–22]. The RLC-approach was exploited to unify the model of PCB interconnect time delay [19,20]. It was also used to optimize the buffered interconnect trees [21]. More accurate algebraic equation for calculating interconnect delay was established [22]. Despite the development of lumped RC- and RLC-model, more explicit analytical investigations of PCB interconnect are necessary at higher frequencies. More advanced interconnect models based on the distributed TL were introduced [23–28]. The distributed model allows one predicting the electronic packaging effects [23] and high-density copper interconnect structures [24]. It also allows one to efficiently estimate the interconnect delay model, including the conductor and dielectric losses [25]. By considering the characteristic impedance and propagation constants, analytical expressions of adequate transfer function [26] and S-parameters [27] of microstrip PCB interconnect were established. In addition to the signal distribution properly through the isolated interconnects, some new and unfamiliar phenomena linked to the group delay might be open questions for the EMC researchers.

In spite of this progress on the interconnect analytical modelling, further studies must be performed for the analytical approach. The tensorial analysis of networks (TAN) formalism can be classified among the unfamiliar methods. The Z-matrix is a key parameter for establishing an equivalent model of interconnected electrical networks [1,2]. It is used to improve different properties of system as the stability assessment [3]. The Z-matrix is regularly used to deal with all system levels from high-voltage power system to microelectronic packaged circuits [4–6]. For the last area, commercial tools and some expensive time-consuming techniques [4–6] are deployed.

In this chapter, we would like to introduce an alternative sharp method to determine the Z-matrix of any multiport network in the present chapter by using the TAN [7–12]. The TAN was initiated by Kron in the 1930s for electrical machine modelling [7]. In the 1990s, Kron's method was developed by Olivier Maurice and his team for solving complex system EMC issues [8–10]. More recently, the method was also used for the S-matrix modelling of multiport structures as PCB interconnections [11] and coupled coaxial cables [12].

For better understanding, this chapter is organized into three different sections. In Section 4.2, the investigation of PCB Y-tree modelling is developed. Section 4.3 is focused on the PCB trace Ψ-tree modelling. The section explains how to use Kron's method for straightforward extraction of frequency-dependent Z-matrix with the four-port proof of concept (POC) represented by a Ψ-tree symmetric network. Last, Section 4.5 summarizes the chapter with a brief conclusion.

4.2 General description of *P*-port system

The present section describes the general introduction of electrical system representing a PCB interconnect. The analytical variables necessary for the development of TAN modelling will be introduced. Then, the methodology of TAN modelling will be presented.

4.2.1 Diagram representation

The planar PCB usually constitutes the main background of electronic devices. During the design phase, the prediction of electrical interconnects on the PCB performance remains one of the most challenging tasks. To highlight the TAN modelling method for the PCB interconnects, we can consider the example of planar structure shown in Figure 4.1. In this configuration, the topological analysis allows one to report that this electrical system is terminated by $P = 10$ ports. This system is assumed as passive network constituted by resistor R, inductor L and capacitor C lumped elements. The

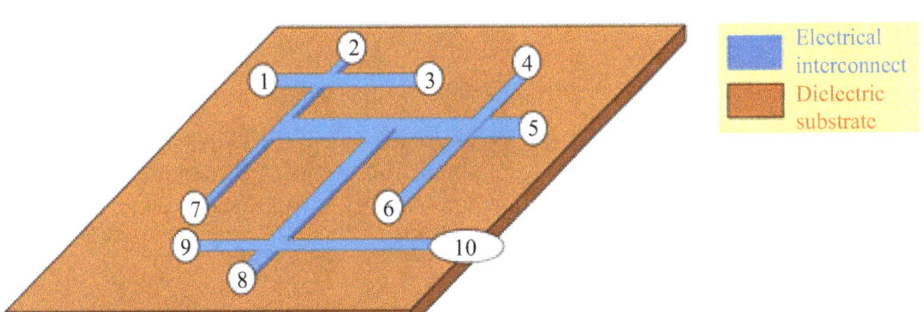

Figure 4.1 Example of planar PCB with electrical interconnects terminated by multiports

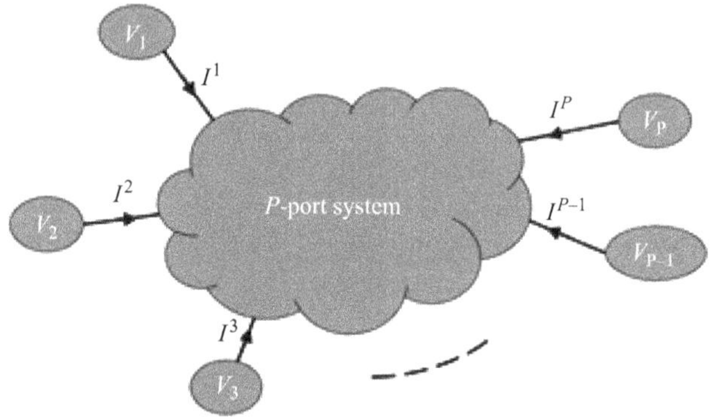

Figure 4.2 General topology of P-port electrical system

electrical operation of this PCB corresponds to the interaction of electrical currents and electrical voltages among different ports.

To study this system in a more fundamental approach, let us consider the ideal and general topology of PCB interconnect as an electrical system depicted in Figure 4.2. This general topology can be assumed as an abstract multiport system defined by the number of access ports $P = 1, 2, 3 \ldots$, which is a strictly positive integer. The determination of this system electrical parameters constitutes the main problem of the present chapter. The key unknowns of our problem to be solved with TAN approach will be defined in the next paragraphs.

4.2.2 Analytical variables constituting PCB electrical interconnections

To analyse a multiport PCB interconnect, each port of the representing system must be fed by access voltage source vector:

$$[V] = \begin{bmatrix} V_1 & V_2 & \cdots & V_P \end{bmatrix} \tag{4.1}$$

During the normal functioning, we will have access currents propagating through each branch:

$$[I] = \begin{bmatrix} I^1 \\ I^2 \\ \vdots \\ I^P \end{bmatrix} \tag{4.2}$$

As the systemic modelling will be performed in the frequency domain, each components V_k and I_k $(k = 1, 2, \ldots, P)$ should be complex number scalars. These two vectors enable the fundamental characterization of multiport system equivalent models with the following:

- impedance ($[Z]$) matrix,
- admittance ($[Y]$) matrix,
- transfer ($[T]$) matrix and
- scattering ($[S]$) matrix.

4.2.2.1 *Z*-matrix definition

The impedance matrix is one of the most fundamental representations of multiport electrical system that depends mainly on the knowledge of access voltage $[V]$. According to the circuit theory, the equivalent matrix impedance is defined by the matrix relationship:

$$[V] = [Z] \times [I] \tag{4.3}$$

By using the vectors defined in (4.1) and (4.2), this relation can be explicitly expressed as

$$\begin{bmatrix} V_1 \\ V_2 \\ \vdots \\ V_P \end{bmatrix} = \begin{bmatrix} Z_{11} & Z_{12} & \cdots & Z_{1P} \\ Z_{21} & Z_{22} & \cdots & Z_{2P} \\ \vdots & \vdots & \ddots & \vdots \\ Z_{P1} & Z_{P2} & \cdots & Z_{PP} \end{bmatrix} \times \begin{bmatrix} I^1 \\ I^2 \\ \vdots \\ I^P \end{bmatrix} \tag{4.4}$$

The impedance matrix must be a square matrix with dimension $(P \times P)$. Each components Z_{km} $(k, m = 1, 2, \ldots, P)$ should be complex number scalars.

4.2.2.2 *Y*-matrix definition

For main cases of parallel structure-based circuits, it would be more practical to consider admittance matrix to analyse certain interconnect PCBs. According to the circuit theory, the equivalent admittance matrix is defined by the matrix relationship:

$$[I] = [Y] \times [V] \tag{4.5}$$

Mathematically speaking, it can be interpreted as the inverse matrix relation of (4.3). By using the vectors defined in (4.1) and (4.2), this relation can be explicitly expressed as

$$\begin{bmatrix} I^1 \\ I^2 \\ \vdots \\ I^P \end{bmatrix} = \begin{bmatrix} Y^{11} & Y^{12} & \cdots & Y^{1P} \\ Y^{21} & Y^{22} & \cdots & Y^{2P} \\ \vdots & \vdots & \ddots & \vdots \\ Y^{P1} & Y^{P2} & \cdots & Y^{PP} \end{bmatrix} \times \begin{bmatrix} V_1 \\ V_2 \\ \vdots \\ V_P \end{bmatrix} \tag{4.6}$$

The admittance matrix must be a square matrix with dimension $(P \times P)$. Each components Z_{km} $(k, m = 1, 2, \ldots, P)$ should be complex number scalars.

4.2.2.3 *T*-matrix definition

One of the most general analytical representations of multiport is the transfer matrix. It consists of representing the couple of voltage–current variables of certain ports as $P_{in} + 1, P_{in} + 2, \ldots, P(P_{in} < P)$ in function of other ports $1, 2, \ldots, P_{in}$. The equivalent transfer matrix is defined by the matrix relationship:

$$\begin{bmatrix} V_{P_0+1} \\ I^{P_0+1} \\ \vdots \\ V_P \\ I^P \end{bmatrix} = [T] \times \begin{bmatrix} V_1 \\ I^1 \\ \vdots \\ V_{P_0} \\ I^{P_0} \end{bmatrix} \tag{4.7}$$

As the present chapter will not have much focus on this $[T]$ matrix, we will not have less explicit representation on it.

4.2.2.4 *S*-matrix definition

The scattering matrix is one of the most useful measurable parameters for the RF and microwave analyses of PCB interconnects. According to the S-parameter theory, the S-matrix can be expressed in function of Z- or Y-matrices with the following matrix relationships:

$$[S] = ([Z] - Z_0[\Lambda_P]) \times ([Z] + Z_0[\Lambda_P])^{-1} \tag{4.8}$$

$$[S] = ([\Lambda_P] - Z_0[Y]) \times ([\Lambda_P] + Z_0[Y])^{-1} \tag{4.9}$$

where the real positive term Z_0 is equal to 50 Ω. In the present chapter, it is the common reference impedance of all access ports. And the term Λ_P is the P size identity matrix:

$$[\Lambda_P] = \begin{bmatrix} 1 & 0 & \cdots & 0 \\ 0 & 1 & \ddots & \vdots \\ \vdots & \ddots & \ddots & 0 \\ 0 & \cdots & 0 & 1 \end{bmatrix} \tag{4.10}$$

This last relation will serve for the S-matrix calculation of the present chapter.

4.2.3 TAN modelling methodology

As reported in [26–32], the TAN formalism is implemented based on the graph topology followed by the problem metric elaboration with tensorial approach.

4.2.3.1 Algorithmic methodological representation

Figure 4.3 indicates the workflow of TAN method applied to PCB trace modelling for solving electrical problems. This describes the successive steps of the analytical processes of the proposed fast S-parameter TAN model. The starting point consists in formulating the problem by defining the initial parameters that can be physical, geometrical or/and electrical variables. The next steps are the classical electrical circuit representation. The medium steps are the branch and mesh space analyses elaborated in the previous paragraphs. The application of this routine algorithm to difference-lumped PCB trace will be investigated in the next two sections.

Before the analytical exploration, the main vector variables in branch and mesh spaces are recalled in the next paragraphs.

4.2.3.2 Topological parameters

This analytical process must start with the topological analysis by identifying the quintuplet integers:

- the network number, Nw,
- the branch number, B,
- the node number, N,

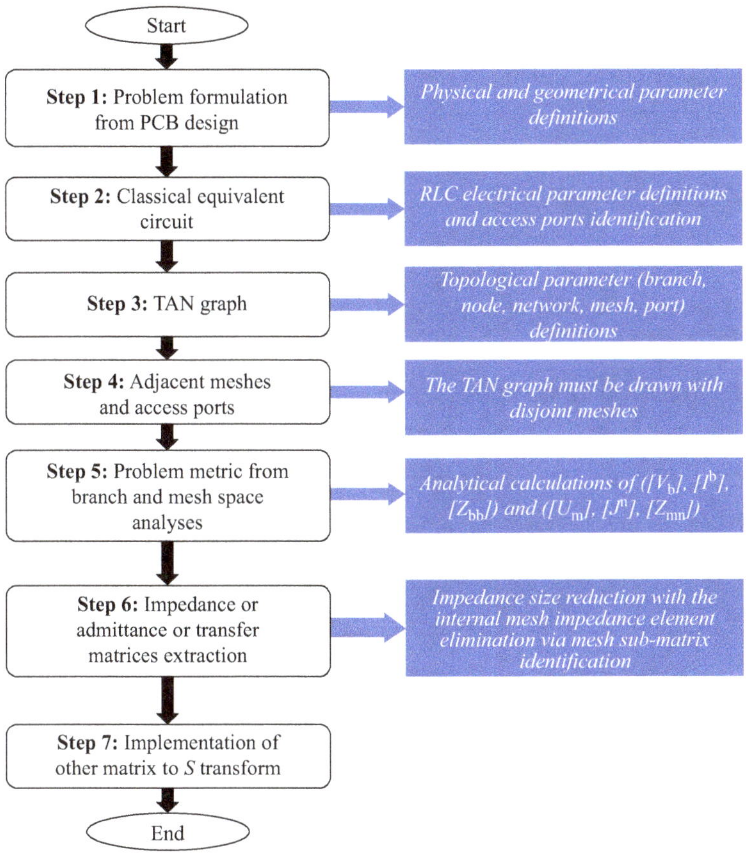

Figure 4.3 Workflow of PCB trace TAN modelling

- the mesh number, $M < B$ and
- the port number, $P < M$.

4.2.3.3 Branch space variables

The branch space analysis of the PCB trace structure consists of the implementation of voltage, current and impedance as follows:

- The covariable V_b represented by the branch source voltages vector:

$$[V_b] = \begin{bmatrix} V_1 & V_2 & \cdots & V_B \end{bmatrix} \tag{4.11}$$

- The contravariable I^b represented by the branch current vector:

$$[I^b] = \begin{bmatrix} I^1 \\ I^2 \\ \vdots \\ I^B \end{bmatrix} \tag{4.12}$$

- And the twice covariable metric represented by the branch impedance matrix that is a systematically diagonal matrix for the lumped circuit with interbranch coupling:

$$[Z_{bb}] = \begin{bmatrix} Z_{11} & 0 & \cdots & 0 \\ 0 & Z_{22} & \ddots & \vdots \\ \vdots & \ddots & \ddots & 0 \\ 0 & \cdots & 0 & Z_{BB} \end{bmatrix} \tag{4.13}$$

The mesh space analysis is derived from these branch space variables via the connectivity matrix $\left[C_b^m\right]$ which will be described in the next paragraph.

4.2.3.4 Mesh space variables

The mesh space analysis consists in expressing:

- The mesh voltage is the covariable U_m represented by the vector:

$$[U_m] = \left[\, U_1 \ U_2 \ \cdots \ U_M \,\right] \tag{4.14}$$

$$[U_m] = \left[C_m^b\right] \times [V_b] \tag{4.15}$$

This matrix relation can be rewritten in Einstein compact notation:

$$U_m = C_m^b V_b \tag{4.16}$$

- The connectivity matrix transposition can be expressed as

$$\left[C_b^m\right] = \left[C_m^b\right]^t \tag{4.17}$$

- The mesh current is the contravariable J^n represented by the vector:

$$[J^n] = \begin{bmatrix} J^1 \\ J^2 \\ \vdots \\ J^M \end{bmatrix} \tag{4.18}$$

$$[J^n] = \left[C_b^n\right] \times \left[I^b\right] \tag{4.19}$$

or in Einstein notation:

$$J^n = C_b^n I^b \tag{4.20}$$

- And the twice covariable metric represented by the mesh impedance matrix:

$$[Z_{mn}] = \begin{bmatrix} Z_{11} & Z_{12} & \cdots & Z_{1M} \\ Z_{21} & Z_{22} & \cdots & Z_{2M} \\ \vdots & \vdots & \ddots & \vdots \\ Z_{M1} & Z_{M2} & \cdots & Z_{MM} \end{bmatrix} \tag{4.21}$$

$$[Z_{mn}] = \left[C_m^b\right] \times [Z_{bb}] \times \left[C_n^b\right] \tag{4.22}$$

$$Z_{mn} = C_m^b Z_{bb} C_n^b \tag{4.23}$$

The fundamental metric of the problem can be formulated by the compact Einstein relation:

$$[J^n] = [Y^{nm}] \times [U_m] \tag{4.24}$$

In Einstein notation, it becomes

$$J^n = Y^{nm} U_m \tag{4.25}$$

where

$$[Y^{nm}] = [Z_{mn}]^{-1}. \tag{4.26}$$

4.2.3.5 Branch and mesh current identity

The present TAN modelling depends mainly on the access port currents. These variables can be obtained from the equality between certain components of branch currents and mesh currents. The analytical expression can be established from disjoint access meshes formulated by

$$\begin{bmatrix} J^1 \\ J^2 \\ \vdots \\ J^P \end{bmatrix} = \begin{bmatrix} I^1 \\ I^2 \\ \vdots \\ I^P \end{bmatrix} \tag{4.27a}$$

This means that the 1 to P column and 1 to P row elements of connectivity matrix can be expressed as the identity matrix:

$$\left[C_b^m(1\ldots P, 1\ldots P)\right] = \Lambda_P \tag{4.27b}$$

The detailed description of the Step 6 analytical calculation will be explained in the next subsection.

4.2.3.6 Calculation of S-matrix

The main original point of the fast developed S-parameter TAN model lies on the extraction of the impedance matrix from the mesh impedance matrix $[Z_{mn}]$ expressed in (4.23). The analytical algebraic mechanism of the Z-matrix extraction from the mesh impedance is explained in the following equations:

$$[Z_{mn}] = \left[\begin{array}{cccc|ccc} Z_{11} & Z_{12} & \cdots & Z_{1p} & Z_{1(p+1)} & \cdots & Z_{1M} \\ Z_{21} & Z_{22} & \cdots & Z_{2p} & Z_{2(p+1)} & \cdots & Z_{2M} \\ \vdots & \vdots & \ddots & \vdots & \vdots & \ddots & \vdots \\ Z_{p1} & Z_{p2} & \cdots & Z_{pp} & Z_{p(p+1)} & \cdots & Z_{pM} \end{array}\right] \tag{4.28}$$

$$\left\{\begin{array}{ll} (\Gamma_1): & U_1 = Z_{11}J^1 + Z_{12}J^2 + \cdots + Z_{1p}J^p + Z_{1(p+1)}J^{p+1} + \cdots + Z_{1M}J^M \\ (\Gamma_2): & U_2 = Z_{21}J^1 + Z_{22}J^2 + \cdots + Z_{2p}J^p + Z_{2(p+1)}J^{p+1} + \cdots + Z_{2M}J^M \\ \vdots & \\ (\Gamma_p): & U_p = Z_{p1}J^1 + Z_{p2}J^2 + \cdots + Z_{pp}J^p + Z_{p(p+1)}J^{p+1} + \cdots + Z_{pM}J^M \\ \hline (\Gamma_{p+1}): & 0 = Z_{(p+1)1}J^1 + Z_{(p+1)2}J^2 + \cdots + Z_{(p+1)p}J^p + Z_{(p+1)(p+1)}J^{p+1} + \cdots + Z_{(p+1)M}J^M \\ \vdots & \\ (\Gamma_M): & 0 = Z_{M1}J^1 + Z_{M2}J^2 + \cdots + Z_{Mp}J^p + Z_{M(p+1)}J^{p+1} + \cdots + Z_{MM}J^M \end{array}\right. \tag{4.29}$$

$$\left.\begin{array}{c} (\Gamma_{(p+1)}) \\ \vdots \\ (\Gamma_M) \end{array}\right\} \Rightarrow -\begin{bmatrix} Z_{(p+1)1} & \cdots & Z_{(p+1)p} \\ \vdots & \ddots & \vdots \\ Z_{M1} & \cdots & Z_{Mp} \end{bmatrix} \times \begin{bmatrix} J^1 \\ \vdots \\ J^p \end{bmatrix} = \begin{bmatrix} Z_{(p+1)(p+1)} & \cdots & Z_{(p+1)M} \\ \vdots & \ddots & \vdots \\ Z_{M(p+1)} & \cdots & Z_{MM} \end{bmatrix} \times \begin{bmatrix} J^{p+1} \\ \vdots \\ J^M \end{bmatrix} \tag{4.30}$$

$$\Leftrightarrow \quad \begin{bmatrix} J^{p+1} \\ \vdots \\ J^M \end{bmatrix} = -\begin{bmatrix} Z_{(p+1)(p+1)} & \cdots & Z_{(p+1)M} \\ \vdots & \ddots & \vdots \\ Z_{M(p+1)} & \cdots & Z_{MM} \end{bmatrix}^{-1} \times \begin{bmatrix} Z_{(p+1)1} & \cdots & Z_{(p+1)p} \\ \vdots & \ddots & \vdots \\ Z_{M1} & \cdots & Z_{Mp} \end{bmatrix} \times \begin{bmatrix} J^1 \\ \vdots \\ J^p \end{bmatrix} \tag{4.31}$$

$$\left.\begin{array}{c} (\Gamma_1) \\ \vdots \\ (\Gamma_p) \end{array}\right\} \Rightarrow \begin{bmatrix} U^1 \\ \vdots \\ U^p \end{bmatrix} = \begin{bmatrix} Z_{11} & \cdots & Z_{1p} \\ \vdots & \ddots & \vdots \\ Z_{p1} & \cdots & Z_{pp} \end{bmatrix} \times \begin{bmatrix} J^1 \\ \vdots \\ J^p \end{bmatrix} + \begin{bmatrix} Z_{1(p+1)} & \cdots & Z_{1M} \\ \vdots & \ddots & \vdots \\ Z_{p(p+1)} & \cdots & Z_{pM} \end{bmatrix} \times \begin{bmatrix} J^{p+1} \\ \vdots \\ J^M \end{bmatrix} \tag{4.32}$$

$[Z_{mn}] =$

$1\ldots P$ $\leftrightarrow$	$P+1\ldots M$ $\leftrightarrow$	
$[Z_A]$	$[Z_B]$	$\updownarrow 1\ldots P$
$[Z_C]$	$[Z_D]$	$\updownarrow P+1\ldots M$

(4.33)

As a matter of fact, the mesh impedance can be subdivided, as illustrated in (4.33). We can see that we can represent it in four submatrices denoted as

- Z_A having size $P \times P$,
- Z_B having size $P \times (M - P)$,
- Z_C having size $(M - P) \times P$ and
- Z_D having size $(M - P) \times (M - P)$.

It yields the impedance matrix expressed as

$$[Z_{matrix}] = [Z_A] - [Z_B] \times [Z_D]^{-1} \times [Z_C] \tag{4.34}$$

Then, the associated S-parameter can be determined from the Z-to-S transform introduced in (4.8).

For the further illustration about the feasibility of the proposed S-parameter TAN model, application cases of a two-port circuit will be discussed in the next section.

4.3 Application study of the TAN method to Y-tree shape PCB trace modelling

The present section is focused on the TAN modelling of Y-shape interconnect based on the RLC passive network.

4.3.1 Y-tree PCB problem description

Figure 4.4 represents the distributed microstrip Y-tree under study. This interconnect network is supposed to be constituted by three different microstrips TLs, $TL_{k=1,2,3}$ (interconnecting nodes M_k and the middle node M_0) with characteristic impedance Z_k and physical length d_k. By denoting the Laplace variable $s = j\omega$, each branch is traversed by port current:

$$[I(s)] = \begin{bmatrix} I^1(s) & I^2(s) & I^3(s) \end{bmatrix} \tag{4.35}$$

The equivalent lumped circuit of the Y-tree is established with Hammerstad–Jensen model [16,17]. Each branch tree is constituted by T-cell built with series impedance and parallel admittance.

$$Z_k(s) = R_k + L_k s \tag{4.36}$$

$$Y_k(s) = C_k s \tag{4.37}$$

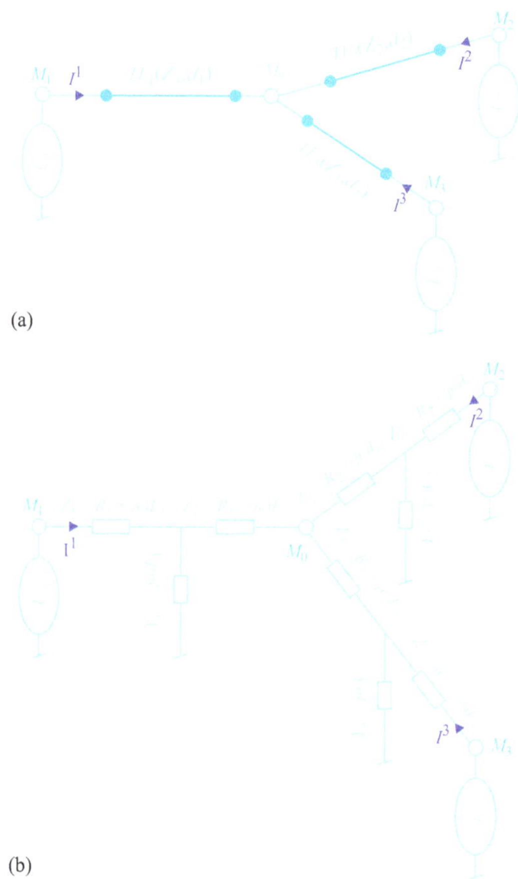

Figure 4.4 (a) Distributed and (b) lumped element equivalent network of the Y-tree circuit under study

4.3.2 TAN modelling of RLC Y-tree

According to the circuit theory, with $a, b = 1, 2, 3$, the equivalent Y-matrix

$$[Y(s)] = Y^{ab}(s) \tag{4.38}$$

is analytically defined by the relationship:

$$[I(s)] = [Y(s)] \times [U(s)] \tag{4.39}$$

where the structure excitation voltage source:

$$[U(s)] = \begin{bmatrix} U_1(s) & U_2(s) & U_3(s) \end{bmatrix} \tag{4.40}$$

with U_k at each node M_k.

4.3.2.1 TAN graph topology

The TAN modelling starts with the elaboration of Kron's branch allowing one to represent both the branch and mesh space variables of the problem. Figure 4.5 depicts Kron's graph equivalent to the circuit shown in Figure 4.4(b).

4.3.2.2 Topological index parameters

The topological analysis of the graph enables one to the following parameters:

1. number of branches: $B = 9$,
2. number of nodes: $N = 5$,
3. number of meshes: $M = 5$ and
4. number of ports: $P = 3$.

The TAN modelling can be implemented in the following four steps must not have an email address marker after their name.

4.3.2.3 Branch space analysis

In the branch space, the problem can be formulated by the covariable voltage $V_b(s)$, the contravariable current $I^b(s)$ and the double covariable branch impedance $Z_{bb}(s)(b = 1, 2, \ldots, B)$ whose extra-diagonal elements are null. With subscripts $a, b = 1, 2, 3$, the Z_{bb} diagonal elements are given by

$$\begin{cases} Z_{b,(a+1+b)}(s) = Z_b(s) \\ Z_{(a+b),(a+b)}(s) = 1/Y_b(s) \\ Z_{b,(a+6+b)}(s) = Z_b(s) \end{cases} \tag{4.41}$$

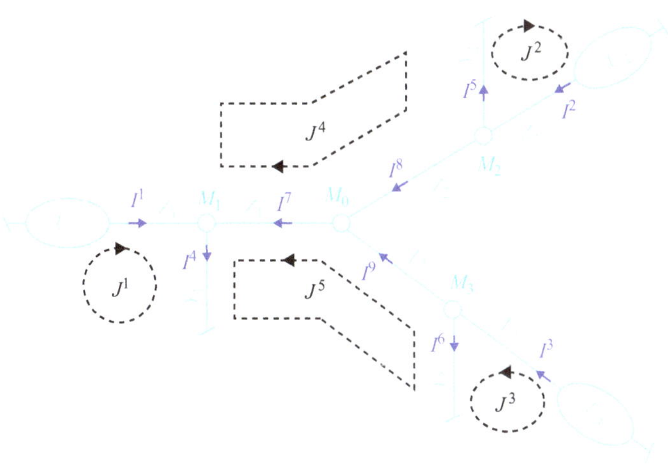

Figure 4.5 Kron's graph equivalent circuit

4.3.2.4 Mesh space analysis

First, the branch-to-mesh connectivity of graph shown in Figure 4.5 is written as

$$C_b^m = \begin{bmatrix} U_{33} & O_2^3 \\ U_{33} & \begin{bmatrix} 1_2 \\ -U_{22} \end{bmatrix} \\ O_{33} & \begin{bmatrix} 1_2 \\ -U_{22} \end{bmatrix} \end{bmatrix} \tag{4.42}$$

with U_{22} and U_{33} are two- and three-dimensional identity matrices, O_2^3 is a two-column-and-three-row zero matrix, O_{33} is a zero three-dimensional square matrix and 1_2 is a one-dimensional unity matrix. In the mesh space, with $m, n = 1, 2, \ldots, M$, the covariable voltage, the contravariable current and the double covariable impedance are linked to the branch space variables by the Einstein notation, respectively:

$$\Lambda_m = C_m^b V_b \tag{4.43}$$

$$J^n = C_b^n I^b \tag{4.44}$$

$$\Omega_{mn} = C_m^b Z_{bb} C_n^b \tag{4.45}$$

It yields the metric equation written in Einstein notation:

$$\Lambda_m = \Omega_{mn} J^n \tag{4.46}$$

4.3.2.5 *Z*-matrix calculation

By identifying with the relationship earlier written in (4.19), the Y-tree Y-matrix can be extracted algebraically from (4.27a) and (4.27b) knowing that

$$\begin{cases} I^{1,2,3} = J^{1,2,3} \\ U^{1,2,3} = \Lambda^{1,2,3} \end{cases} \tag{4.47}$$

By explicating relationship (4.23) with subtensors defined by subscripts $\alpha, \mu = 1, 2, \ldots, P$ and $\beta, \nu = P + 1, \ldots, M$, we have

$$\begin{bmatrix} U_\mu \\ O_\nu \end{bmatrix} = \begin{bmatrix} \Omega_{\mu\alpha} & \Omega_{\mu\beta} \\ \Omega_{\nu\alpha} & \Omega_{\nu\beta} \end{bmatrix} \begin{bmatrix} J^\alpha \\ J^\beta \end{bmatrix} \tag{4.48}$$

with O_ν is the zero $M - P$ size 1-rank tensor. Substituting the solution of second line of (4.43) into the first line, we have the Z-matrix:

$$[Z] = \Omega_{\mu\alpha} - \Omega_{\mu\beta}(\Omega_{\nu\beta})^{-1}\Omega_{\nu\alpha} \tag{4.49}$$

4.3.2.6 *Y*-matrix calculation

The calculation of the Y-matrix can be realized from (4.49) by the matrix inversion:

$$[Y] = \left\{\Omega_{\mu\alpha} - \Omega_{\mu\beta}(\Omega_{\nu\beta})^{-1}\Omega_{\nu\alpha}\right\}^{-1} \tag{4.50}$$

4.3.2.7 *S*-matrix extraction

With the reference impedance R_0, the Y-tree S-matrix model can be extracted from Z-matrix from (4.49) with the Z-to-S matrix transform:

$$[S(s)] = \{[Z(s)] - R_0 U_{33}\} \times \{[Z(s)] + R_0 U_{33}\}^{-1} \tag{4.51}$$

This S-matrix can also be determined from the Y-matrix from (4.50) with the Y-to-S matrix transform [15], we have:

$$[S(s)] = \{U_{33} - R_0\,[Y(s)]\} \times \{U_{33} + R_0\,[Y(s)]\}^{-1} \tag{4.52}$$

4.3.3 Validation result with SPICE simulations

To validate the Y-tree TAN model, comparisons between SPICE AC simulations have been performed.

4.3.3.1 POC description

To validate the TAN model, as POC, the microstrip Y-tree shown in Figure 4.6 was designed and simulated in Momentum environment of ADS®. The physical and electrical constituting TL parameters are listed in Table 4.1.

4.3.3.2 Discussion on computed results

After MATLAB® programing of previous TAN model, comparisons between calculated ("Calc.") and simulated ("Simu.") results from 0 to 0.5 GHz are presented in Figure 4.7. The calculated S-parameters from TAN model are in good agreement with ADS simulation. The slight discrepancies above 0.3 MHz are mainly due to the skin effect and momentum numerical solver inaccuracies.

4.3.3.3 Partial conclusion

In summary of the present section, an efficient S-matrix modelling of multiport interconnect networks is investigated. The proposed model is based on the unfamiliar

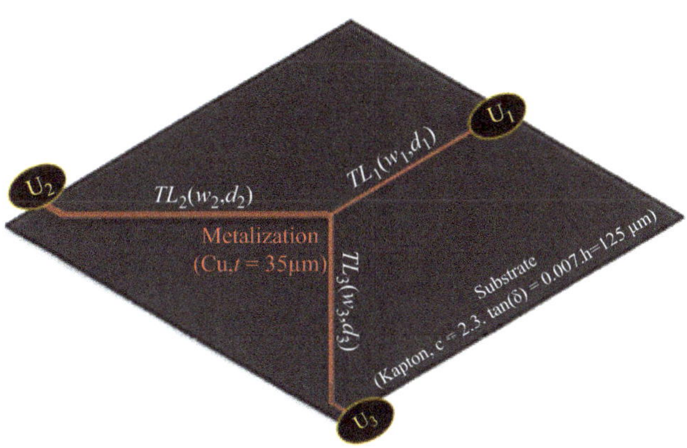

Figure 4.6 Three-dimensional design of Y-tree POC

Table 4.1 Y-tree PCB parameters

TL	Parameters		
	Length (mm)	Width (μm)	*RLC*
TL_1	$d_1 = 50$	$w_1 = 0.6$	$R_1 \approx 43\,m\Omega$, $L_1 \approx 4.25\,nH$, $C_1 \approx 9\,pF$
TL_2	$d_2 = 60$	$w_2 = 1.43$	$R_2 \approx 22\,m\Omega$, $L_2 \approx 2.55\,nH$, $C_2 \approx 2.28\,pF$
TL_3	$d_3 = 70$	$w_3 = 0.6$	$R_3 \approx 59\,m\Omega$, $L_3 \approx 11.9\,nH$, $C_3 \approx 12.66\,pF$

TAN formalism. The TAN methodology from the graph representation and tensorial analysis of the problem are described. The compact Einstein notation of the problem is established. The model is validated with a microstrip *Y*-tree. A very good agreement between the calculation, commercial tool simulation and measurement is realized from 0 to 0.5 GHz.

4.4 Application study to ψ-shape microstrip interconnect

The present section is focused on the TAN modelling of ψ-shape microstrip interconnect based on the *RLC* passive network.

4.4.1 Analytical investigation on the TAN modelling of ψ-tree PCB

After the Ψ-tree topological description, Kron's method analytical modelling is introduced.

4.4.1.1 Problem formulation of ψ-tree PCB trace

Figure 4.8 shows the Ψ-network under study. It is composed of *LC* π-cell-based four branches connecting access nodes M_k (k=1,2,3,4) to central node M_0. By denoting s the Laplace variable:

- the branches $M_0M_2 \Leftrightarrow M_0M_3$ are coupled and formulated by mutual inductance:

$$Z_{M_0M_2 \Leftrightarrow M_0M_3}(s) = K_2 s \tag{4.53}$$

- the branches $MM_3 \Leftrightarrow MM_4$ are coupled with mutual inductance:

$$Z_{M_0M_3 \Leftrightarrow M_0M_4}(s) = K_3 s \tag{4.54}$$

To determine the Z-matrix $[Z]$, this network can be fed by voltage sources U_p that generates access branch currents I^q (p,q=1,2,3,4). The problem associated with this

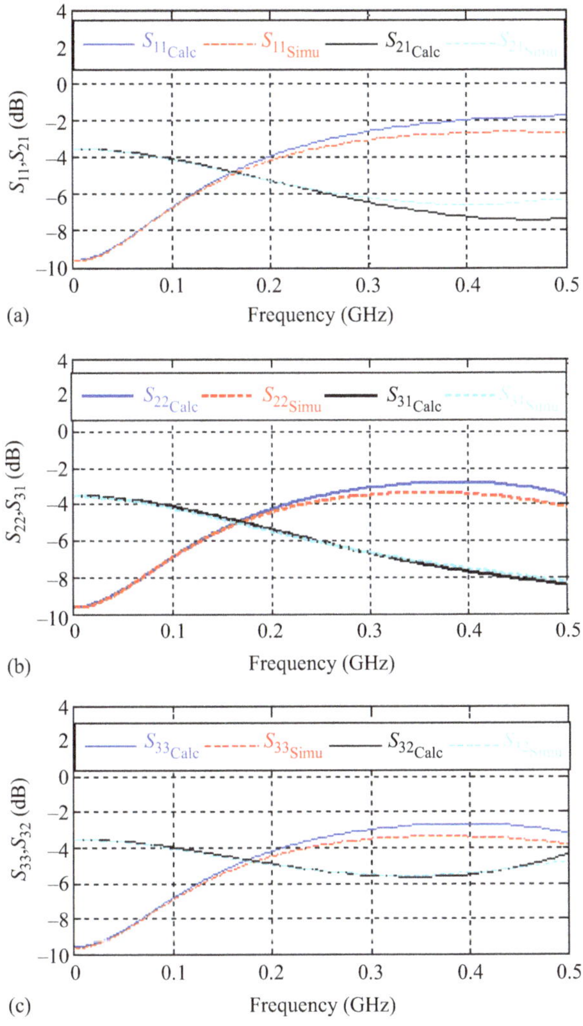

Figure 4.7 Calculated and simulated: (a) $S_{11,21}$, (b) $S_{22,31}$ and (c) $S_{33,22}$

network can be formulated by the determination of $Z_{pq}(s)$ that verifies the matrix relation:

$$U_p(s) = Z_{pq}(s)I^q(s) \tag{4.55}$$

4.4.1.2 Topological index parameters

The analysis of this graph reveals that the tensorial parameters:

- $B = 13$ branches,
- $N = 6$ nodes,

- $R = 1$ network,
- $M = 8$ meshes (which can be verified with Euler–Poincaré characteristic topological relation [13]), and
- $P = 4$ ports.

4.4.1.3 TAN graph topology

Figure 4.9 represents the TAN equivalent graph of the circuit introduced in Figure 4.8. The analytical implementation of the model will be elaborated in the next paragraph.

4.4.1.4 Branch space analysis

In the branch space, this graph can be defined by

- 1-rank covariant tensor voltage with $a = 1, \ldots, B$:

$$V_a = \begin{bmatrix} U_1 & \cdots & U_4 & 0 & \cdots & 0 \end{bmatrix} \tag{4.56}$$

- 1-rank contravariant tensor current with $b = 1, \ldots, B$:

$$I^b = \begin{bmatrix} I^1 \\ I^2 \\ \vdots \\ I^B \end{bmatrix} \tag{4.57}$$

- 2-rank covariant double tensor branch impedance Σ_{ab} essentially with the diagonal element Σ_{aa} equal to $Z(s)$ or $1/Y(s)$ and non-zero non-diagonal four elements:

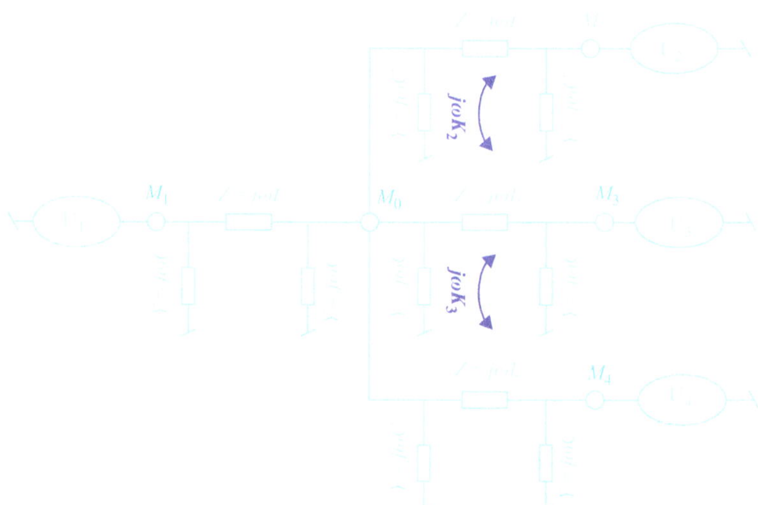

Figure 4.8 Ψ*-Tree network under study*

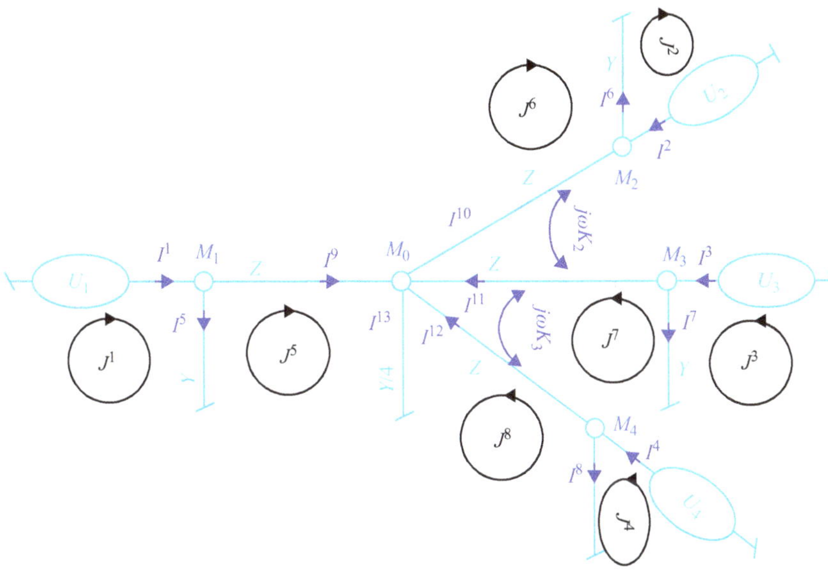

Figure 4.9 Kron's graph equivalent of Ψ-network

$$\begin{cases} \Sigma_{10,11}(s) = \Sigma_{11,10}(s) = -K_2 s \\ \Sigma_{11,12}(s) = \Sigma_{12,11}(s) = -K_3 s \end{cases} \tag{4.58}$$

4.4.1.5 Connectivity

We can express mixed 2-rank hybrid tensor connectivity Θ^b_m between the branch and mesh variables with respect to Einstein notation relationship:

$$I^b = \Theta^b{}_m J^m \tag{4.59}$$

which can be expressed as

$$\Theta^m_b = \begin{bmatrix} 0^4_4 & -\Lambda^4_4 & \Lambda^4_4 & 1^4 \\ \Lambda^4_4 & \Lambda^4_4 & 0^4_4 & 0^4 \end{bmatrix} \tag{4.60}$$

where

- 0^r_c represents the zero matrix with r rows and c columns,
- 1^r_c represents the one matrix with r rows and c columns, and
- Λ^4_4 represents the four-dimensional identity matrix.

4.4.1.6 Mesh space analysis

In the mesh space, our graph can be described by

- 1-rank covariant tensor mesh voltage with $m = 1, \ldots, M$:

$$X_m = \Theta^a{}_m V_a \tag{4.61}$$

- 1-rank contravariant tensor current with $n = 1, \ldots, M$:

$$J^n = \Theta^n{}_b I^b \tag{4.62}$$

- 2-rank double covariant tensor branch impedance (4.60) which yields

$$\Omega_{mn} = \Theta^a{}_m \Sigma_{ab}\, \Theta^b_n \tag{4.63}$$

The explicit expression of the mesh impedance will be

$$\Omega_{mn} = \begin{bmatrix} \frac{\Lambda_4^4}{Y} & \frac{-\Lambda_4^4}{Y} \\ \frac{-\Lambda_4^4}{Y} & \begin{bmatrix} \Xi & \frac{1}{4Y} & \frac{1}{4Y} & \frac{1}{4Y} \\ \frac{1}{4Y} & \Xi & \Delta_2 & \frac{1}{4Y} \\ \frac{1}{4Y} & \Delta_2 & \Xi & \Delta_3 \\ \frac{1}{4Y} & \frac{1}{4Y} & \Delta_3 & \Xi \end{bmatrix} \end{bmatrix} \tag{4.64}$$

with

$$\begin{cases} \Xi = Z + 5/(4Y) \\ \Delta_{2,3} = K_{2,3}s + 1/(4Y) \end{cases} \tag{4.65}$$

Now, our Ψ-network problem is completely tensorialized by the Einstein notation metric:

$$X_m = \Omega_{mn} J^n \tag{4.66}$$

with mesh current J^n as main unknown.

1. **Z-matrix calculation**

The Ψ-network Z-matrix can be extracted by splitting metric (4.23) into two sub-equations with unknowns:

$$J^{1\ldots P} = I^{1\ldots P} \tag{4.67}$$

and $J^{P+1,\ldots,M}$. Therefore, we have equation system (4.24). Substituting the intermediate solution $J^{P+1\ldots M}$ of second equation into the first one, we have the algebraic solution expressed in (4.25):

$$\begin{cases} \left[\Omega_{1\ldots P,1\ldots P}\right] \times \left[J^{1\ldots P}\right] + \left\{ \begin{array}{c} \left[\Omega_{1\ldots P,P+1\ldots M}\right] \\ \times \left[J^{P+1\ldots M}\right] \end{array} \right\} = \left[U_{1\ldots P}\right] \\ \\ \left[\Omega_{P+1\ldots M,1\ldots P}\right] \times \left[J^{1\ldots P}\right] + \left\{ \begin{array}{c} \left[\Omega_{P+1\ldots M,P+1\ldots M}\right] \\ \times \left[J^{P+1\ldots M}\right] \end{array} \right\} = \left[0^{M-P+1}\right] \end{cases} \tag{4.68}$$

$$Z_{pq} = \left[\Omega_{1\ldots P,1\ldots P}\right] - \left\{ \begin{array}{c} \left[\Omega_{1\ldots P,P+1\ldots M}\right] \\ \times \left[\Omega_{P+1\ldots M,P+1\ldots M}\right]^{-1} \\ \times \left[\Omega_{P+1\ldots M,1\ldots P}\right] \end{array} \right\} \tag{4.69}$$

To validate this Z-matrix Kron's model, numerical application compared with commercial tool simulations will be presented in the next section.

4.4.2 Validation results with SPICE simulations

The previous analytical equations were implemented as a MATLAB routine program for this numerical application. A Ψ-network POC constituted by *LC*-network was simulated with the commercial tool ADS from Keysight Technologies® to verify the relevance of the developed fast model.

4.4.2.1 POC description

As POC, the arbitrary values of inductor, capacitor C and mutual inductance K_2 varied from 0.1 to 5 nH, with fixed $K_3 = 3$ nH. It is to be noticed that Table 4.2 gives an overview of the different test cases considered in this work. The schematic of simulated *LC* Ψ-network POC is shown in Figure 4.10. The 50 Ω reference impedance will not affect the Z-matrix by using the S-to-Z ADS function.

Table 4.2 Test cases (definition of inputs)

Input type	**Case 1**	**Case 2**	**Case 3**
Inductance L (in nH)	10	10	10
Capacitor C (in nF)	0.1	0.1	2
Inductance K_2 (in nH)	0.1	5	5
Inductance K_3 (in nH)	3	3	3

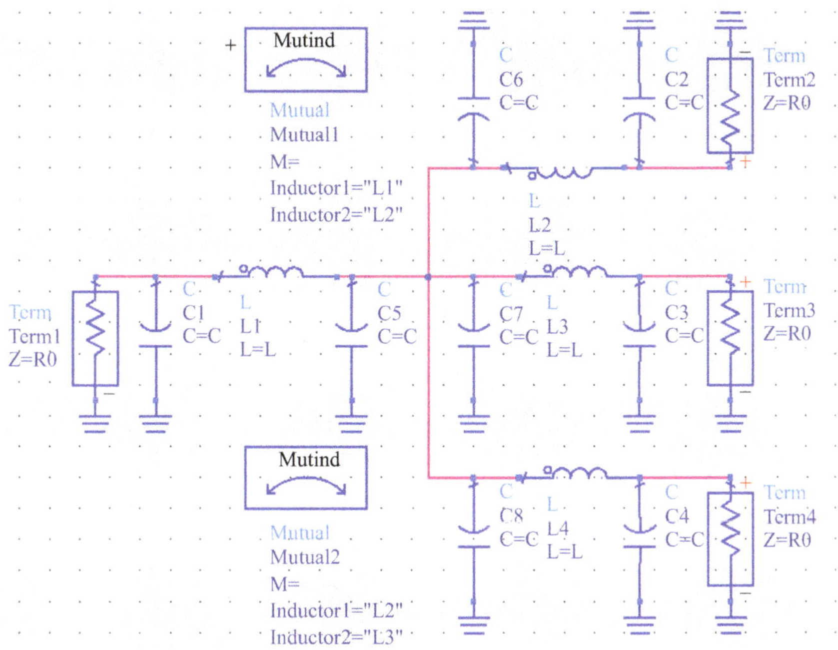

Figure 4.10 Schematic of simulated Ψ-network proof of concept

4.4.2.2 Discussion on computed results

The numerical computations and simulations have been run in the frequency band defined from 1 to 400 MHz. The simulated Z-matrix was extracted from S-parameters via S-to-Z function. Figure 4.11 displays the plot of impedances Z_{11} and Z_{12}. As expected, excellent correlation between Kron's model and the ADS simulations is obtained. The proposed model presents a considerable advantage for simulating complex structures that can be represented with multi-rank tensors and also the fast computation time.

4.4.2.3 Partial conclusion

A fast modelling method of Z-impedance is investigated. The methodology was described with tensorial algebra. The effectiveness of the method was confirmed with simulation of LC ψ-network with excellent agreement. Moreover, the TAN calculation method provides an analytical framework allowing to predict the S-parameter

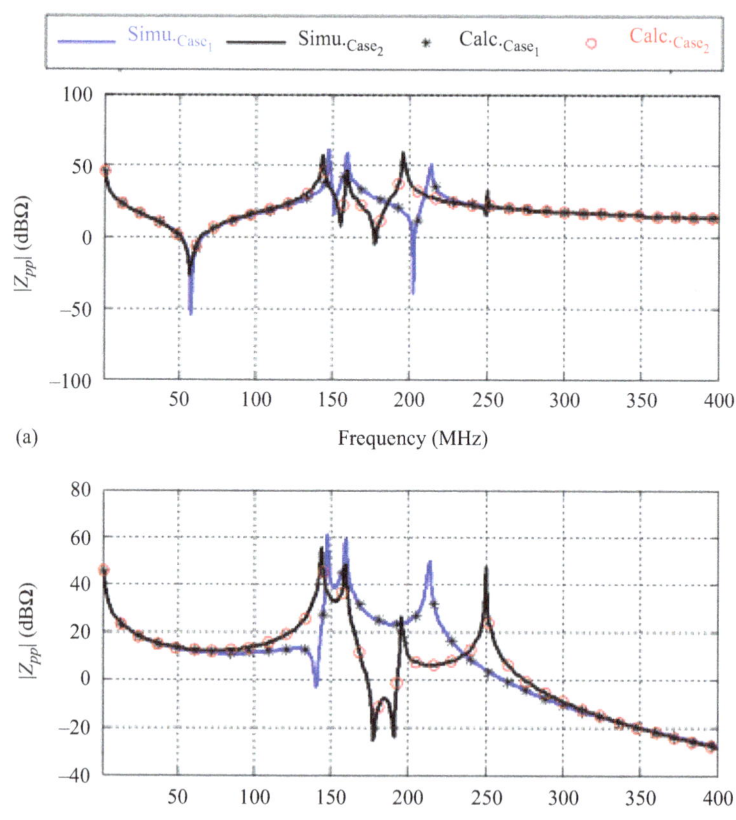

Figure 4.11 Calculated and simulated: (a) reflection Z_{pp} and (b) transmission Z_{pq} impedances with $p,q = 1,2,3,4$

responses of typical four-port electrical network. The developed method enables to demonstrate with a fast calculation the influence of physical parameters in a very wide frequency band from 0 Hz to several gigahertz. In the future, the TAN method can be exploited to emulate a relevant tool to generate a deep knowledge of the equivalent system and the particular effects behind as the resonances.

4.5 Conclusion

A TAN modelling applied to PCB trace was investigated. For better understanding, the PCB trace as interconnect structures was assumed by their R, L and C equivalent lumped circuits. The methodology of the TAN formalism allowing one to determine the Z, Y and S matrices is elaborated. The different steps of the routine algorithm are described. The feasibility of the TAN approach is illustrated with two different PCB trace examples of Y- and ψ-tree structures. The TAN model is particularly fast to implement, accurate and enables one to understand the physical mechanism behind the electrical signals interacting with the PCB traces.

References

[1] R. Archambeault, C. Brench, and S. Connor, "Review of printed-circuit-board level EMI/EMC issues and tools," IEEE Trans. EMC, vol. 52, no. 2, pp. 455–461, 2010.

[2] F. Jun, Y. Xiaoning, J. Kim, B. Archambeault, and A. Orlandi, "Signal integrity design for high-speed digital circuits: Progress and directions," IEEE Trans. EMC, vol. 52, no. 2, pp. 392–400, 2010.

[3] J. Kim and E. Li, "Special issue on PCB level signal integrity, power integrity, and EMC," IEEE Trans. EMC, vol. 52, no. 2, pp. 246–247, 2010.

[4] C. Schuster and W. Fichtner, "Parasitic modes on printed circuit boards and their effects on EMC and signal integrity," IEEE Trans. EMC, vol. 43, no. 4, pp. 416–425, 2001.

[5] J. F. Buckwalter, "Predicting microwave digital signal integrity," IEEE Trans. Adv. Packag., vol. 32, no. 2, pp. 280–289, 2009.

[6] R. M. Averill, K. G. Barkley, M. A. Bowen, *et al.*, "Chip integration methodology for the IBM S/390 G5 and G6 custom microprocessors," IBM J. Res. Dev., vol. 43, no. 5/6, pp. 681–706, 1999.

[7] A. E. Ruehli and A. C. Cangellaris, "Progress in the methodologies for the electrical modeling of interconnects and electronic packages," Proc. IEEE, vol. 89, no. 5, pp. 740–771, 2001.

[8] A. Ruan, J. Yang, L. Wan, B. Jie, and Z. Tian, "Insight into a generic interconnect resource model for Xilinx Virtex and Spartan series FPGAs," IEEE Trans. Circuits Syst. II, Express Briefs, vol. 60, no. 11, pp. 801–805, 2013.

[9] W. J. R. Hoefer, "The transmission-line matrix method – theory and applications," IEEE Trans. MTT, vol. 33, no. 10, pp. 882–893, 1985.

[10] M. Ney, “Method of moments as applied to electromagnetic problems”, IEEE Trans. MTT, vol. 33, no. 10, pp. 972–980, 1985.
[11] J. Jin, The Finite Element Method in Electromagnetics. New York, USA: John Wiley & Sons, 1993.
[12] K. S. R. Krishna, J. L. Narayana, and L. P. Reddy, “ANN models for microstrip line synthesis and analysis,” Int. J. Electr. Syst. Sci. Eng., vol. 1, pp. 196–200, 2008.
[13] W. C. Elmore, “The transient response of damped linear networks,” J. Appl. Phys., vol. 19, pp. 55–63, 1948.
[14] L. Wyatt, Circuit Analysis, Simulation and Design. North-Holland. The Netherlands: Elsevier Science, 1978.
[15] J. Rubinstein, P. Penfield, Jr., and M. A. Horowitz, “Signal delay in RC tree networks,” IEEE Trans. CAD, vol. 2, no. 3, pp. 202–211, 1983.
[16] R. Mita, G. Palumbo, and M. Poli, “Propagation delay of an RC-chain with a ramp input,” IEEE Trans. Circuits Syst. II, Express Briefs, vol. 54, no. 1, pp. 66–70, 2007.
[17] J. J. Cong and K.-S. Leung, “Optimal wiresizing under Elmore delay model,” IEEE Trans. CADICS, vol. 14, no. 3, pp. 321–336, 1995.
[18] L. Vandenberghe, S. Boyd, and A. El Gamal, “Optimizing dominant time constant in RC circuits,” IEEE Trans CAD, vol. 17, no. 2, pp. 110–125, 1998.
[19] A. B. Kahng and S. Muddu, “An analytical delay model of RLC interconnects,” IEEE Trans. CAD, vol. 16, pp. 1507–1514, 1997.
[20] R. Venkatesan, J.A. Davis, and J.D. Meindl, “Compact distributed RLC interconnect models – Part IV: Unified models for time delay, crosstalk, and repeater insertion,” IEEE Trans. Electron Devices, vol. 50, no. 4, pp. 1094–1102, 2003.
[21] S. L. Wang and Y. W. Chang, “Delay modeling for buffered RLY/RLC trees,” Proc. of 2005 IEEE Int. Symp. on VLSI Design, Automation and Test (VLSI-TSA), Hsinchu, Taiwan, pp. 237–240, 25–27 Apr. 2005.
[22] S. Roy and A. Dounavis, “Efficient delay and crosstalk modeling of RLC interconnects using delay algebraic equations,” IEEE Trans. VLSI, vol. 19, no. 2, pp. 342–346, 2011.
[23] Z. A. Khan, “A novel transmission line structure for high-speed high-density copper interconnects,” IEEE Trans. CPMT, vol. 6, no. 7, pp. 1077–1086, 2016.
[24] K. M. C. Branch, J. Morsey, A. C. Cangellaris, and A. E. Ruehli, “Physically consistent transmission line models for high-speed interconnects in lossy dielectrics,” IEEE Trans. Adv. Packag., vol. 25, no. 2, pp. 129–35, 2002.
[25] B. Ravelo, “Delay modelling of high-speed distributed interconnect for the signal integrity prediction,” Eur. Phys. J. Appl. Phys. (EPJAP), vol. 57, no. 31002, pp. 1–8, 2012.
[26] M. Choi, J.-Y. Sim, H.-J. Park, and B. Kim, “An approximate closed-form transfer function model for diverse differential interconnects,” IEEE Trans. Circuits Syst. I, Reg. Pap., vol. 62, no. 5, pp. 1335–1344, 2015.
[27] T. Eudes, B. Ravelo, and A. Louis, “Experimental validations of a simple PCB interconnect model for high-rate signal integrity,” IEEE Trans. EMC, vol. 54, no. 2, pp. 397–404, 2012.

[28] M. S. Maza and M. L. Aranda "Analysis of clock distribution networks in the presence of crosstalk and ground bounce," Proc. of the 8th IEEE International Conference on Electronics, Circuits and Systems, 2001 (ICECS 2001), vol. 2, Malta, Malta, pp. 773–776, 2–5 Sep. 2001.
[29] B. Ravelo, "Theory on coupled line coupler-based negative group delay microwave circuit," IEEE Trans. MTT, vol. 64, no. 11, pp. 3604–3611, 2016.
[30] R. Das, Q. Zhang, and H. Liu, "Lossy coupling matrix synthesis approach for the realization of negative group delay response," IEEE Access, vol. 6, pp. 1916–1926, 2017.
[31] Z. Wang, Y. Cao, T. Shao, S. Fang, and Y. Liu, "A negative group delay microwave circuit based on signal interference techniques," IEEE Microwave Wireless Compon. Lett., vol. 28, no. 4, pp. 290–292, 2018.
[32] S. L. March, "Analyzing lossy radial-line stubs," IEEE Trans. MTT, vol. 33, no. 3, pp. 269–271, 1985.

Chapter 5

Fast S-parameter Kron–Branin's modelling of rectangular wave guide (RWG) structure via mesh impedance reduction

Blaise Ravelo[1] and Olivier Maurice[2]

Abstract

The present chapter elaborates an innovative Kron–Branin's model of WG S-parameters. The graph topology is drawn from the equivalent electrical 1D circuit of the wave guide. The branch and mesh space analyses are introduced to determine the main unknowns of the problem represented by the contravariant mesh currents. Then, the mesh impedance reduction method is originally developed. Then, the rigorous tensorial equations enabling one to rapidly calculate the S-parameters are presented. Application examples are explored to validate the fast S-parameters modelling.

Keywords: TAN approach, Kron–Branin's model, frequency-domain analysis, modelling methodology, wave guide (WG) theory

5.1 Introduction to Chapter 5

The diversity of wireless communicating IoT objects forecasted for 5G technology raises questions on electromagnetic (EM) pollution. Behind this constraint, the printed circuit boards (PCBs) must operate under protection against EM interferences (EMIs). Therefore, different EM shielding techniques [1–3] have been deployed.

However, further improvements are still necessary to face up ultra-wideband (UWB) radiated EM emissions. Among the existing EM shielding techniques, the cavities with metallic enclosure are particularly popular [4–6]. The assessment of EMI effect in function of the real enclosure shielding effectiveness (SE) requires further study [5,6]. An improvement of shielding technique based on rectangular metallic enclosure with aperture by using extra shielding wall was proposed [4]. In general, the SE experimental approach of enclosure-shielding is usually expensive [6,7]. Nowadays, to study the EM shielding with enclosures, different simulation

[1]IRSEEM/ESIGELEC, Rouen, France
[2]ArianeGroup, Paris, France

and numerical modelling methods are developed. In terms of modelling aspect of particular shapes as parallelepiped metallic enclosures, the short-circuited rectangular wave guide (RWG) constitutes a relevant approach.

Since the 1990s, the electromagnetic compatibility (EMC) engineering on PCB shielding evolves with the use of emerging computer-aided designers. These EM simulators are developed based on numerical computation solvers. 3D full wave simulation methods have been deployed to predict the SE of enclosures, including the associated apertures [8,9]. The simulators use the solvers based on numerical iterative calculation methods with discrete integration of elementary 1D/2D/3D cells as FDTD [10,11], MoM [12] and also hybrid methods of MoM/FEM [13] and MoM/FDTD [14]. However, the full wave models are excessively time-consuming and do not enable us to comprehend the phenomenology of the computed SE.

Alternative modelling methods based on the equivalent circuit approach of resonant cavity representing the shielding enclosure have been developed [1–3,15]. The circuit models are generally based on the equivalent *RLC* resonant [16] or transmission line (TL) [17,18] networks. The circuit approach is advantageous for the comprehensive analysis. Furthermore, the EM and circuit hybrid model can also be performed as proposed in [19] by considering finite element time-domain computation. Nevertheless, the classical circuit approach does not enable us to realize critical EM coupling.

To overcome this approach limitation, an unfamiliar SE modelling based on the TAN using Kron's method was proposed [20]. The TAN approach consists in analysing EM problems via a graph topology [20–22]. The TAN formalism offers outstanding possibilities of representing any physical phenomena and could be a potential candidate to solve universal engineering problems [23–26]. The TAN allows one to model in relevant way diverse physical interactions via the tensor products and additions of object primitive elements [27]. Since the 1990s, the TAN concept was suggested by Maurice and his team [21,22] to solve the EMC problems of complex electronic, RF and microwave systems. Associated with Branin's method, the TAN allows one to integrate the propagation aspect as the TLs into the Kron–Branin (KB) formalism as proposed in [28].

In the present chapter, the unfamiliar TAN approach will be exploited to model originally RWGs under UWB excitation. The chapter is organized into four sections. Section 4.2 presents the problem formulation about the RWG S-parameters modelling. Section 4.3 is focused on the TAN modelling of RWG assumed as two-port symmetric circuit. The KB model is established in function of the RWG parameters. Section 4.4 discusses the comparison of the TAN-computed RWG model with simulations from a commercial tool. Then, Section 4.5 is the final conclusion of the chapter.

5.2 Problem formulation

This section formulates the problem related to the RWG modelling. The 3D structure is assumed as a two-port passive device. The necessity to establish the S-matrix black box is underlined.

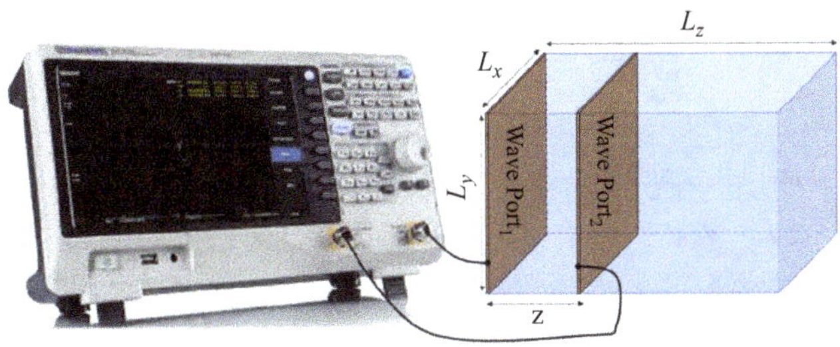

Figure 5.1 RWG connected to two-port VNA

5.2.1 Structural description

The central object of the present chapter is an RWG defined with

- physical dimensions ($L_x \times L_y \times L_z$),
- filled by lossless medium that is a vacuum material known with characteristics: absolute permittivity $\varepsilon_0 = 1/(36\pi\, 10^9)$ SI and absolute permeability $\mu_0 = 4\pi\, 10^{-7}$ SI,
- excited by plane wave sources represented by wave Port$_1$ ($z = 0$) from the left side and short-circuited ($z = L_z$) at the right side,
- consideration only of transversal modes propagating from $z = 0$ to the arbitrary position z.

The EM problem related to this RWG is the estimation of the EM fields or the EM powers at the different position inside this parallelepiped object. To characterize experimentally this RWG, we proceed regularly with S-parameters measurement as illustrated in Figure 5.1.

Because of symmetry, this RWG can be assumed as a two-port passive circuit. Under this approach, the most significant physical parameter is the longitudinal one represented by the variable z in Figure 5.1. Therefore, the problem can be transformed as a 1D equivalent circuit with fixed wave Port$_1$ (positioned at $z = 0$) and variable position wave Port$_2$ as illustrated in Figure 5.1.

5.2.2 Representation of S-matrix black box

In the electrical point of view, the RWG problem can be transformed as a two-port passive circuit. By taking into account the wave port excitation, the problem can be reformulated as an electrical network as depicted in Figure 5.2. The wave port $n = 1,2$ is represented as voltage sources E_n with internal reference resistive load R_0.

In this case, the RWG can be represented by a two-port S-matrix:

$$[S(j\omega)] = \begin{bmatrix} S_{11}(j\omega) & S_{12}(j\omega) \\ S_{21}(j\omega) & S_{22}(j\omega) \end{bmatrix} \tag{5.1}$$

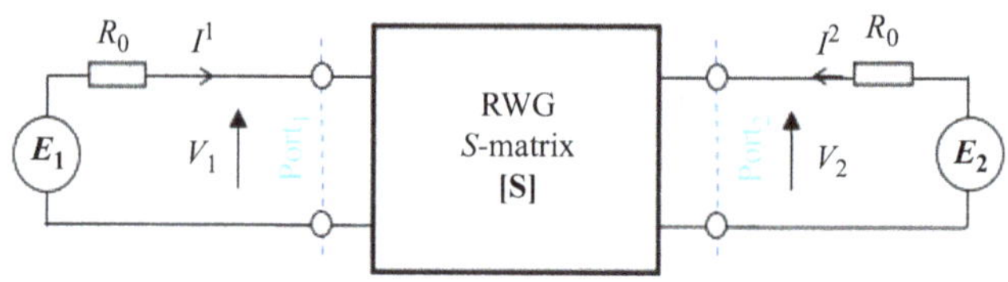

Figure 5.2 S-matrix black box equivalent model of RWG

The S-matrix can be analytically determined from the Z- or Y-matrices. To do this, the relation between the port voltage and port current vectors should be made:

$$\begin{cases} [V(j\omega)] = \begin{bmatrix} V_1(j\omega) & V_2(j\omega) \end{bmatrix} \\ [I(j\omega)] = \begin{bmatrix} I^1(j\omega) \\ I^2(j\omega) \end{bmatrix} \end{cases} \tag{5.2}$$

Based on this electrical representation, the detailed analytical investigation of this RWG will be explored in the next section.

5.3 KB theorization of RWG S-matrix

The analytical theory on RWG is recalled in this section. Then, the electrical circuit analogy between the RWG and TL will be elaborated. The KB modelling of RWG will be developed by considering the KB model investigated in Chapter 4.

5.3.1 Recall on RWG and TL theory

Before the elaboration of the unfamiliar KB model, the RWG theory will be introduced in the following section.

5.3.1.1 RWG characterization

The familiarity to the KB modelling may depend on the simplicity of application example. In the present chapter, we limit the study to the simple case of RWG manifested only by a single mode. In this optic, for the sake of mathematical simplicity, along the chapter, we assume that the RWG as a square section WG:

$$L_x = L_y = a \tag{5.3}$$

In this case, by denoting the speed of light in the vacuum c, the RWG cut-off frequency of fundamental propagating mode TE_{01} or TE_{10} is given by

$$f_c = \frac{\omega_c}{2\pi} = f_{cTE01=TE10} = \frac{c}{2a} \tag{5.4}$$

With the operating frequency,

$$f = \frac{\omega}{2\pi} \tag{5.5}$$

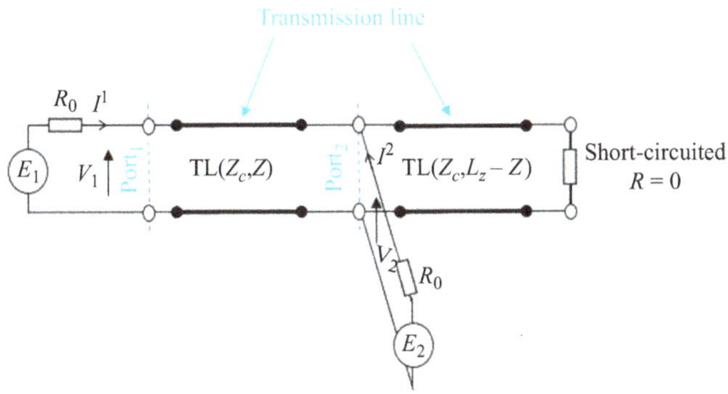

Figure 5.3 TL-based electrical equivalent circuit of the RWG introduced in Figure 5.1

this characteristic frequency enables us to distinguish the propagative mode ($f > f_c$) and the evanescent mode ($f < f_c$). The segregation of these fundamental regimes enables one to consider the electrical equivalent behaviour of the RWG as elaborated in the following section.

5.3.1.2 TL equivalent circuit of the RWG structure

By considering the EM wave propagating from Port_m to Port_n (for our RWG, $m,n = 1,2$), the RWG introduced in Figure 5.1 is equivalent to a TL as depicted in Figure 5.3. The electrical network equivalence can be understood with connections of excitation Port_1 and Port_2. The main circuit comprises two pieces of TLs:

- The first one TL(Z_c,z) connected between Port_1 and Port_2, which is defined with characteristic impedance Z_c and physical length z.
- And the other one TL(Z_c,$L_z - z$) connected with Port_2 and terminated by short-circuit (resistive load $R = 0$) with the same characteristic impedance and physical length $L_z - z$.

The constituting TL parameters as characteristic impedance depending on the air characteristic impedance:

$$Z_{air} = \sqrt{\frac{\mu_0}{\varepsilon_0}} \tag{5.6}$$

and propagation constant depends on the wave propagation regime.

5.3.1.3 TL equivalent circuit of RWG with the first propagative mode

With the first propagative mode ($\omega > \omega_c$), the equivalent TL of the RWG should have its specific frequency-dependent parameters:

- Characteristic impedance:

$$Z_c(\omega) = Z_p(\omega) = \frac{Z_{air}}{\sqrt{1 - (\omega_c^2/\omega^2)}} \tag{5.7}$$

- Constant phase:

TRIAL RESTRICTION (5.8)

- Propagation constant:

TRIAL RESTRICTION (5.9)

As these parameters change with the EM wave regime inside the RWG, the next section will recall the case of evanescent mode.

5.3.1.4 TL equivalent circuit of RWG under the evanescent mode

With the evanescent mode ($\omega < \omega_c$), the equivalent TL of the RWG frequency-dependent parameters are changed as

- Characteristic impedance:

$$Z_c(\omega) = Z_e(\omega) = \frac{Z_{air}}{\sqrt{(\omega_c^2/\omega^2) - 1}} \tag{5.10}$$

- Attenuation constant:

$$\alpha(j\omega) = \alpha_e(j\omega) = -\frac{\sqrt{\omega_c^2 - \omega^2}}{c} \tag{5.11}$$

- Propagation constant:

$$\gamma(j\omega) = \gamma_e(j\omega) = -\frac{\sqrt{\omega_c^2 - \omega^2}}{c} \tag{5.12}$$

5.3.2 KB modelling of RWG

After simplification of the equivalent circuit proposed in Figure 5.3, the KB modelling of the RWG will be developed in the present.

5.3.2.1 TL equivalent circuit

Based on the TL theory, the input impedance of TL(Z_c,$L_z - z$) connected at Port$_2$ can be written as

$$Z_z(j\omega) = Z_c(\omega)x(j\omega) \tag{5.13}$$

where

$$x(j\omega) = \tanh[-\gamma(j\omega)]. \tag{5.14}$$

With this impedance Z_z, the equivalent circuit of Figure 5.3 can be simplified as shown in Figure 5.4. The equivalent graph of this circuit enables us to define the KB modelling of the RWG.

5.3.2.2 KB graph topology equivalent to the RWG

Figure 5.5 represents the equivalent graph inspired from the reduced circuit proposed previously. In this graph, the TL termination figures out the Branin sources B_1 and B_2. The topological parametric indices are

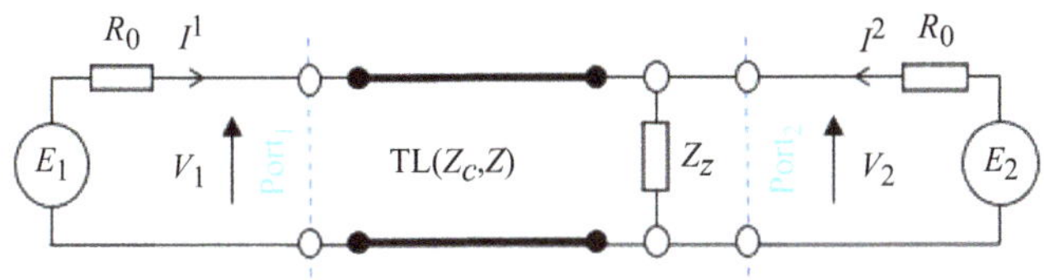

Figure 5.4 Reduced equivalent circuit of the RWG

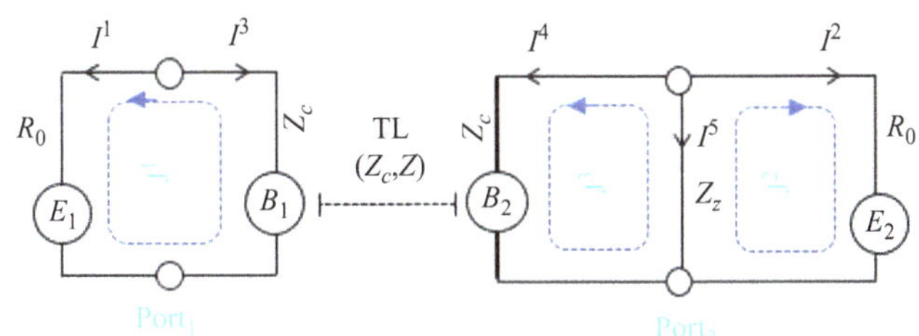

Figure 5.5 Equivalent graph of the RWG

- number of branches: $B = 5$,
- number of meshes: $M = 3$,
- number of ports: $P = 2$.

The branch space analysis enables to define the following:

- The covariant voltage:

$$[V_b] = \begin{bmatrix} E_1 & E_2 & xZ_cI^4 & xZ_cI^3 & 0 \end{bmatrix} \tag{5.15}$$

- The contravariant current:

$$[I^b] = \begin{bmatrix} I^1 \\ I^2 \\ I^3 \\ I^4 \end{bmatrix} \tag{5.16}$$

- The double covariant branch impedance, including the Branin coupling impedance:

$$[Z_{ab}] = \begin{bmatrix} R_0 & 0 & 0 & 0 & 0 \\ 0 & R_0 & 0 & 0 & 0 \\ 0 & 0 & Z_c & -xZ_c & -xZ_z \\ -xR_0 & 0 & -xZ_c & Z_c & 0 \\ 0 & 0 & 0 & 0 & Z_z \end{bmatrix} \tag{5.17}$$

The branch-to-mesh connectivity is given by

$$\left[I^b\right] = \left[C_m^b\right]\left[J^m\right] \Rightarrow \left[C_b^m\right] = \begin{bmatrix} 1 & 0 & 0 \\ 0 & 1 & 0 \\ -1 & 0 & 0 \\ 0 & 0 & 1 \\ 0 & -1 & -1 \end{bmatrix} \tag{5.18}$$

The mesh space analyses enabling one to determine

$$[V_m] = \begin{bmatrix} U^1 & U^2 & U^3 \end{bmatrix} \tag{5.19}$$

$$[J^m] = \begin{bmatrix} J^1 \\ J^2 \\ J^3 \end{bmatrix} \tag{5.20}$$

can be performed following the methodology established in Chapter 4.

5.3.2.3 *S*-matrix analytical expression of the RWG

The branch and mesh space analyses have been performed to determine the equivalent Z-matrix of the RWG seen from Port_1 and Port_2. With the Z-to-S transform, we established the following S-matrix coefficients:

$$S_{11}(j\omega) = \frac{\begin{matrix}[Z_c(\omega) - R_0]\{R_0[Z_c(\omega) + Z_z(j\omega)] + Z_c(\omega)Z_z(j\omega)\}x^2(j\omega) \\ -[R_0 + Z_c(\omega)]\{R_0[Z_c(\omega) - Z_z(j\omega)] + Z_c(\omega)Z_z(j\omega)\}\end{matrix}}{X(j\omega)} \tag{5.21}$$

$$S_{12}(j\omega) = S_{21}(j\omega) = \frac{4R_0Z_c(\omega)\,x(j\omega)Z(j\omega)}{X(j\omega)} \tag{5.22}$$

$$S_{22}(j\omega) = \frac{\begin{matrix}[R_0 + Z_c(\omega)]\{Z_c(\omega)Z_z(j\omega) - R_0[Z_c(\omega) + Z_z(j\omega)]\}x^2(j\omega) \\ -[R_0 - Z_c(\omega)]\{R_0[Z_c(\omega) - Z_z(j\omega)] + Z_c(\omega)Z_z(j\omega)\}\end{matrix}}{X(j\omega)} \tag{5.23}$$

with:

$$X(j\omega) = \left\{\begin{matrix}[R_0 + Z_c(\omega)]\{R_0[Z_c(\omega) + Z_z(j\omega)] + Z_c(\omega)Z_z(j\omega)\}x^2(j\omega) \\ +[R_0 - Z_c(\omega)]\{R_0[Z_c(\omega) - Z_z(j\omega)] + Z_c(\omega)Z_z(j\omega)\}\end{matrix}\right\} \tag{5.24}$$

The effectiveness of this theoretical approach will be checked in the next section with the application example analyses with parametric variations of the RWG section a and position z.

5.4 Validation results with parametric analyses

To verify the feasibility of the RWG KB modelling elaborated previously, illustrative applications will be examined in this section. An electronic circuit simulator

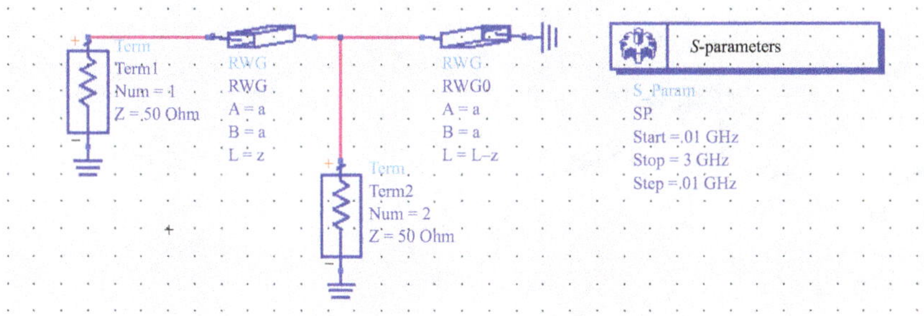

Figure 5.6 ADS schematic of RWG simulated circuit

Table 5.1 RWG POC parameters

Description	Parameter	Minimal value (cm)	Maximal value (cm)
Side	a	1	10
Length	L_z	12	
Depth	z	1	5

was used for validation. Simulations with the commercial tool ADS® from Keysight Technologies will be carried out. Comparisons between the simulated and theoretical calculated frequency responses will be discussed. Parametric analyses with variation of the RWG parameters will be explored.

5.4.1 Description of RWG POC

After description of the proof of concept (POC), the computation methods and simulation methods will be described in this subsection.

5.4.1.1 RWG POC description

Figure 5.6 represents the simulated circuit in the SPICE schematic environment of ADS. Two components of RWG available in the ADS library were considered:

- The component 'RWG' connected between Term1 and Term2 with section a and depth z,
- And the component 'RWG0' connected to the ground plane with section a and depth $L_z - z$.

The RWG POC is a square section structure with fixed length. The parameters of the POC are summarized in Table 5.1.

The S-parameter simulations have been performed in the frequency domain from $f_{\min} = 10$ MHz to $f_{\max} = 3$ GHz with sampling step $f_s = 10$ MHz.

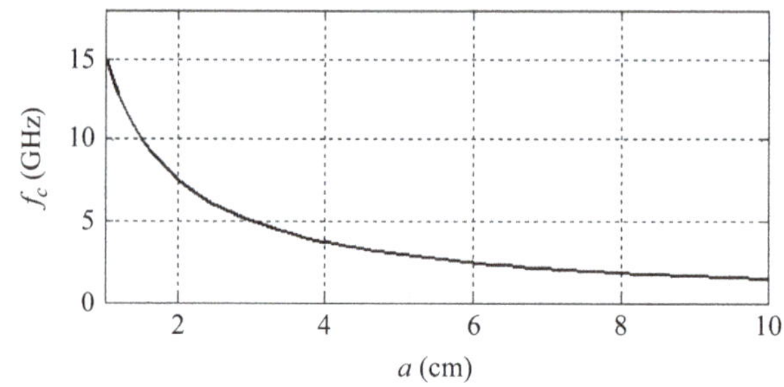

Figure 5.7 Variation of cut-off frequency f_c versus a

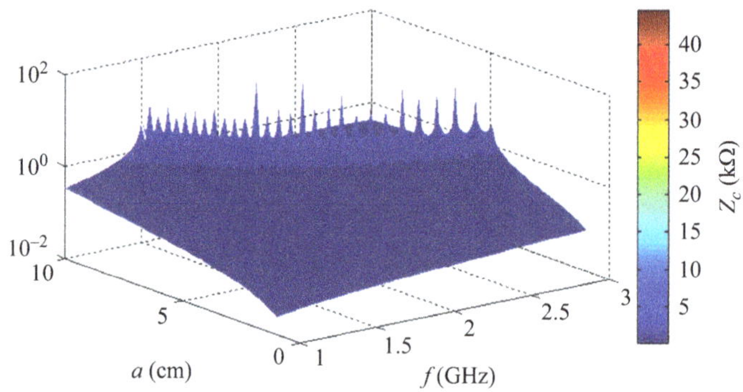

Figure 5.8 RWG equivalent TL characteristic impedance

5.4.1.2 Equivalent TL parameters

As pointed out in Section 4.3, the equivalent TL of the RWG depends on the later section side a. Therefore, the plot of the cut-off frequency in function of a is displayed in Figure 5.7. The cut-off frequency is inversely proportional to a by decreasing from 15 to 1.5 GHz.

The mapping cartography in function of the frequency f and the RWG side a of

- The variation of the characteristic impedance is represented in Figure 5.8. The characteristic impedance is along the line $f = f_c$.
- The variation of the phase constant is represented in Figure 5.9.

5.4.1.3 Routine algorithm of the RWG KB modelling

The RWG KB modelling was computed with MATLAB® by implementing the S-matrix coefficients expressed in (5.21)–(5.23). The programming can be divided into three parts:

- Part 1: Definition of the RWG parameters and the choice of the frequency bandwidth

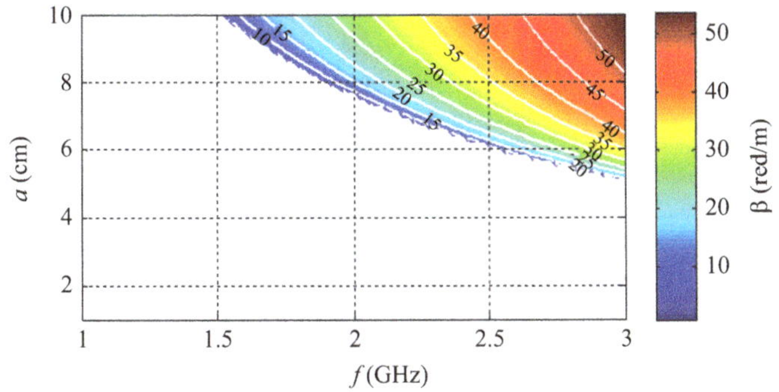

Figure 5.9 RWG equivalent TL phase constant

- Part 2: Definition of the equivalent TL parameters
- Part 3: Computation of S_{11}, $S_{12} = S_{21}$ and S_{22}

The following sections discuss the obtained results.

5.4.2 Discussion on RWG simulation results

Parametric analyses of S-parameter coefficients with respect to the RWG physical sizes have been performed. Then, comparison between KB model computed with MATLAB and ADS simulations has been realized to validate the concept. The present analyses are performed in the frequency band from $f_{min} = 10$ MHz and $f_{max} = 3$ GHz.

1. **S-parameter parametric analyses with respect to RWG side a**

The parametric analyses from (5.21) to (5.23) were performed with respect to the RWG side a varied from $a_{min} = 1$ cm to $a_{max} = 10$ cm. Figure 5.10 displays the mapping of the S-parameter magnitudes.

It can be underlined that the RWG operates with very significant reflections (S_{11} and S_{22}) which are above -1 dB as illustrated by Figure 5.10(a) and (c). It can be understood from Figure 5.10(b), the transmission coefficient confirms this behaviour with very high attenuation more than 20 dB. The high-pass filter behaviour is confirmed by this result. The attenuation is inversely proportional to a. Figure 5.11 presents the phases corresponding to the S-parameters S_{11}, S_{21} and S_{22}.

5.4.2.1 S-parameter parametric analyses versus position z

Similarly, to the previous section, parametric analyses with respect to the RWG depth z were carried out from the S-parameters modelled with the KB approach. The physical depth z is increased from $z_{min} = 1$ cm to $z_{max} = 5$ cm. Figure 5.12 presents the mapping of S-parameter magnitudes S_{11}, S_{21} and S_{22}. The corresponding phases are displayed in Figure 5.13.

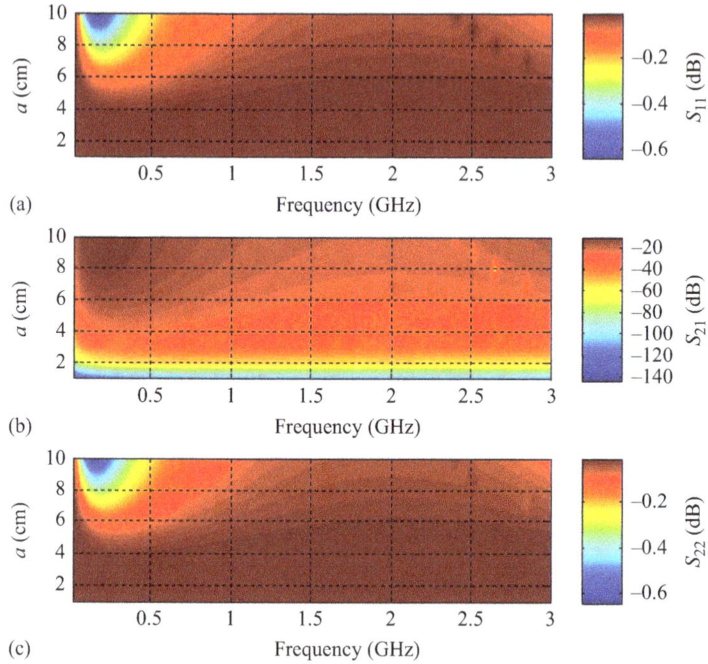

Figure 5.10 Mapping of KB computed S-parameter magnitudes versus (f,a): (a) S_{11}, (b) S_{21} and (c) S_{22}

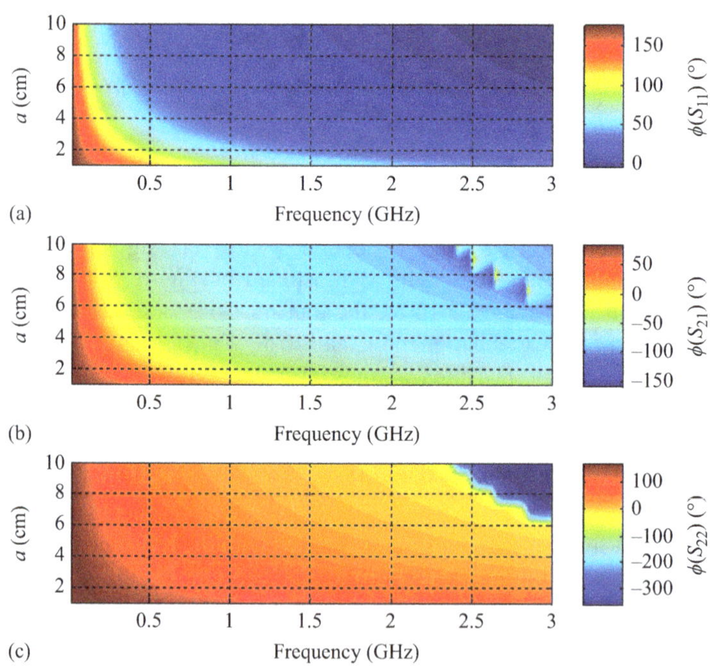

Figure 5.11 Mapping of KB computed S-parameter phases versus (f, a): (a) S_{11}, (b) S_{21} and (c) S_{22}

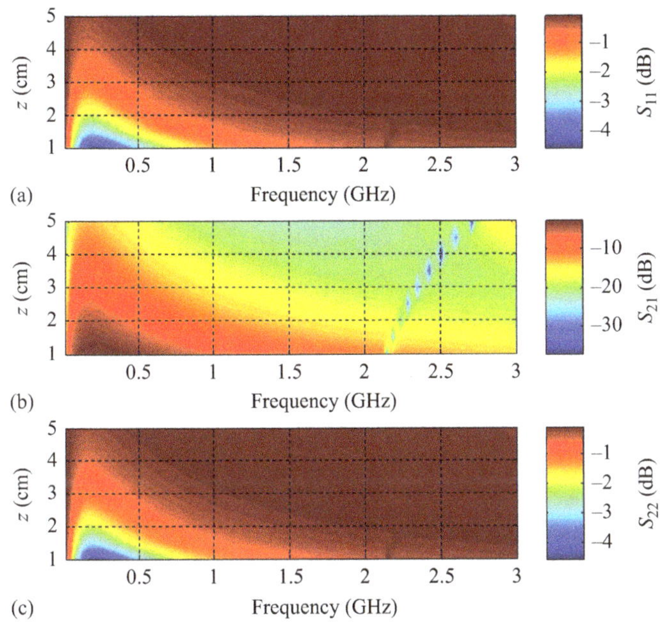

Figure 5.12 KB computed S-parameters versus z: (a) S_{11}, (b) S_{21} and (c) S_{22}

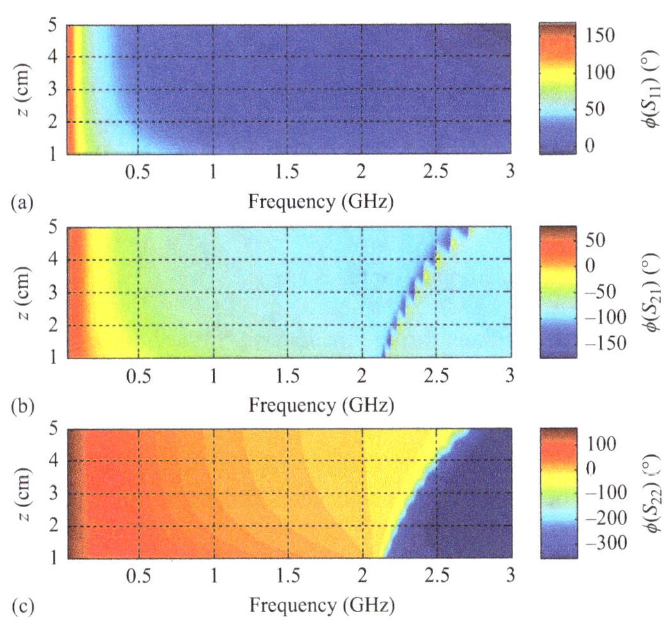

Figure 5.13 KB computed S-parameter phases versus z: (a) S_{11}, (b) S_{21} and (c) S_{22}

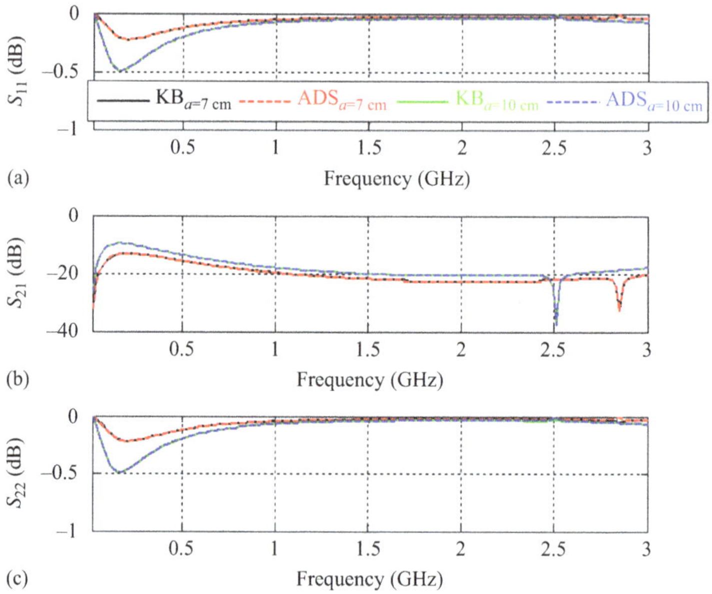

Figure 5.14 Comparisons of calculated and simulated S-parameter magnitudes for $a = \{7, 10\ cm\}$: (a) S_{11}, (b) S_{21} and (c) S_{22}

5.4.2.2 Comparisons between *S*-parameter KB-computed and ADS-simulated results

To confirm in more rigorous way, the effectiveness of the KB modelling applied to RWG, comparisons between the MATLAB computations and ADS simulations have been realized. Figure 5.14 plots the magnitudes of S_{11}, S_{21} and S_{22} for arbitrary chooses $a = \{7, 10\ \text{cm}\}$. Figure 5.15 represents the phase variation in the considered frequency band. It can be emphasized that a good correlation between KB calculations and ADS simulations.

The proposed KB modelling advantages are

- an EM computation without losing time in terms of design pre-processing,
- the possibilities of analytical understanding of EM field inside the RWG,
- the fast computation techniques of RWG responses in very wide frequency bands.

The correlation between the KB model and ADS simulation was also confirmed with the analyses based on the depth z arbitrary with chosen values $\{3, 5\ \text{cm}\}$. Figure 5.16 plots the magnitudes of S_{11}, S_{21} and S_{22} and Figure 5.17 displays the associated phase variation in the considered frequency band.

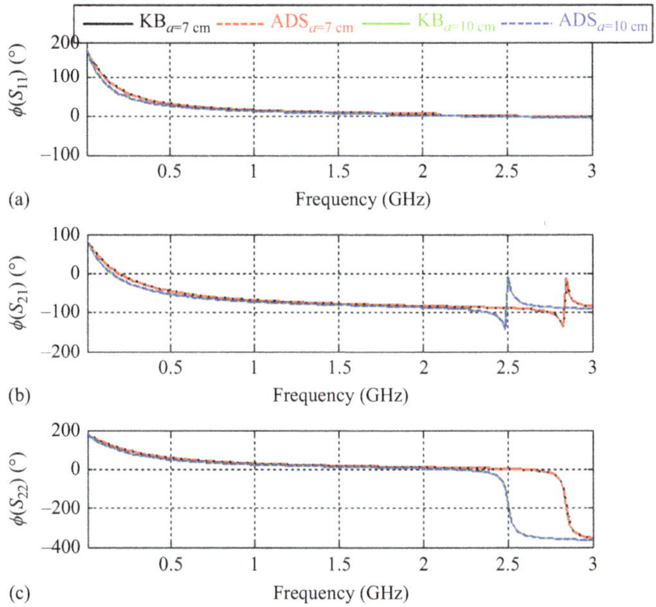

Figure 5.15 Comparisons of calculated and simulated S-parameter phases for $a = \{7, 10\ cm\}$: (a) S_{11}, (b) S_{21} and (c) S_{22}

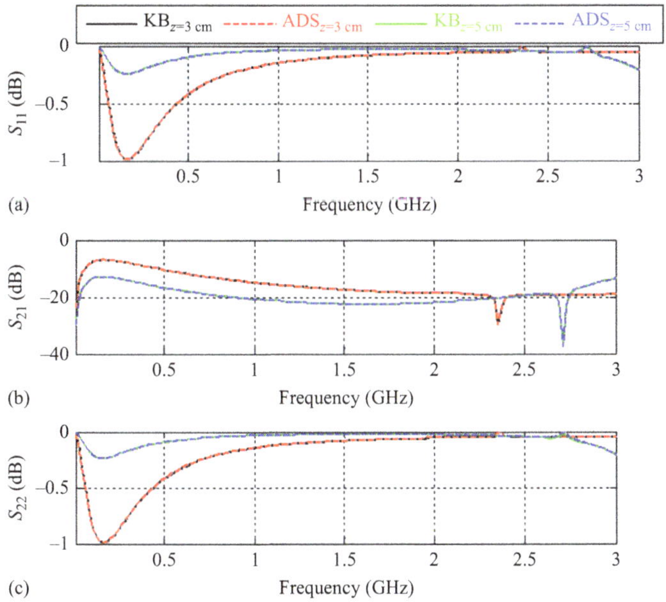

Figure 5.16 Comparisons of calculated and simulated S-parameter magnitudes for $z = \{3, 5\ cm\}$: (a) S_{11}, (b) S_{21} and (c) S_{22}

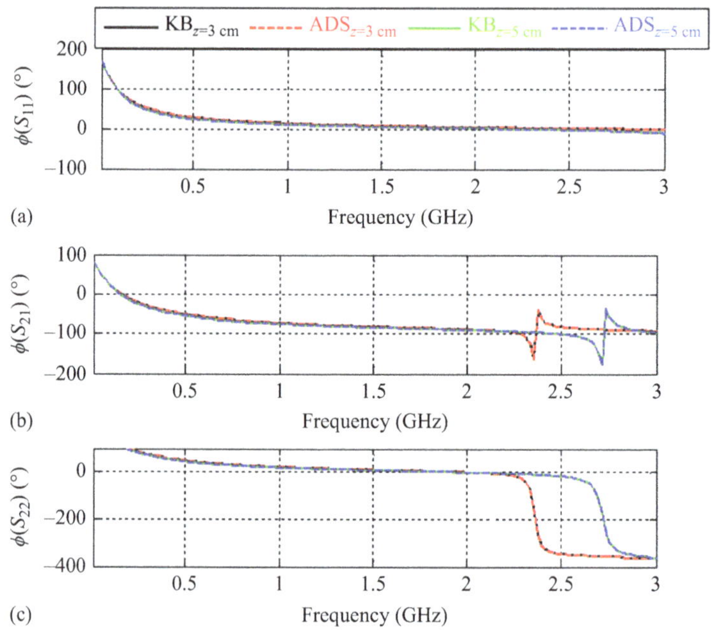

Figure 5.17 Comparisons of calculated and simulated S-parameter phases for $z = \{3, 5\ cm\}$: (a) S_{11}, (b) S_{21} and (c) S_{22}

5.5 Conclusion

A KB modelling method of RWG is developed. The methodology of the RWG modelling is described by assuming as equivalent to electric circuit constituted by TL. The graph topology of the RWG model is elaborated. The expressions of the S-matrix coefficients are presented.

The relevance of the RWG KB model is illustrated with numerical applications with an ideal POC. Comparisons between computed results and simulations from 10 MHz to 3 GHz are discussed.

In the future, the proposed KB model can be useful for the following:

- the EMC analyses of PCB by taking into account the enclosure effects,
- to calculate the EM field induced in the cavities or enclosures,
- to model the SE in UWB by taking into account the evanescent waves at low frequencies.

References

[1] B.-L. Nie, P.-A. Du, Y.-T. Yu, and Z. Shi, "Study of the shielding properties of enclosures with apertures at higher frequencies using the transmission line modeling method," IEEE Transactions on Electromagnetic Compatibility, vol. 53, no. 1, pp. 73–81, 2011.

[2] K. Li, G.-H. Wei, and X.-D. Pan, "Study on shielding effectiveness of rectangular metal enclosure with slots," Journal of Microwaves, vol. 29, no. 4, pp. 48–52, 2013.
[3] P.-Y. Hu and X.-Y. Sun, "Study of the calculation method of shielding effectiveness of rectangle enclosure with an electrically large aperture," Progress in Electromagnetics Research M, vol. 61, pp. 85–96, 2017.
[4] M. Bahadorzadeh and M. N. Moghaddasi, "Improving the shielding effectiveness of a rectangular metallic enclosure with aperture by using extra shielding wall," Proc. of 12th Int. Symp. on Antenna Technology and Applied Electromagnetics and Canadian Radio Sciences Conference, 2006, Montreal, QC, Canada, pp. 1–4, 17–19 July 2006.
[5] R. De Smedt, J. De Moerloose, S. Criel, *et al.*, "Assessment of the shielding effectiveness of a real enclosure," Proc. of Int. Symp. On EMC, Rome, Italy, pp. 248–253, 14–18 September 1998.
[6] F. Olyslager, E. Laermans, D. De Zutter, *et al.*, "Numerical and experimental study of the shielding effectiveness of a metallic enclosure," IEEE Transactions on Electromagnetic Compatibility, vol. 41, no. 3, pp. 202–213, 1999.
[7] M. Li, J. L. Drewniak, S. Radu, *et al.*, "An EMI estimate for shielding-enclosure evaluation," IEEE Transactions on Electromagnetic Compatibility, vol. 43, no. 3, pp. 295–304, 2001.
[8] T. Martin, M. Backstrom, and J. Loren, "Semi-empirical modeling of apertures for shielding effectiveness simulations," IEEE Transactions on Electromagnetic Compatibility, vol. 45, no. 2, pp. 229–237, 2003.
[9] L. G. Garcia-Perez, A. J. Lozano-Guerrero, J. M. Blazquez-Ruiz, *et al.*, "Time-domain shielding effectiveness of enclosures against a plane wave excitation," IEEE Transactions on Electromagnetic Compatibility, vol. 59, no. 3, pp. 789–796, 2017.
[10] C. Jiao, L. Li, X. Cui, and H. Li, "Subcell FDTD analysis of shielding effectiveness of a thin-walled enclosure with an aperture," IEEE Transactions on Magnetics, vol. 42, no. 4, pp. 1075–1078, 2006.
[11] J. Chen and J. Wang, "A three-dimensional semi-implicit FDTD scheme for calculation of shielding effectiveness of enclosure with thin slots," IEEE Transactions on Electromagnetic Compatibility, vol. 49, no. 2, 2007.
[12] R. Araneo and G. Lovat, "An efficient MoM formulation for the evaluation of the shielding effectiveness of rectangular enclosures with thin and thick apertures," IEEE Transactions on Electromagnetic Compatibility, vol. 50, no. 2, pp. 294–304, 2008.
[13] S. Yenikaya, "Electromagnetic analysis and shielding effectiveness of rectangular enclosures with aperture using hybrid MoM/FEM," Iranian Journal of Electrical and Computer Engineering, vol. 10, no. 2, pp. 70–76, 2011.
[14] G. Marrocco and F. Bardati, "Combined time and frequency-domain modeling of electromagnetic radiation from apertures on resonant cavities by FDTD-MOM method," Journal of Electromagnetic Waves and Applications, vol. 16, pp. 523–539, 2012.
[15] J. Shim, D. G. Kam, J. H. Kwon, and J. Kim, "Circuital modeling and measurement of shielding effectiveness against oblique incident plane wave on

apertures in multiple sides of rectangular enclosure," IEEE Transactions on Electromagnetic Compatibility, vol. 52, no. 3, pp. 566–577, 2010.

[16] S. Leman, B. Demoulin, O. Maurice, M. Cauterman, and P. Hoffmann, "Use of the circuit approach to solve large EMC problems," Comptes Rendus Physique, vol. 10, no. 1, pp. 70–82, 2009.

[17] M. P. Robinson, J. D. Turner, D. W. P. Thomas, *et al.*, "Shielding effectiveness of a rectangular enclosure with a rectangular aperture," Electronics Letters, vol. 32, no. 17, pp. 1559–1560, 1996.

[18] M. P. Robinson, T. M. Benson, C. Christopoulos, *et al.*, "Analytical formulation for the shielding effectiveness of enclosures with apertures," IEEE Transactions on Electromagnetic Compatibility, vol. 40, no. 3, pp. 240–248, 1998.

[19] P. Li and L. J. Jiang, "A hybrid electromagnetics-circuit simulation method exploiting discontinuous Galerkin finite element time domain method," IEEE Microwave and Wireless Components Letters, vol. 23, no. 3, pp. 113–115, 2013.

[20] S. Leman, A. Reineix, F. Hoeppe, *et al.*, "Kron's method applied to the study of electromagnetic interference occurring in aerospace systems," Proc. of 2012 ESA Workshop on Aerospace EMC, Venice, Italy, pp. 1–6, 21–23 May 2012.

[21] O. Maurice, "La compatibilité électromagnétique des systems complexes (in French)," Lavoisier, 2007.

[22] O. Maurice, A. Reineix, P. Durand, and F. Dubois, "Kron's method and cell complexes for magnetomotive and electromotive forces," International Journal of Applied Mathematics, vol. 44, no. 4, pp. 183–191, 2014.

[23] G. Kron, "Electrical engineering problems and topology," Matrix and Tensor Quarterly, 1951.

[24] G. Kron, "Solving extremely large and complicated physical systems in easy stages," Matrix and Tensor Quarterly, 1953.

[25] G. Kron, "Diakoptics – a gateway into universal engineering," Electrical Journal (London), 1956.

[26] G. Kron, "A set of principles to interconnect the solutions of physical systems," Journal of Applied Physics, 1953.

[27] B. Ravelo and O. Maurice, "Kron–Branin modeling of Y-Y-tree interconnects for the PCB signal integrity analysis," IEEE Transactions on Electromagnetic Compatibility, vol. 59, no. 2, pp. 411–419, 2017.

[28] B. Ravelo, O. Maurice and S. Lalléchère, "Asymmetrical 1:2 Y-tree interconnects modelling with Kron-Branin formalism," Electronics Letters, vol. 52, no. 14, pp. 1215–1216, 2016.

Chapter 6
Time-domain TAN modelling of PCB-lumped system with Kron's method

Zhifei Xu, Yang Liu[1]*, Blaise Ravelo*[1]*, and Olivier Maurice*[2]

Abstract

The time domain (TD) modelling of the lumped components-based low and medium speed printed circuit board (PCB) is examined in the present chapter. The TD translation of the physical-classical electric circuit-tensorial analysis of networks (TAN) graph dictionary will be established. Then, the workflow of the methodology describing the routine algorithm of the TD Kron's model will be introduced. By considering pulse input signals, the computation methods of the TAN method resolution based on the time-different iterative discrete solver are established. The feasibility of the TAN TD model will be verified with examples of PCB systems excited by the previously cited test signals. Discussion will be made about the accuracy, advantages and limits of the TAN TD computation methods.

Keywords: TAN approach, Kron's method, Kron–Branin's model, frequency-domain analysis, modelling methodology, low and medium speed PCB, SI analysis, PI analysis, EMC analysis

6.1 Introduction

As pointed in [1], with respect to the printed circuit board (PCB) design technology progress, progressive analysis techniques and methodologies are necessary. Various approaches and simulation tools have been developed to predict the behaviour of multilayer PCBs in a holistic perspective. Currently, the existing tools operate with different solvers which can be categorized into circuit analysis, 3D full wave approaches and hybridized methods. However, global simulations of the complete multilayer PCB complete layouts with 3D full wave electromagnetic (EM) solvers will consume huge

[1]IRSEEM/ESIGELEC, Rouen, France
[2]ArianeGroup, Paris, France

time in consequence of the meshes. Based on the transmission line (TL) RC- [2], RLC- [3] and RLCG- [4] network models, analytical computational approaches of planar PCB microstrip symmetric tree interconnects were proposed. However, these approaches are mathematically too complicated for higher level asymmetric trees [5] because of the successive Z-, Y- and T-matrix transforms. For this reason, modelling and simulation methods of multilayer interconnect structures constitute one of the hottest topics for component packaging technology design engineers [6,7]. Nowadays, the prediction of high density interconnect (HDI) as multilayer PCB trace and signal distribution buses constitutes on of the current open challenges [6]. Further research work is expected to elaborate a technique allowing us to reduce the cost of signal integrity (SI) and electromagnetic compatibility (EMC) characterizations for novel multilayer PCBs with acceptable performances [8,9].

To face up this challenge, the unfamiliar tensorial analysis of networks (TAN) was recently introduced for analysing the SI and power integrity (PI) of multilayer interconnect structures [10,11]. Simple tree interconnect structures with single-input and double-outputs were considered [10,11]. The proof-of-concept (POC) structures were designed following the standard described in [12]. The POCs were mainly composed of elementary TLs, via, pads and anti-pads. The TAN model enables one to compute rapidly the multilayer interconnect S-parameters and voltage transfer function. The model integrates the TL and via characteristics as proposed, respectively, in [13] and [14,15]. The TAN model presents an advantage of flexibility to be hybridized with various models as the different shapes of via introduced in [16,17]. In difference with the previous works [10,11], the present paper investigates the TAN modelling of multilayer hybrid PCB, including surface mounted device (SMD) components and the innovative TAN direct time-domain (TD) computation. In addition, the TAN model is validated experimentally with fabricated 3D six-layer PCB prototype.

Before developing the main content of the paper, it is worth to get clear insight about the TAN model which is unfamiliar for most of electrical and electronic engineers. Let us describe the state-of-the-art on the TAN computational method. The TAN was initiated by the General Electric Research Engineer, Kron in the 1930s [18,19]. Inspired with the concept, Kron established electrical machine modelling based on the tensor calculus. He defined the TAN as modern physics computational method. In the 1950s, he envisioned to extend the TAN computational method for the analyses of universal complex systems. To do this, he proposed an innovative method to analyse complex system based on the decomposition method called 'diakoptics' approach. In the 1950s, few research groups as the tensor society of Japan were explored Kron's formalism for the electrical and electronics engineering [20]. In this way, in the 1960s, a complementary approach of Kron's formalism has been initiated by Branin for the TL transient analysis [21]. The Branin model enables one to calculate the signal propagation through TLs by considering voltage source coupling between the TL terminations [21,22]. Since the 1990s, Maurice and his group have introduced an extension of Kron's formalism by creating modified Kron's method for the EMC complex system analyses [23,24]. By combining Kron's and Branin's models, they develop more general TAN computational method for electronic systems, including the PCBs. More recently, the Kron–Branin (KB) formalism was applied to the tree interconnects of single-layer planar PCB [25,26]. It was found that the KB formalism enables one to analyse rapidly and accurately the tree interconnect SI analyses. In the continuation, the general theory on the TAN model for the 3D multilayer PCB modelling is developed in the present chapter.

This chapter is organized into main sections: Section 6.2 focuses on the basic methodology of TD implementation of TAN modelling. After the PCB problem formulation, the graph topology equivalent to the structure under study is elaborated. Then, the tensor metrics abstracting the system is introduced. The validation results with TT and Y-tree topologies of lumped element networks are examined in Section 6.3. Finally, Section 6.4 addresses the conclusion.

6.2 Basic definitions and general methodology of the innovative direct TD TAN modelling of PCBs

This subsection is focused on the general algebraic approach of the innovative direct TD TAN modelling. It is necessary to define the basic parameters before the elaboration of the TD TAN methodology. The specific differences between the frequency domain and TD methods will be pointed out in this section. Then, the primitive element representations, the general TD TAN methodology and the specific expressions of the unknown variable analytical calculations will be introduced.

6.2.1 Representation of TAN topology in the TD

Let us denote m and n integers, the number of input and output parameters which depends on function of the time-dependent continuous variable t. The particularity of TD analysis is the necessity to consider the discretization of input signal.

6.2.1.1 Excitation signal description

To do this, the plot of discrete signal as shown in Figure 6.1 must be defined correctly in function of the PCB characteristics.

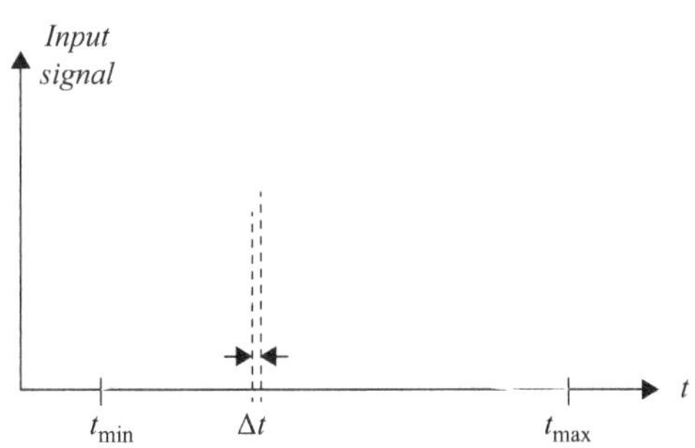

Figure 6.1 Transient signal plot and the associated time parameters

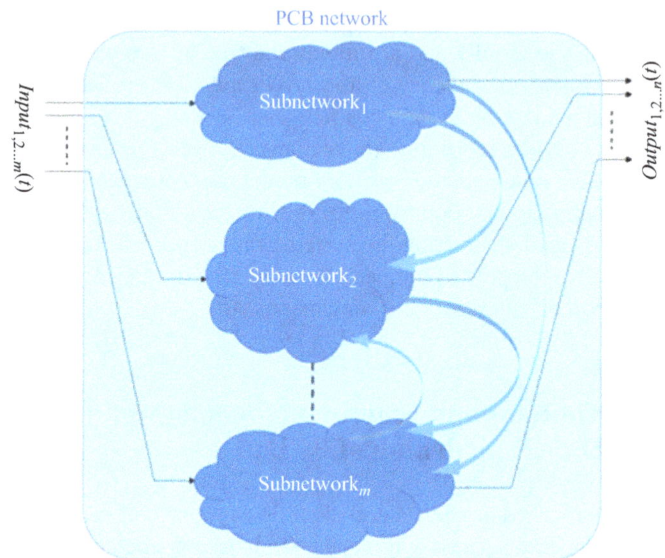

Figure 6.2 General representation of abstracted network with m-inputs and n-outputs

6.2.1.2 Block diagram representation

Figure 6.2 represents a general representation of the graph topology constituting the PCB network. This equivalent graph of PCB was defined by Kron as the universe of the problem. The first level of the PCB structure abstraction would be the description of the overall graph. This later can be constituted by several interactive or coupled subnetworks. The overall graph network can be analysed in function of the m-dimension inputs $Input_{1,2,\ldots,m}(t)$ as excitation and n-dimension outputs $Output_{1,2,\ldots,n}(t)$ as unknowns. In algebraic approach, these inputs and outputs can be assumed as 1-rank tensor or generally vector objects. The interaction between these mathematical objects is traduced from the physical laws governing the system. With the TAN approach, this interaction can be compactly formulated as tensorial metric of the problem. The ultimate objective of the proposed TD TAN model is to solve this problem metric via the suitable analytical way to determine the time-dependent unknown outputs $Output_{1,2,\ldots,n}(t)$.

6.2.2 Key parameters of TD implementation of TAN approach

As argued previously, the TD TAN modelling consists in determining the time-dependent unknown currents and voltages ($Output_{1,2,\ldots,m}$) of the PCB supposed excited by transient signals ($Input_{1,2,\ldots,n}$). In this point of view, the network can be assumed as a linear time invariant transfer system. In this chapter, the SI and PI analysis study is limited to the case of excitations represented by inputs and unknown outputs as electrical currents and voltage.

Similarly, to the frequency domain, this direct TD must be realized following the methodology indicated in Figure 6.5. Nevertheless, in difference to the frequency domain, the TD modelling must respect the following golden rules:

- Instantaneous and continuous signal processing: First, the TD approach is the analysis of physical phenomenon in the time window delimited by the instant times t_{min} (generally referenced as zero) and t_{max}. This time duration in addition to the rise and fall times of the excitations constitute the nature of electrical and EM interactions to take into account during the SI and PI analyses. Then, the modern HS PCB analyses necessitate the consideration of microwave phenomena in the frequency ranges from DC up to several GHzs.
- Discrete signal processing: The resolution of the TAN metric in the TD must be performed with the discretization approach. The discretization consists in representing the phenomena with successive discrete times t_k for $k = \{1,2,\ldots,k_{max}\}$, with $t_1 = 0$ s by assuming that the time step is equal to constant:

$$\Delta t = t_{k+1} - t_k \tag{6.1}$$

  for all values of the index k. This time step must be chosen with respect to the speed of the electrical and EM phenomena and also the geometrical parameters constituting the PCB network.

- Initial conditions: Because of the successive implicit delays $Delay_{1,2,\ldots}$ induced by the reactive and distributed elements constituting the TAN graph, the TD metric must be defined with initial conditions between the input variables. These initial conditions can be defined as 1-rank tensor having indices from 1 to the natural integer:

$$k_0 = \max\left[\frac{Delay}{\Delta t}\right] \tag{6.2}$$

  with $[x]$ is the superior integer part of the real x.

- Time invariant topology: Similar to the fundamental physical laws as Maxwell's equations, in the present chapter, we adopt that the PCB network transfer system does not change the behaviour in function of t. Such PCB networks belong to the hypothesis of time invariant system. In other word, the TD tensor metrics are assumed as time invariant tensorial operations.
- Time-difference operations: It acts as iterative difference equations between the input/output currents and voltages in the considered time window. Knowing the initial conditions, the resolution of the TD tensor metrics can be implemented as routine algorithms of iterative calculations by varying the time index k from k_0 defined in (6.2) to

$$k_{max} = \left[\frac{t_{max}}{\Delta t}\right] \tag{6.3}$$

- Stability condition: The time-difference computations require critical condition in term of stability. The condition allows us to ensure the calculability of difference equations between the parameters at the instant time $t = t_k$ and the previous instants $t = t_{1,2,...,(k-1)}$ for k incremented from k_0 to $k_{\max}$.
- Causality: One of the golden rules of fundamental physics in the TD is the causality. Translated in the area of SI and PI engineering, the causality means that the state of $Output_{1,2...,n}(t_k)$ appears after the previous states corresponding to the indices $\{1,2,\ldots,k-1\}$. The illustrative general analytical expression of this causality will be given in the next section.

6.2.3 TAN TD primitive elements

To model a PCB with the direct TD TAN method, we need the tensorial definitions of TD primitive elements. Acting as an electrical object, the primitive elements are defined in function of the input/output currents and voltage operating with the PCB network. The analytical equation governing the dipole corresponds to the physical law taking into account the instantaneous voltage $v(t)$ and current $i(t)$. The TD TAN primitive elements are defined in the following sections (Figure 6.3).

6.2.3.1 TD TAN general transfer equation

Therefore, the associated TD graph must be associated with the time-dependent tensorial objects $[v_{in}(t)]$, $[i_{in}(t)]$, $[v_{out}(t)]$ and $[i_{out}(t)]$. In systemic representation, the primitive elements of PCB network, including the input and output variables in the TD, can be represented in Figure 6.4.

After the graph elaboration, the TAN mathematization must be proceeded with tensorial metric formulation. By denoting the transfer tensors $[\psi_{in}]$, $[\psi_{out}]$ and

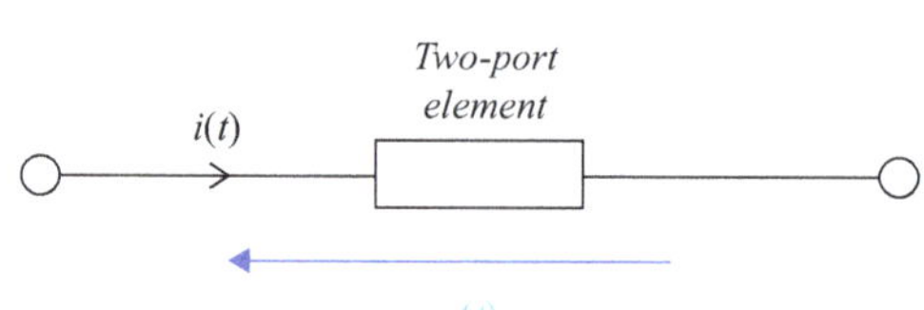

Figure 6.3 Two-port element of PCB

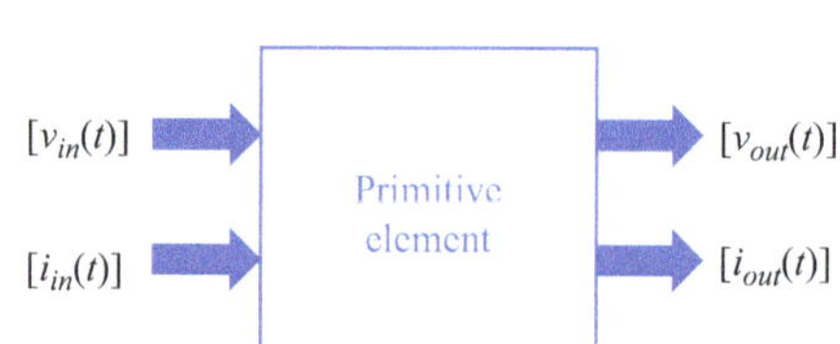

Figure 6.4 Diagram of primitive element representation in the TD

$[\psi_{in-out}]$ associated with the PCB network, the instantaneous laws and time-difference operations can be intuitively formulated by

$$[\psi_{out_0}(t_k)] \odot \begin{bmatrix} [v_{out}(t_k)] \\ [i_{out}(t_k)] \end{bmatrix} = \left\{ \begin{array}{l} [\psi_{in-out}(t_k)] \odot \begin{bmatrix} [v_{in}(t_k)] \\ [i_{in}(t_k)] \end{bmatrix} + \\ \sum_{l=1}^{k-1} \left\{ \begin{array}{l} [\psi_{out}(t_l)] \odot \begin{bmatrix} [v_{out}(t_l)] \\ [i_{out}(t_l)] \end{bmatrix} \\ +[\psi_{in}(t_l)] \odot \begin{bmatrix} [v_{out}(t_l)] \\ [i_{out}(t_l)] \end{bmatrix} \end{array} \right\} \end{array} \right\} \tag{6.4}$$

The tensorial product figures out in this expression in order to take into account the TD integro-differential operators.

6.2.3.2 Dictionary of TD TAN modelling

During the implementation of the TD metric, the primitive elements can be defined following the dictionary of TD tensorial objects introduced in Tables 6.1 and 6.2. This representation concerns the TD TAN models of the *RLC*-lumped elements. In the present chapter, the TL, via, pads and other elements constituting the PCB can also be implemented as *RLC*-lumped elements based on the Telegraphers equation equivalence circuit.

The present paragraph is focused on the detailed TD expressions of tensorial objects associated with the primitive elements addressed in Tables 6.1 and 6.2. The proposed tensorial implementation of primitive elements depends on the time-difference formulation of the TD TAN.

6.2.3.3 Inductive element time-difference tensorial expression

The integro-differential equations of lumped element TD TAN model depend on the properties of the lumped via-pad inductors L_k and capacitors C_k with $k = 1, 2, \ldots$. Accordingly, the time-dependent operator of the inductor L is defined by

$$J(t) = \frac{\Delta t}{L}\left[V(t) + \frac{L}{\Delta t}J(t - \Delta t)\right] \tag{6.5}$$

In this expression, Δ_t is the time step, the rise and fall times require minimum 10 time steps, this defines the time steps, Δ_t/L is the inverse of impedance operator in TD, V is the extrinsic source term, $J(t - \Delta_t)L/\Delta_t$ is an intrinsic source term, generated by the inductance load which should appear in the Wronskian vector W.

6.2.3.4 Capacitive element time-difference tensorial expression

The capacitor TD operator is defined by the instantaneous charge Q with respect to the number of meshes p and the following formula:

$$Q^p(t) = Q^p\,(t - \Delta t) + \Delta t J^p\,(t) \tag{6.6}$$

Table 6.1 Primitive elements expression in time domain

Elements	Time domain	Discrete	Kron mesh impedance tensor	Kron Wronskian tensor	Kron source tensor	Initial condition
Resistor	$v(t) = Ri(t)$	$v(t) = Ri(t)\ t = \{0, \infty\}$	$Z_{mn} = [R]$	N/A	$E_m = [v(t)]\ t = \{0, \infty\}$	$if\ t \leq \Delta t$ $i = 0$
Inductor	$v(t) = L\dfrac{di(t)}{dt}$	$\dfrac{L}{\Delta t}i(t) = \left[v(t) = \dfrac{L}{\Delta t}i(t - \Delta t)\right]$ $t - \{\Delta t, \infty\}$	$Z_{mn} = \left[\dfrac{L}{\Delta t}\right]$	$W = \left[\dfrac{L}{\Delta t}i(t - \Delta t)\right]$	$E_m = [v(t)]\ t = \{0, \infty\}$	$if\ t \leq \Delta t$ $i_0 = 0$
Capacitor	$i(t) = C\dfrac{dv(t)}{dt}$	$Q(t) = Q(t - \Delta t) + i(t)$ $\dfrac{\Delta t}{C}i(t) = \dfrac{\Delta t}{C}Q(t - \Delta t)$	$Z_{mn} = \left[\dfrac{\Delta t}{C}\right]$	$W = \left[\dfrac{\Delta t}{C}Q(t - \Delta t)\right]$	N/A	$if\ t \leq \Delta t$ $\begin{cases} i = 0 \\ Q = 0 \end{cases}$

Table 6.2 Primitive elements expression in time domain

Structure	Time domain	Kron mesh impedance tensor	Kron Wronskian tensor	Kron source tensor	Initial condition
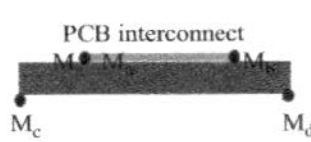	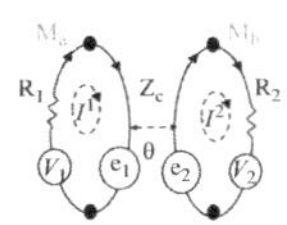	$Z_{mn} = \begin{bmatrix} Z_C & 0 \\ 0 & -Z_C \end{bmatrix}$	$W = \begin{bmatrix} -V_2(t-\tau) + Z_C I^2(t-\tau) \\ -V_1(t-\tau) - Z_C I^1(t-\tau) \end{bmatrix}$	$E_{mn} = \begin{bmatrix} V_1(t) \\ V_2(t) \end{bmatrix}$	$V_2(t) = 0$ *and if* $t \le \Delta t, \begin{cases} V_1(t-\tau) = 0, V_2(t-\tau) = 0 \\ I^1(t-\tau) = 0, I^2(t-\tau) = 0 \end{cases}$
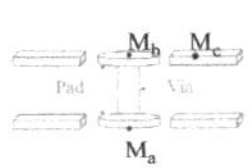	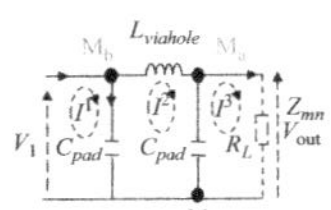	$= \begin{bmatrix} \frac{\Delta t}{C_{pad}} & -\frac{\Delta t}{C_{pad}} & 0 \\ -\frac{\Delta t}{C_{pad}} & 2\frac{\Delta t}{C_{pad}} + \frac{L_{viahole}}{\Delta t} & -\frac{\Delta t}{C_{pad}} \\ 0 & -\frac{\Delta t}{C_{pad}} & \frac{\Delta t}{C_{pad}} + R_L \end{bmatrix}$	$W = \begin{bmatrix} -\frac{\Delta t}{C_{pad}} Q_1(t-\Delta t) + \frac{\Delta t}{C_{pad}} Q_2(t-\Delta t) \\ \begin{Bmatrix} -\frac{\Delta t}{C_{pad}} Q_1(t-\Delta t) - 2\frac{\Delta t}{C_{pad}} Q_2(t-\Delta t) \\ -\frac{\Delta t}{C_{pad}} Q_3(t-\Delta t) + \frac{L_{viahole}}{\Delta t} I^2(t-\Delta t) \end{Bmatrix} \\ \frac{\Delta t}{C_{pad}} Q_2(t-\Delta t) + \frac{\Delta t}{C_{pad}} Q_3(t-\Delta t) \end{bmatrix}$	$E_m = \begin{bmatrix} V_1(t) \\ 0 \\ 0 \end{bmatrix}$	*if* $t \le \Delta t,$ $\begin{cases} Q_n(n = 1, 2, 3) = 0 \\ I^2(t-\Delta t) = 0 \end{cases}$

The mesh current $J(t)$ is calculated at each instant time step with the capacitor C_k integral operator:

$$\frac{\Delta t}{C_k}\sum_{t=0}^{t=n} J^p(t) = \frac{\Delta t}{C_k} Q^m(t-\Delta t) \tag{6.7}$$

where the left-side quantity $\Delta t/C_k$ is the impedance operator in TD and the right-side quantity is the element of Wronskian internal source vector W.

6.2.4 Methodology of PCB trace modelling with TAN TD approach

Knowing the previous TD implementation golden rules and the specifications of primitive elements, a general methodology is necessary to apply the TD TAN. This TD modelling is organized as multi-interactions of different steps. Figure 6.5 illustrates the workflow of the routine algorithm to treat a PCB SI and PI problem.

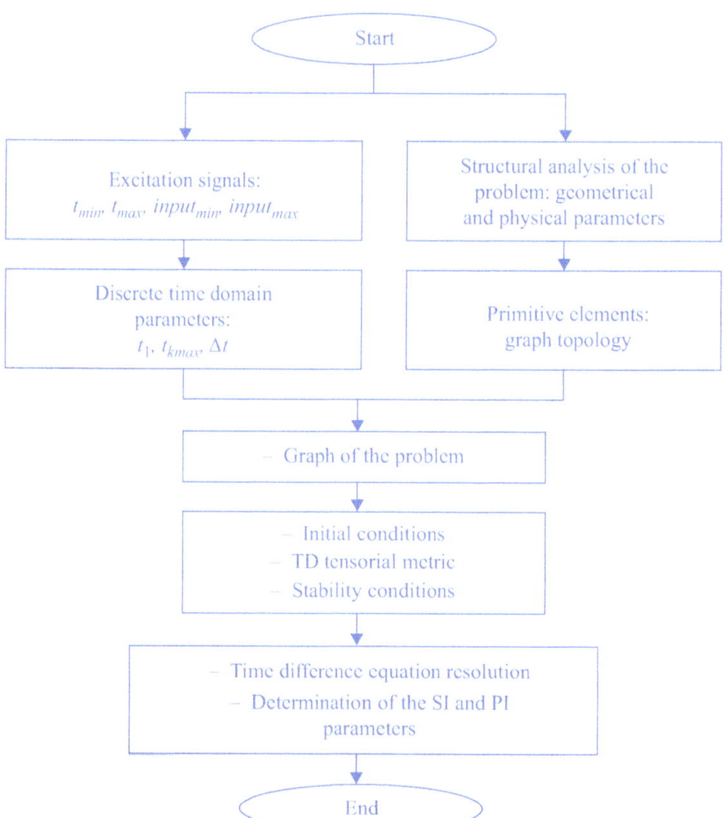

Figure 6.5 Workflow of the proposed TD TAN method

This routine can be chronologically described in following steps:

- Step 1: The parallel actions about the analyses of the excitation signals and the PCB structures must be performed. The initial time, physical and geometrical parameters of the problem are defined in this step.
- Step 2: The following actions are the implementations of the graph primitive elements by taking into account the time discrete parameters. At this stage, the implicit initial conditions associated with the delayed elements must be respected.
- Step 3: The equivalent graph of the problem constitutes the main action of this step. This topology induces the basic understanding of the electrical and EM parameter interactions.
- Step 4: In this step, the previous topology must be mathematized with a compact tensorial metric. The initial and the probable calculability conditions must be associated with the established metric. At this stage, the problem is already analytically transformed as an analytical expression.
- Step 5: It consists in the resolution of the posed tensorial metric of the previous step. This resolution is generally performed with iterative computations.

The relevance of this theoretical approach will be checked in the next section with the application example analyses of lumped circuits.

6.3 Application to two port LC circuits

To verify the feasibility of the TD TAN modelling previously defined, illustrative applications will be examined. Doing this, TT-tree and Y-tree circuits are considered as POCs. Transient simulations with the electronic circuit simulator tool ADS® from Keysight Technologies® will be carried out. Comparisons between the simulated and theoretical calculated TAN TD responses will be discussed.

6.3.1 TD TAN application with TT LC circuit

The first TD application results are discussed in the present subsection. Following the methodology described earlier in Figure 6.5, the modelling will be based on the graph topology, branch and mesh space TAN, and TD metric.

6.3.1.1 Description of the TT-topology circuit

Figure 6.6 presents the schematic of the TT-tree circuit. It can be assumed as a two-port system. The circuit is constituted by identical series resistor R and shunt capacitor C. The voltage source U_0 connected at input M_{in} is considered as excitation. The aim of this investigation is to plot the output voltage at node M_{out}.

6.3.1.2 Equivalent of the TT-topology circuit

The TAN equivalent graph is drawn in Figure 6.7. The topological index parameters of this graph are

- number of branches: $B = 5$,
- number of nodes: $N = 3$,

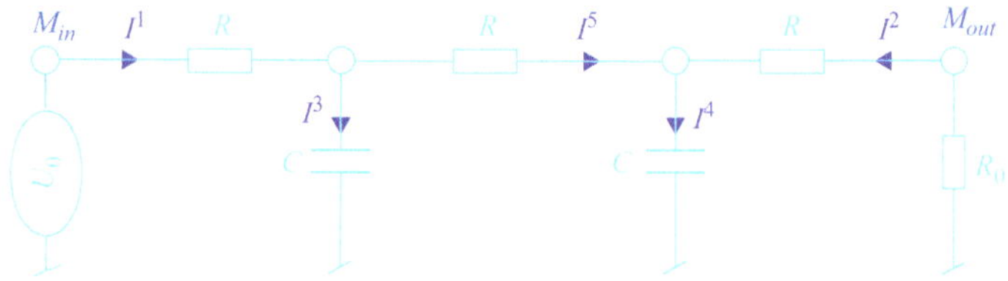

Figure 6.6 Schematic of classical TT-circuit

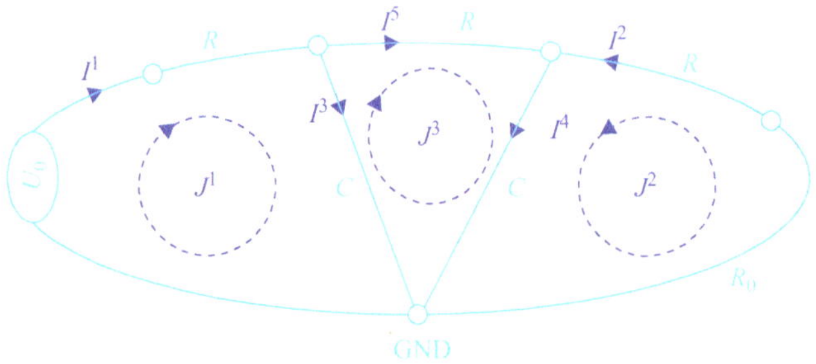

Figure 6.7 Equivalent graph topology of circuit

- number of meshes $M = 3$,
- number of ports $P = 2$.

6.3.1.3 Branch space analysis

The branch space analysis enables us to write the branch source voltage vector, the branch current vector and the branch impedance matrix given by, respectively,

$$[V_b] = \begin{bmatrix} U_0 & 0 & 0 & 0 & 0 \end{bmatrix} \tag{6.8}$$

$$\left[I^b\right] = \begin{bmatrix} I^1 \\ I^2 \\ I^3 \\ I^4 \\ I^5 \end{bmatrix} \tag{6.9}$$

$$[Z_{bb}(s)] = \begin{bmatrix} Z_1(s) & 0 & 0 & 0 & 0 \\ 0 & Z_2(s) & 0 & 0 & 0 \\ 0 & 0 & Z_3(s) & 0 & 0 \\ 0 & 0 & 0 & Z_4(s) & 0 \\ 0 & 0 & 0 & 0 & Z_5(s) \end{bmatrix} \tag{6.10}$$

with

$$\begin{cases} Z_1(s) = Z_5(s) = R \\ Z_2(s) = R + R_0 \\ Z_3(s) = 1/(Cs) \end{cases} \tag{6.11}$$

6.3.1.4 Mesh space analysis

The connectivity matrix is expressed as

$$\left[C_b^m\right] = \begin{bmatrix} 1 & 0 & 0 \\ 0 & 1 & 0 \\ 1 & 0 & -1 \\ 0 & 1 & 1 \\ 0 & 0 & 1 \end{bmatrix} \tag{6.12}$$

The mesh voltage derived from (6.8) is transformed as

$$[V_m] = \begin{bmatrix} U_0 & 0 & 0 \end{bmatrix} \tag{6.13}$$

The mesh impedance is given in (6.10) and connectivity (6.12) can be written as

$$[Z_{mn}(s)] = \begin{bmatrix} R + \frac{1}{Cs} & 0 & \frac{-1}{Cs} \\ 0 & R + R_0 + \frac{1}{Cs} & 0 \\ \frac{-1}{Cs} & \frac{1}{Cs} & R + \frac{2}{Cs} \end{bmatrix} \tag{6.14}$$

Therefore, the mesh current is equal to:

$$[J^n(s)] = \begin{bmatrix} \frac{U_0\left[R(R+R_0)(Cs)^2+(3R+2R_0)Cs+1\right]}{D(s)} \\ \frac{-U_0}{D(s)} \\ \frac{U_0[(R+R_0)Cs+1]}{D(s)} \end{bmatrix} \tag{6.15}$$

with

$$D(s) = (R + R_0)(RCs)^2 + (4R + 3R_0)RCs + 3R + R_0 \tag{6.16}$$

6.3.1.5 Expression of VTF

The voltage transfer function (VTF) of the TT-circuit is defined by

$$VTF(s) = \frac{-Z_2 J^2(s)}{U_0(s)} \tag{6.17}$$

Knowing the mesh current expressed in (6.16), the VTF can be explicitly written as

$$VTF(s) = \frac{1}{\zeta_2 s^2 + \zeta_1 s + \zeta_0} \tag{6.18}$$

with:

$$\begin{cases} \zeta_0 = \frac{3R+R_0}{R+R_0} \\ \zeta_1 = \frac{(4R+3R_0)RC}{R+R_0} \\ \zeta_0 = (RC)^2 \end{cases} \tag{6.19}$$

6.3.1.6 Discrete expression of the output

The present discretization is realized by considering the time sampling Δt, and the instant time

$$t = k\Delta t \tag{6.20}$$

The TD expression of the output is extracted from the VTF knowing that

$$VTF(s) = \frac{V_{out}(s)}{V_{in}(s)} \Rightarrow V_{in}(s) = \zeta_0 V_{out}(s) + \zeta_1 s V_{out}(s) + \zeta_2 s^2 V_{out}(s) \tag{6.21}$$

From inverse Laplace, we have

$$v_{in}(t) = \zeta_0 v_{out}(t) + \zeta_1 \frac{dv_{out}(t)}{dt} + \zeta_2 \frac{d^2 v_{out}(t)}{dt^2} \tag{6.22}$$

As introduced in the second subsection, an initial condition must be defined to solve the following equation:

$$\begin{cases} v_{in}(t=0) = 0 \\ v_{out}(t=0) = 0 \\ \frac{dv_{out}(t=0)}{dt} = 0 \end{cases} \tag{6.23}$$

The time discretization must be performed substituting the continuous time by

$$\begin{cases} v_{in}(t) \Rightarrow v_{in}(k) \\ \frac{dv_{out}(t=0)}{dt} \Rightarrow \frac{v_{in}(k)-v_{in}(k-1)}{\Delta t} \end{cases} \tag{6.24}$$

Therefore, we have the TD metric:

$$\begin{bmatrix} v_{out}(k) \\ dv_{out}(k) \\ d^2 v_{out}(k) \end{bmatrix} = [Z_t(k)] \begin{bmatrix} v_{in}(k) \\ 0 \\ 0 \end{bmatrix} + [W(k)] \begin{bmatrix} v_{out}(k-1) \\ dv_{out}(k-1) \\ d^2 v_{out}(k-1) \end{bmatrix} \tag{6.25}$$

6.3.1.7 Computed results

The validation is based on the comparison between transient simulation from the electronic and RF/microwave circuit simulator ADS from Keysight Technologies. Figure 6.8 depicts the POC representing the TT-cell of the lumped two-port circuit.

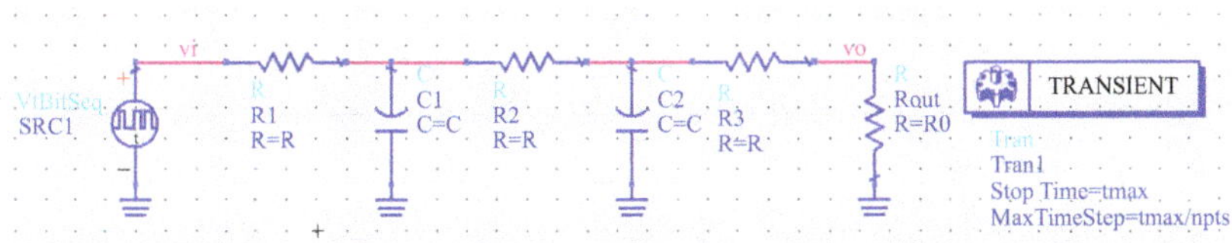

Figure 6.8 Proof-of-concept of SPICE circuit schematic

Table 6.3 Parameters of the proof-of-concept circuit

Parameter	Nature	Value
–	Resistor	$R = \{5\ \Omega, 10\ \Omega\}$
–	Capacitor	$C = \{1\ \text{nF}, 2\ \text{nF}\}$
Load	Resistor	$R_0 = 1\ \text{k}\Omega$
Z_1	Resistor	R
Z_2	Resistor	$R + R_0$
Z_3	Capacitor	C
Z_4	Capacitor	C
Z_5	Resistor	R

It is mainly constituted by R and C components. The branch impedance element parameters are addressed in Table 6.3.

As numerical application, the routine algorithm of the TAN model inspired from the theoretical equations developed in the previous sections was implemented in MATLAB® program.

After programming previous TAN model, comparisons between calculated ('Calc.') and simulated ('Simu.') results from 0 to 200 ns are carried out with different values of TT-circuit R and C parameters.

The present transient analyses have been performed with source U_0 assumed as a sequence of 8-bit data '0101100'. The data rate is 40 Mbps with rise and fall times equal to 2.5 ns. Figure 6.9 plots the obtained results from the proposed TAN computations and the ADS simulations. It can be emphasized that for the three cases of arbitrary R and C parameters, the TAN computation and simulation results are in excellent correlation.

6.3.1.8 Partial conclusion

The present subsection reports how to implement the TAN approach in the TD. The modelling was applied to two-port system representing a PCB with lumped circuits. The TAN methodology from the graph representation and tensorial analysis of the problem are recalled. The time discrete representation of TAN formulas is established. The model is validated with a microstrip TT-circuit. A very good agreement between the TAN calculation and commercial tool simulation is realized.

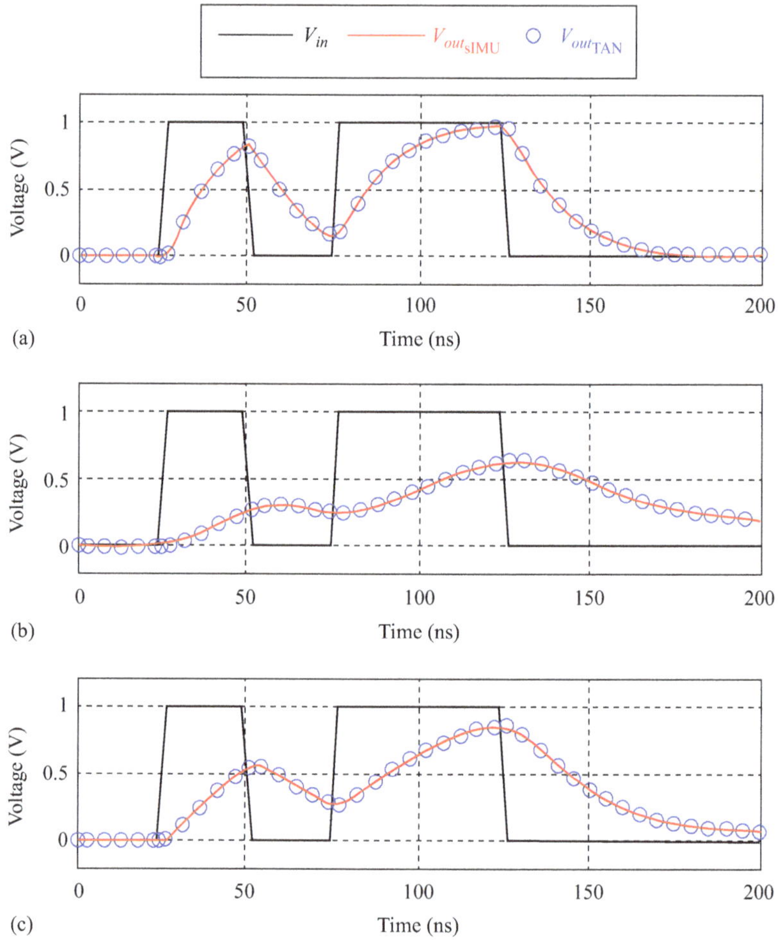

Figure 6.9 Comparison of TAN computed and simulated TT-circuit transient responses: (a) {R = 5 Ω, C = 1 nF}, (b) {R = 5 Ω, C = 2 nF} and (c) {R = 10 Ω, C = 1 nF}

6.3.2 TD TAN application with Y-tree LC circuit

In contrast to the previous application case, the present one treats a case of three-port PCB based on R, L and C lumped elements.

6.3.2.1 Recall on the mesh impedance of *RLC*-based Y-tree network

Figure 6.10 shows the classical schematic of the Y-tree fed by U_0 at its input port M_1 and loaded by resistance R_0 at its outputs M_2 and M_3. The branches M_kM_0 are constituted by R_k, L_k and C_k components for $k = 1, 2, 3$.

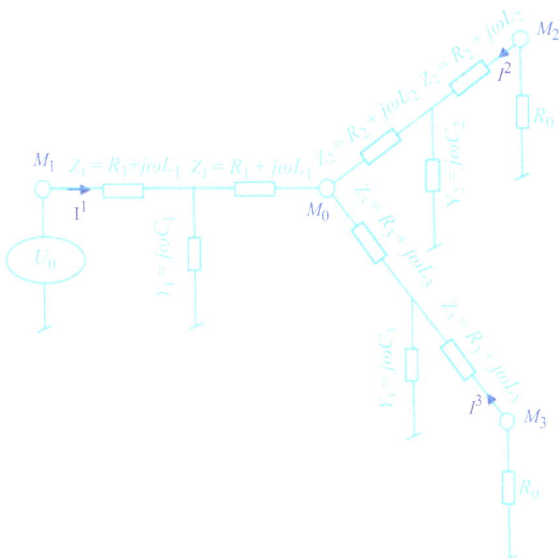

Figure 6.10 Y-tree classical circuit

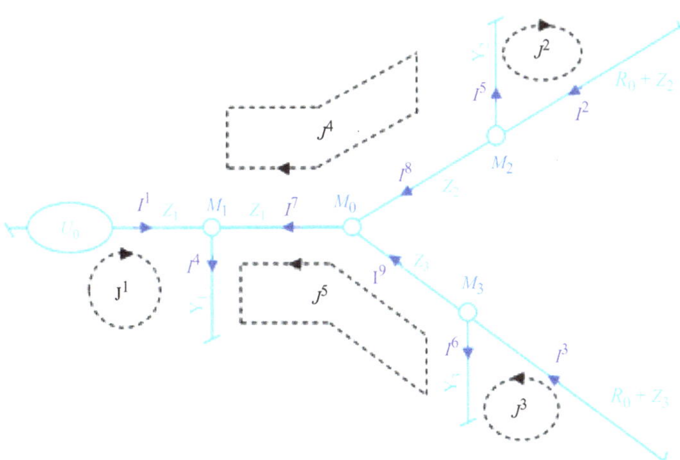

Figure 6.11 Y-tree equivalent graph

6.3.2.2 TAN modelling of *RLC*-based Y-tree network

The equivalent graph of Y-tree introduced in the previous section is represented in Figure 6.11. Contrary to the Y-tree graph of Chapter 4, the loads are supposed to be equal to resistor R_0.

The mesh voltage derived from (6.8) is transformed as

$$[V_m] = \begin{bmatrix} U_0 & 0 & 0 & 0 & 0 \end{bmatrix} \tag{6.26}$$

The mesh impedance is given in (6.10) and connectivity (6.12) can be written as

$$[Z_{mn}(s)] = \begin{bmatrix} Z_{11} & 0 & 0 & \frac{1}{C_1 s} & \frac{1}{C_1 s} \\ 0 & Z_{22} & 0 & \frac{-1}{C_2 s} & 0 \\ 0 & 0 & Z_{33} & 0 & \frac{-1}{C_3 s} \\ \frac{1}{C_1 s} & \frac{-1}{C_2 s} & 0 & Z_{44} & R_1 + L_1 s + \frac{1}{C_1 s} \\ \frac{1}{C_1 s} & 0 & \frac{-1}{C_3 s} & R_1 + L_1 s + \frac{1}{C_1 s} & Z_{55} \end{bmatrix} \tag{6.27}$$

with

$$\begin{cases} Z_{11} = R_1 + L_1 s + \frac{1}{C_1 s} \\ Z_{22} = R_0 + R_2 + L_2 s + \frac{1}{C_2 s} \\ Z_{33} = R_0 + R_3 + L_3 s + \frac{1}{C_3 s} \end{cases} \tag{6.28}$$

$$\begin{cases} Z_{44} = R_1 + R_2 + (L_1 + L_2)s + \frac{1}{C_1 s} + \frac{1}{C_2 s} \\ Z_{55} = R_1 + R_3 + (L_1 + L_3)s + \frac{1}{C_1 s} + \frac{1}{C_3 s} \end{cases} \tag{6.29}$$

Therefore, the mesh currents are equal to

$$[J^n(s)] = \begin{bmatrix} J^1(s) \\ J^2(s) \\ J^3(s) \\ J^4(s) \\ J^5(s) \end{bmatrix} \tag{6.30}$$

6.3.2.3 VTFs of Y-tree

Across the two output ports $M_{k=2,3}$, the VTFs of the Y-tree are defined by

$$VTF_k(s) = \frac{-(R_0 + Z_k)J^k(s)}{U_0(s)} \tag{6.31}$$

Knowing the mesh current expressed in (6.16), the VTF can be explicitly written as

$$VTF_k(s) = \frac{1}{\zeta_{8k} s^8 + \cdots + \zeta_{1k} s + \zeta_{0k}} \tag{6.32}$$

6.3.2.4 Time-domain metric

Similar to the previous subsection, the TD expression of the output is extracted from the VTF from inverse Laplace, we have

$$v_{in}(t) = \zeta_0 v_{out}^k(t) + \zeta_1 \frac{dv_{out}^k(t)}{dt} + \cdots + \zeta_8 \frac{d^8 v_{out}^k(t)}{dt^8} \tag{6.33}$$

As introduced in the second subsection, an initial condition must be defined to solve the following equation:

$$\begin{cases} v_{in}(t=0)=0 \\ v_{out}^{k}(t=0)=0 \\ \frac{d^{(m)}v_{out}^{k}(t=0)}{dt^{m}}=0 \end{cases} \tag{6.34}$$

The time discretization $t = m\Delta t$ must be performed substituting the continuous time by taking into account the eighth-order derivative of the output. Therefore, we have the following TD metric:

$$\begin{bmatrix} v_{out}^{k}(m) \\ dv_{out}^{k}(m) \\ \vdots \\ d^{8}v_{out}^{k}(m) \end{bmatrix} = [Z_{tk}(m)] \begin{bmatrix} v_{in}^{k}(m) \\ 0 \\ \vdots \\ 0 \end{bmatrix} + [W_{k}(m)] \begin{bmatrix} v_{out}^{k}(m-1) \\ dv_{out}^{k}(m-1) \\ \vdots \\ d^{8}v_{out}^{k}(m-1) \end{bmatrix} \tag{6.35}$$

6.3.2.5 Computed results

Figure 6.12 displays the ADS schematic of the simulated Y-circuit POC. It acts as the same microstrip Y-tree proposed in Chapter 4. The equivalent *RLC*-tree parameters were calculated from Hammerstad–Jensen microstrip line model. The physical and electrical constituting TL parameters are listed in Table 6.4.

In this case of study, the present transient analyses have been performed with source U_0 assumed as a sequence of 8-bit data '0101100'. The data rate is 0.5 Gbps with rise and fall times equal to 0.2 ns.

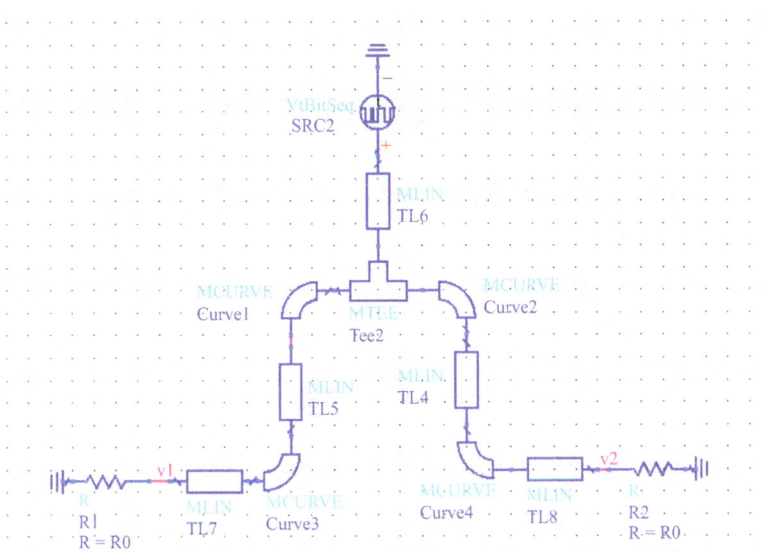

Figure 6.12 Design of Y-tree POC

Table 6.4 Y-tree PCB parameters

TL	Parameters		
	Length (mm)	**Width (μm)**	***RLC* parameters**
TL_1	$d_1 = 50$	$w_1 = 0.6$	$R_1 \approx 43$ mΩ, $L_1 \approx 4.25$ nH, $C_1 \approx 9$ pF
TL_2	$d_2 = 60$	$w_2 = 1.43$	$R_2 \approx 22$ mΩ, $L_2 \approx 2.55$ nH, $C_2 \approx 2.28$ pF
TL_3	$d_3 = 70$	$w_3 = 0.6$	$R_3 \approx 59$ mΩ, $L_3 \approx 11.9$ nH, $C_3 \approx 12.66$ pF

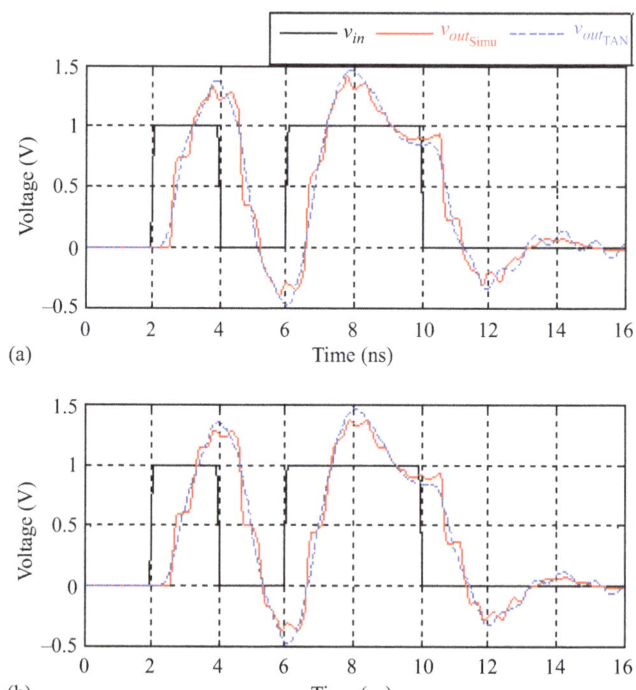

Figure 6.13 Comparisons between calculated and simulated Y-tree transient responses: (a) port M_2 and (b) port M_3 outputs

After MATLAB programing of the previous TAN model, comparisons between calculated ('Calc.') and simulated ('Simu.') transient results from 0 to 16 ns are presented in Figure 6.13. A notable signal-to-noise ratio between the computed and simulated signals of about 17 dB is observed. Despite the noise discrepancies, once again, the calculated TD responses from TAN model are in good agreement with ADS simulation. The slight discrepancy is mainly due to the limited order of the equivalent lumped circuits.

6.3.2.6 Partial conclusion

An efficient TD modelling of three-port interconnect network is investigated in this subsection. The TAN methodology from the graph representation and tensorial analysis of the lumped equivalent network is described. The model is validated with a microstrip Y-tree. The results of TAN computation and commercial tool simulation are in good agreement.

6.4 Conclusion

A TD theory on TAN modelling for PCB system without using Fourier transform is developed in this chapter. The PCB system is represented by electrical lumped circuits. The particular conditions on the time parameters in function of the PCB circuit parameters are defined. The main methodology of the TAN TD is presented. The TAN routine algorithm was implemented as a MATLAB program for the result computations.

To illustrate the feasibility of TAN direct TD computation, two cases of application are investigated. The first one is based on fully lumped circuit with TT-topology. The different steps of TAN modelling from the graph topology to the TAN metric in the TD are established. The validity of the TD modelling was confirmed with a commercial tool simulation based on SPICE environment. It was shown that the TAN approach enables us to predict the SI of the lumped circuit PCB outputs.

The second POC is based on Y-tree which was considered in the frequency domain analysis in the previous chapter. The equivalent lumped circuit constituted by *RLC* elements is established from Hammerstad–Jensen microstrip line model. The implementation of the TAN model is globally described. Transient simulations have been carried out to validate the TAN computations. As expected, the TAN results present a good trend of the microstrip Y-tree outputs.

References

[1] A. E. Ruehli, and A. C. Cangellaris, "Progress in the methodologies for the electrical modeling of interconnects and electronic packages," Proceedings of the IEEE, vol. 89, no. 5, pp. 740–771, 2001.

[2] D. Standley, and Jr. J. L. Wyatt, "Improved Signal Delay Bounds for RC Tree Networks," VLSI Memo, No. 86–317, MIT, Cambridge, MA (USA), May 1986.

[3] A. B. Kahng, and S. Muddu, "An analytical delay model of RLC interconnects," IEEE Transactions on Computer-Aided Design of Integrated Circuits and Systems, vol. 16, no. 12, pp. 1507–1514, 1997.

[4] T. Eudes, and B. Ravelo, "Analysis of multi-gigabits signal integrity through clock H-tree," International Journal of Circuit Theory and Applications, vol. 41, no. 5, pp. 535–549, 2013.

[5] B. Ravelo, "Modelling of asymmetrical interconnect T-tree laminated on flexible substrate," The European Physical Journal Applied Physics (EPJAP), vol. 72, no. 2, pp. 1–9, 2015. 20103.

[6] S. Wen, J. Zhang, and Y. Lu, "Modeling and quantification for electromagnetic radiation of power-bus structure with multilayer printed circuit board," IEEE Transactions on Components, Packaging and Manufacturing Technology, vol. 6, no. 1, pp. 79–86, 2016.

[7] D.-H. Kim, S.-J. Joo, D.-O. Kwak, and H.-S. Kim, "Warpage simulation of a multilayer printed circuit board and microelectronic package using the anisotropic viscoelastic shell modeling technique that considers the initial warpage," IEEE Transactions on Components, Packaging and Manufacturing Technology, vol. 6, no. 11, pp. 1667–1676, 2016.

[8] H. Tang, J.-X. Chen, H. Chu, G.-Q. Zhang, Y.-J. Yang, and Z.-H. Bao, "Integration design of filtering antenna with load-insensitive multilayer balun filter," IEEE Transactions on Components, Packaging and Manufacturing Technology, vol. 6, no. 7, pp. 1408–1416, 2016.

[9] C.-L. Cho, H.-L. Kao, L.-C. Chang, Y.-H. Wu, and H.-C. Chiu, "Inkjet-printed multilayer bandpass filter using liquid crystal polymer system-on-package technology," IEEE Transactions on Components, Packaging and Manufacturing Technology, vol. 6, no. 4, pp. 622–629, 2016.

[10] Z. Xu, Y. Liu, B. Ravelo, and O. Maurice, "Modified Kron's TAN Modeling of 3D Multilayer PCB," in Proc. 11th International Workshop on Electromagnetic Compatibility of Integrated Circuits, EMC Compo 2017, St. Petersburg, Russia, 4–8 July 2017, pp. 242–247.

[11] Z. Xu, Y. Liu, B. Ravelo, and O. Maurice, "Multilayer Power Delivery Network Modeling With Modified Kron's Method (MKM)," in Proc. of 16th Int. Symposium on Electromagnetic Compatibility (EMC) Europe 2017, Angers, France, 4–8 September 2017, pp. 1–6.

[12] T. Nikola, "Standard 4-and 6-Layer PCB stackups," White Paper, Data Response, Available Online, 2013, http://www.bitweenie.com/listings/standard-4-and-6-layer-pcb-stackups/.

[13] M. E. Goldfarb, and R. A. Pucel, "Modeling via hole grounds in microstrip," IEEE Microwave and Guided Wave Letters, vol. 1, no. 6, pp. 135–137, 1991.

[14] Y. Shang, C. Li, and H. Xiong, "One method for via equivalent circuit extraction based on structural segmentation," International Journal of Computer Science Issues, vol. 10, no. 3, pp. 18–22, 2013.

[15] S. Liu, W. Tang, W. Zhuang, G. Wang, and Y. L. Chow, "Capacitance and conductance of through silicon vias with consideration of multilayer media and different shapes," IEEE Transactions on Components, Packaging and Manufacturing Technology, vol. 6, no. 1, pp. 256–264, 2016.

[16] S. Huang, and L. Tsang, "Fast broadband modeling of traces connecting vias in printed circuit boards using broadband Green's function method," IEEE Transactions on Components, Packaging and Manufacturing Technology, vol. 7, no. 8, pp. 1343–1355, 2017.
[17] E. Laermans, J. De Geest, D. De Zutter, F. Olyslager, S. Sercu, and D. Morlion, "Modeling complex via hole structures," IEEE Transactions on Advanced Packaging, vol. 25, no. 2, pp. 206–214, 2002.
[18] G. Kron, Tensor Analysis of Networks, Wiley Publishers, New York, Chapman & Hall, London, 1939.
[19] G. Kron, IEEE Transactions on Circuit Theory, vol. 15, no. 3, p. 174, 1968.
[20] K. Kondo, and Y. Ishhizuka, "Geometrical Aspects of Kron's Non-Riemannian Electrodynamics," Memoirs of the unifying study of the basic problems in engineering science by means of geometry, vol. I, Published by Unified Study Group, G. B. Fukyu-Kai, Tokyo, Japan, pp. 195–239, 1955.
[21] F. H. Branin, "Transient analysis of lossless transmission lines," Proceedings of the IEEE, vol. 55, pp. 2012–2013, 1967.
[22] J. Dobes, and L. Slama, "A Modified Branin Model of Lossless Transmission Lines," in Proc. 52nd IEEE Int. Midwest Symp. on CAS 2009, Cancun, Mexico, 2–5 August 2009, pp. 236–239.
[23] O. Maurice, A. Reineix, P. Durand, and F. Dubois, "Kron's method and cell complexes for magnetomotive and electromotive forces," International Journal of Applied Mathematics, vol. 44, no. 4, pp. 183–191, 2014.
[24] O. Maurice, A. Reineix, P. Hoffmann, B. Pecqueux, and P. Pouliguen, "A formalism to compute the electromagnetic compatibility of complex networks," Advances in Applied Science Research, vol. 2, no. 5, pp. 439–448, 2011.
[25] B. Ravelo, O. Maurice, and S. Lalléchère, "Asymmetrical 1:2 Y-tree interconnects modelling with Kron–Branin formalism," Electronics Letters, vol. 52, no. 14, pp. 1215–1216, 2016.
[26] B. Ravelo, and O. Maurice, "Kron–Branin modeling of Y-Y-tree interconnects for the PCB signal integrity analysis," IEEE Transactions on Electromagnetic Compatibility, vol. 59, no. 2, pp. 411–419, 2017.

Chapter 7
Direct time-domain analysis with TAN method for distributed PCB modelling

Zhifei Xu[1], *Blaise Ravelo, Jonathan Gantet*[2], *Nicolas Marier*[2], *and Olivier Maurice*[3]

Abstract

The time domain (TD) tensorial analysis of networks (TAN) modelling of high-speed printed circuit board (PCB) system will be developed in the present chapter. Similar to the previous chapter, the dictionary of the TD TAN primitive elements based on distributed elements will be presented. The TD translation of the Kron–Branin method integrating the signal propagator operator will be used in the present case. Then the workflow of the routine algorithm illustrating how to analyse the high-speed PCB system will be described. By considering pulse waveform, discrete mathematical solvers will be developed. The feasibility of the TAN TD model will be verified with examples of SI and PI analyses of high-speed PCB systems excited by the previously cited test-signal waveforms. Discussion will be made about the accuracy, advantages and limits of the TAN TD computation methods.

Keywords: TAN approach, Kron's method, Kron–Branin's model, frequency domain analysis, modelling methodology, high speed PCB system, SI analysis, PI analysis, EMC analysis

7.1 Introduction

Since the speed of electronics is highly increased, digital signals in time domain (TD) constitute a crucial step of printed circuit board (PCB) signal integrity (SI) and power integrity (PI) analyses. The TD investigation requires generally the assessment of certain parameters as overshoot, undershoot, rise/fall times, jitters and so on. Working in the frequency domain can be efficient for the engineer and the solvers in the narrow band but more difficult from DC. The frequency domain parameters can

[1]IRSEEM/ESIGELEC, Rouen, France
[2]AFSCET (Association Française de Science des Systèmes), Paris, France
[3]ArianeGroup, Paris, France

also be extracted from TD by applying Fourier transform. Nevertheless, the Fourier transform can be limited by the bandwidth and time-frequency sampling.

During the digital PCB operation, the transitions between low and high logic can induce undesirably EMI problems, and with increase in speed, the adjacent coupling will increase by the fast switching time. In difference to the frequency domain, the TD analysis enables to extract the tested structure instantaneous responses. The TD computation allows to adapt the analysis to the signal shapes and time-range. More precisely, it enables to materialize the time-parameters as rise- and fall-times and steady-state responses. When the time step largely decreases in modern electronics because of the fast switching time, rise/fall times, the precision of time step, the resolution of the solution of the integrodifferential equations and the boundary conditions become important.

This chapter is focused on the direct TD Kron–Branin (KB) modelling with tensorial analysis of networks (TAN) concept based on finite difference time method, applied to 3D multilayer PCB. The multilayer technology constitutes the ultimate solution for the design of HD PCB. Challenging modelling method is always required to predict the SI of multilayer PCB in TD. Prior to detail the model for the multilayer structure in TD, some TD parameters are needed to be characterized [1].

7.1.1 Branin's TD expression

Here we can develop the expression of Branin's model in frequency domain to TD by replacing the frequency domain operators of (7.1) and by denoting the propagation delay τ.

$$\begin{bmatrix} V(t) - (R_2 - Z_c)J^1(t-\tau) \\ V(t-\tau) + (Z_c - R_1)J^2(t-\tau) \end{bmatrix} = \begin{bmatrix} R_1 + Z_c & 0 \\ 0 & R_2 + Z_c \end{bmatrix} \begin{bmatrix} J^1(t) \\ J^2(t) \end{bmatrix} \tag{7.1}$$

Where (7.2) in (7.1) is the Wronskian vector of the Branin's model

$$W_m = \begin{bmatrix} -(R_2 - Z_c)J^1(t-\tau) \\ V(t-\tau) + (Z_c - R_1)J^2(t-\tau) \end{bmatrix} \tag{7.2}$$

Equation (7.3) is the source vector of the Branin's model.

$$E_m = \begin{bmatrix} V(t) \\ 0 \end{bmatrix} \tag{7.3}$$

It can be solved with equation as follows and initial conditions:

$$W + E_m = Z_{mn}J^n \tag{7.4}$$

$$if \quad t \le \tau \quad \begin{cases} V(t-\tau) = 0 \\ J^1(t-\tau) = 0 \\ J^2(t-\tau) = 0 \end{cases} \tag{7.5}$$

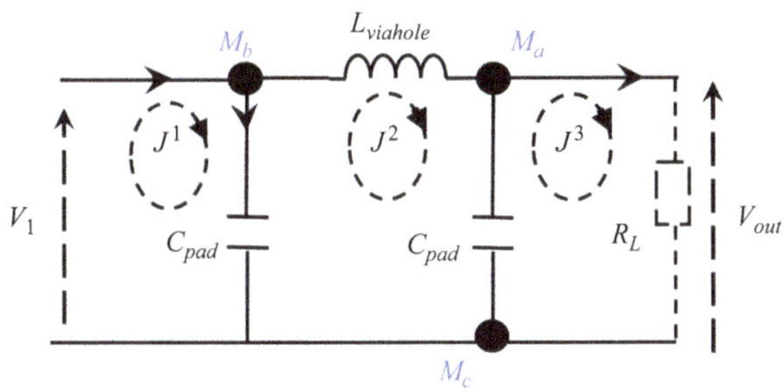

Figure 7.1 TD via representation with source and load

The Wronskian tensor representing the internal sources and the parameter, τ, is the delay of the transmission line (TL) in Branin's model.

7.1.2 Via's TD expression

The via's model can be completed with the load and source as in Figure 7.1. The impedance, source and Wronskian tensors in mesh space in discrete TD can be obtained:

$$Z_{mn} = \begin{bmatrix} \frac{\Delta t}{C_{pad}} & -\frac{\Delta t}{C_{pad}} & 0 \\ -\frac{\Delta t}{C_{pad}} & 2\frac{\Delta t}{C_{pad}} + \frac{L_{viahole}}{\Delta t} & -\frac{\Delta t}{C_{pad}} \\ 0 & -\frac{\Delta t}{C_{pad}} & \frac{\Delta t}{C_{pad}} + R_L \end{bmatrix} \tag{7.6}$$

$$W_m = \begin{bmatrix} -\frac{\Delta t}{C_{pad}} Q_1(t-\Delta t) + \frac{\Delta t}{C_{pad}} Q_2(t-\Delta t) \\ -\frac{\Delta t}{C_{pad}} Q_1(t-\Delta t) - 2\frac{\Delta t}{C_{pad}} Q_2(t-\Delta t) - \\ \frac{\Delta t}{C_{pad}} Q_3(t-\Delta t) + \frac{L_{viahole}}{\Delta t} J^2(t-\Delta t) \\ \frac{\Delta t}{C_{pad}} Q_2(t-\Delta t) + \frac{\Delta t}{C_{pad}} Q_3(t-\Delta t) \end{bmatrix} \tag{7.7}$$

where Q_n represents the charge of different meshes in the via model.

$$E_m = \begin{bmatrix} V_1(t) \\ 0 \\ 0 \end{bmatrix} \tag{7.8}$$

Then by applying the equation in (7.4), the mesh current in TD can be easily obtained for this simple via structure.

7.2 Application example of TD TAN modelling

As proof of concept (POC) of the developed TD TAN model, let us consider a multi-layer PCB. It acts as a hybrid structure which can be assumed as a multiport system. It

Table 7.1 Interconnect lines parameters in mm

Width	TL_1	TL_2	TL_3	TL_4	TL_5	TL_6	TL_7
0.2	10.47	14.30	18.90	7.20	47.86	27.20	29.20

Table 7.2 Via and SMD components parameters

Via diameter	Pad diameter	Capacitor and package	Resistor and package
0.4 mm	0.6 mm	1 nF – 0603 mm	49.9 Ω – 0603

Table 7.3 PCB layout parameters

Board width	Board length	Board height	Substrate permittivity	Loss tangent	Metal thickness
30 mm	40 mm	1.6 mm	4.5	0.012	0.035 mm

is mainly constituted by the subnetworks composed of interconnect lines, vias, pads and anti-pads. Each subnetwork is transcribed in its equivalent electrical scheme.

The structure is excited by the voltage source v_{in} and loaded by resistor R_0. The perspective 3D view of the six-layer structure is depicted in Figure 7.3(a). It acts as a three-port complex 3D multilayer structure with surface mounted devices (SMDs).

The electrical network configuration can be referenced with the nodes $M_n(n = 1, 2, \ldots, 15)$. The parameters can be found in Tables 7.1–7.3.

7.2.1 Graph topology of the 3D multilayer hybrid PCB

The equivalent graph in Figure 7.2 is described from the electrical diagram. The considered structure basic subnetworks must be defined as initial objects. The combination of each subnetwork model becomes the associated graph. As aforementioned, this KB model must begin with the structural segmentation analysis. It is necessary to elaborate the 3D multilayer hybrid PCB equivalent graph. The basic elements constituting the primitive graph are as follows:

- $R_{1,2,3} = R_0$ are the resistive loads connected to the interconnects on each port;
- Vias, pads and anti-pads are modelled by LC lossless lumped network; and

- interconnect TLs are parametrized by their characteristic impedances Z_k and their propagation constants θ_k with $k = 1, 2, \ldots, 9$.

These characteristics have been defined differently in function of the implementation technology as microstrip and strip lines constituting the multilayer PCB.

The covariables internal source E_m with $m = 1, 2, \ldots, n$ represents the Branin's electromotive force (EMF) coupling sources. From these primitive elements, the KB equivalent graph topology of the multilayer structure is highlighted in Figure 7.2. This graph is essentially composed of $B = 36$ branches, $M = 23$ meshes and nodes connecting different components. The graph includes the external excitation voltage source V_{in}.

7.2.2 Integration of the innovative direct TD method

7.2.2.1 Characteristic matrix in mesh space

To extract the contravariant mesh current, the analytical implementation of KB needs to change the branch space into the mesh space. The branch space impedance matrix is denoted $Z_{ab}(t)$. Accompany with the previous section of lumped element TD expression, the branch impedance tensor can be established.

$$Z_{ab}(t) = \begin{bmatrix} Z_{b1}(t) & 0 & 0 & 0 & 0 \\ 0 & \ddots & 0 & 0 & 0 \\ 0 & 0 & Z_{b10}(t) & 0 & 0 \\ 0 & 0 & 0 & \ddots & 0 \\ 0 & 0 & 0 & 0 & Z_{bn}(t) \end{bmatrix} \tag{7.9}$$

where $a, b = 1, 2, \ldots, B$ and $m, n = 1, 2, \ldots, M$. $Z_{bn}(t)$ is the branch impedance expression. In this configuration, the connection matrix is defined from the relationship between branch and mesh currents in (7.10). The connectivity matrix compact form in (7.11) is developed with the coefficients of mesh current relationship expressed in (7.10)

$$\begin{aligned} I^1 &= 1J^1 + 0J^2 + \cdots + 0J^{23} \\ &\vdots \\ I^4 &= 0J^1 + 0J^2 + 1J^3 + 1J^4 + 0J^5 + \cdots + 0J^{23} \\ I^5 &= 0J^1 + 0J^2 + 0J^3 + 1J^4 + 0J^5 + \cdots + 0J^{23} \\ &\vdots \\ I^{36} &= 0J^1 + \cdots + J^{22} + J^{23} \end{aligned} \tag{7.10}$$

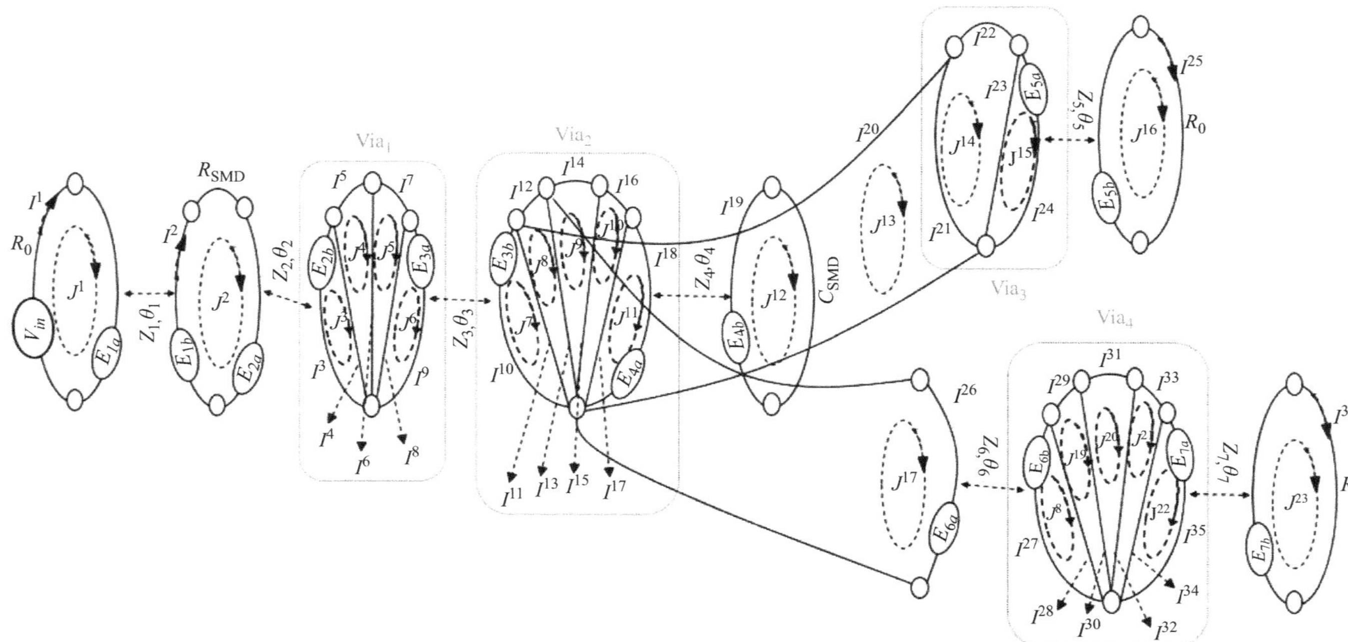

Figure 7.2 KB equivalent graph topology of the 3D multilayer hybrid structure [1]

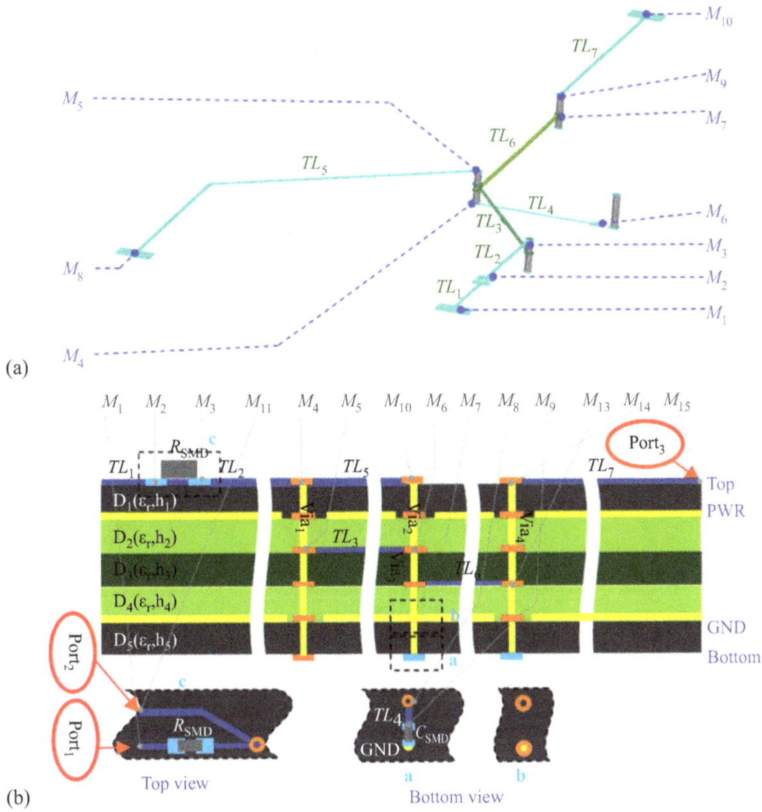

Figure 7.3 Illustrative configuration of (a) perspective and (b) profile view of the structure [1]

$$
C = \begin{bmatrix} 1 & 0 & 0 & \cdots & \cdots & 0 & 0 \\ \vdots & \vdots & \vdots & \vdots & \vdots & \vdots & \vdots \\ 0 & 0 & 1 & -1 & 0 & \cdots & 0 \\ 0 & 0 & 0 & 1 & 0 & \cdots & 0 \\ \vdots & \vdots & \vdots & \vdots & \vdots & \vdots & \vdots \\ 0 & 0 & 0 & 0 & 0 & \cdots & \cdots \end{bmatrix} \tag{7.11}
$$

Following the definitions in branch space, the branch excitation sources are denoted as E_a. The external covariable voltage sources are written in mesh space as follows:

$$
E_m(t) = C_m^a E_a(t) = \begin{bmatrix} V_1(t) & 0 \cdots & 0 & 0 & 0 & 0 & 0 \end{bmatrix} \tag{7.12}
$$

7.2.2.2 TD KB modelling principle

In difference with the frequency domain, the TD KB model is defined by the iterative time-difference resolution of discretized integrodifferential equations.

The TD analysis of the electrical network drawn in Figure 7.2 needs more exhaustive parameter analyses. The internal Branin's EMF sources must be represented by the voltage source vectors. All the EMFs in the considered network can be implemented into the internal coupling matrix in mesh space. Following Branin's model, this matrix size will be the same as the impedance comatrix or rank 2 tensors without internal Branin's sources. Based on the KB equations, the mesh tensor impedance and covariable sources of the Branin's sources are, respectively, expressed as

$$Z'_{mn}(t) = \begin{bmatrix} [Z_{e_1}(t)] & & & & \\ & \ddots & & & \\ & & [Z_{e_k}(t)] & & \\ & & & \ddots & \\ & & & & [Z_{e_7}(t)] \end{bmatrix} \tag{7.13}$$

$$E'_m(t) = [0 \quad V_1(t - D_0) \quad \cdots \quad 0 \quad 0 \quad 0 \quad 0] \tag{7.14}$$

with D_0 is the TL induced delay. The mesh tensor formulation leads to the generalized Ohm's law of the whole topology:

$$E_m(t) + E'_m(t) = \left\{[Z_{mn}(t)] + \left[Z'_{mn}(t)\right]\right\} \odot J^n(t)] \tag{7.15}$$

This relation represents the abstraction of the initial physical posed-problem sketched in Figure 7.3(a) into a purely mathematical problem. The mesh contravector currents $J^n(t)$ are the posed problem unknowns. In a nutshell, the problem can be summarized with the KB characteristic matrix metric:

$$[M_{KB}(t)] = [Z_{mn}(t)] + \left[Z'_{mn}(t)\right] \tag{7.16}$$

By considering the internal excitation sources, the previous equation becomes

$$T(t) = E(t) + W(t) = Z(t) \odot J(t) \tag{7.17}$$

where $E(t)$ is the vector of the extrinsic sources and $W(t)$ is the Wronskian vector of the intrinsic sources. The TD extrinsic source vector is defined by

$$E(t) = \left[e_0(t) \quad e_0(t - D_1) \quad \cdots \quad 0 \quad 0\right] \tag{7.18}$$

The TD impedance comatrix is derived from these operators. Accordingly, the innovative Wronskian covariable of the intrinsic TD sources of the 3D multilayer hybrid PCB shown in Figure 7.3 is expressed as

$$
W(t) = \begin{bmatrix}
-(R + Z_1 + Z_2) J[(t - D_1), 1] + Z_2 J[(t - D_1 - D_2), 2] - \frac{\Delta t}{C_1} Q_1 (t - D_1 - D_2) + \frac{\Delta t}{C_1} Q_2 (t - D_1 - D_2) \\
(Z_1 - R_1) J[(t - D_1), 0] + Z_2 J[(t - D_2), 2] - \frac{\Delta t}{C_1} Q_1 (t - D_2) + \frac{\Delta t}{C_1} Q_2 (t - D_2) \\
\vdots \\
\frac{\Delta t}{C_2} Q_2 (t - \Delta t) - \frac{\Delta t}{C_2} Q_3 (t - \Delta t) - \frac{\Delta t}{C_3} Q_3 (t - \Delta t) + \frac{L_2}{\Delta t} J[(t - \Delta t), 4] + \frac{\Delta t}{C_3} Q_4 (t - \Delta t) \\
\vdots \\
\frac{\Delta t}{C_1} Q_{12} (t - \Delta t) - \frac{\Delta t}{C_1} Q_{13} (t - \Delta t) - (R_2 - Z_5) J[(t - D_5), 15] \\
\frac{\Delta t}{C_1} Q_{12} (t - D_5) - \frac{\Delta t}{C_1} Q_{13} (t - D_5) + Z_5 J[(t - D_2), 14] \\
\vdots \\
\frac{\Delta t}{C_1} Q_{20} (t - \Delta t) - \frac{\Delta t}{C_1} Q_{21} (t - \Delta t) - (R_3 - Z_7) J[(t - D_7), 22] \\
\frac{\Delta t}{C_1} Q_{20} (t - D_7) - \frac{\Delta t}{C_1} Q_{21} (t - D_7) + Z_7 J[(t - D_7), 21]
\end{bmatrix} \quad (7.19)
$$

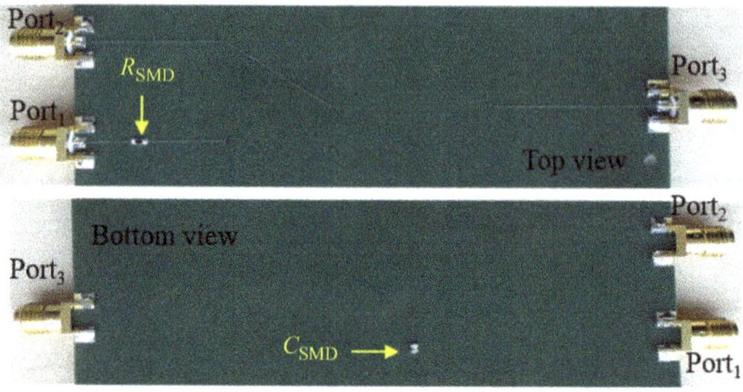

Figure 7.4 Photographs of the multilayer PCB top and bottom views [1]

where D_k is the induced delay from the TL_k for $k = 0, 1, \ldots, 7$ and Δt is the defined time step based on rise time of the signal. To confirm the relevance of the proposed direct TD circuit theory, validation results with a 3D multilayer-based hybrid PCB prototype are examined in the next subsection.

7.2.3 Validation results

7.2.3.1 Prototype design and fabrication

The tested circuit prototype is built with the FR4 epoxy dielectric with relative permittivity $\varepsilon_r = 4.5$, loss tangent $\tan(\delta) = 0.012$. The fabricated PCB metallization parameters are followed with Tables 7.2 and 7.3. The interconnect TLs present the same physical width and different lengths.

The signal, power, ground planes, layer interconnects, anti-pads and vias are constituted by Cu-metallization. The test signals are obtained via identical SubMiniature version A (SMA) connectors placed at ports 1, 2 and 3. This layout figures out the placement of the lumped SMD capacitor and resistor. The photographs of top and bottom view of the fabricated PCB are displayed in Figure 7.4.

7.2.3.2 TD experimental results

TD experimentation has been performed to validate the developed direct TD unfamiliar KB model. The present results correspond to the circuit configuration with output ports 2 and 3 loaded by $R_0 = 50\ \Omega$. The performed TD experimental setup is shown in Figure 7.5. It consists in injecting the transient voltage from the pattern generator to the 3D multilayer hybrid PCB. The input and output voltage plots are compared via the signal analyser visualizations. The pattern generator Infiniium DSA90404A provided by Agilent® presents a maximum 600 MHz data rate. As seen in Figure 7.6, the TD outputs present similar behaviour. Therefore, only the output signal from port 2 is displayed in this part. The TD results have been visualized and recorded with 4 GHz Agilent digital signal analyser with 20 GSa/s.

As illustrated in Figure 7.6, the test sequence is a voltage signal having risen time of about 1 ns, fall-time of about 1.5 ns and amplitude of about 2.5 V. This transient signal was injected to input port 1 of the circuit under test. The experimental

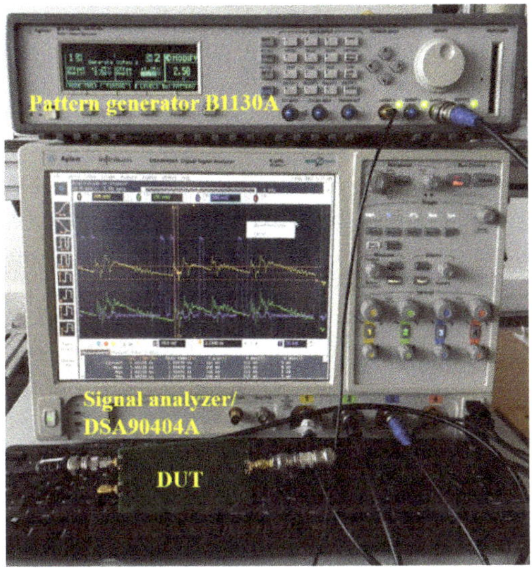

Figure 7.5 TD experimental setup [1]

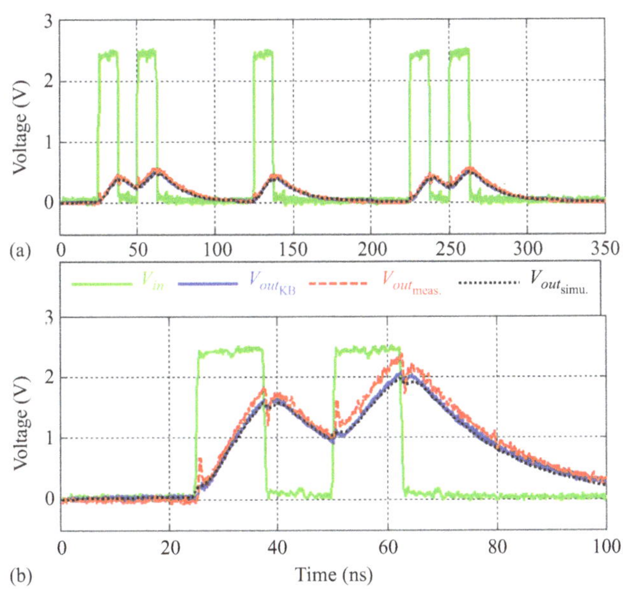

Figure 7.6 TD results of 80 Mbps rate input sequence with (a) real and (b) four-time scaled output plots

results denoted by $v_{out_{meas}}$ and the KB TD calculated results denoted by $v_{out_{KB}}$ are performed with time-step Δt=75 ps with 80 Mbps rate input. Figure 7.6(a) displays the comparison between the KB computation, SPICE transient simulation ("simu.") and measurement ("meas.") results. Figure 7.6(b) is plotted in $4 \times v_{out_{KB}}$. The outputs of the tested PCB are, respectively, displayed in solid and dashed lines. The applicative test configurations enable to verify the KB model efficiency. The TL losses have been integrated into the KB model via the propagation constants. As expected, the KB computations are well-correlated to the simulations but slightly different to the measurements.

The KB model and measurement results are in good correlation. Significant distortions, delays and attenuation are observed because of the hybrid multilayer interconnect structure effects and the real-time measurement. As the main part of the test signal energy is focused in the frequency band lower than 1 GHz, the higher frequency effects are not significant in TD. The slight differences between the KB and measured output signal spikes and distortions are caused mainly by the numerical inaccuracies and the time steps. The difference between the model and measurement results can be predicted with the transmission coefficients. Moreover, the measurement results were generated with periodical sequences which charge the tested PCB capacitors in difference to the calculations performed with the initial condition $v_{out_{KB}}(t = 0) = 0$.

In order to quantify these differences, the error vector magnitude (EVM) modelled between calculation and measurement results are evaluated via the relation

$$EVM(t_k) = \frac{\sqrt{\sum_{k=0}^{k_{\max}} \left[v_{KB}^2(t_k) - v_{meas}^2(t_k)\right]}}{\sqrt{\sum_{k=0}^{k_{\max}} v_{KB}^2(t_k)}} \tag{7.20}$$

with $t_k = k\Delta t$ represents the sampling time varying from 0 to $k_{\max} = 1,001$. The EVMs related to 80 Mbps is, respectively, lower than 10%. The performed KB modelling computation speed is of about 10 ms by using a PC equipped a single-core processor Intel® Core i7-3120M CPU @ 2.50 GHz and 16 GBytes physical RAM with 64-bits Windows 10.

7.3 Conclusion

The basic elements TD expressions of the multilayer PCB have been detailed with KB formalism. Then the direct TD modelling methodology has been presented through an example of a multilayer hybrid PCB. To validate the developed direct TD KB model, a six-layer PCB is designed and fabricated. As expected, KB computed transient outputs voltages are in good agreement with measurements. The TD experimentation has been performed with 80 Mbps rate pulse square wave signal for the multilayer PCB. A very good prediction of the transient signal behaviours is verified by simulations and experiments of both outputs of tested PCB.

Reference

[1] Xu, Z., Liu, Y., Ravelo, B., Gantet, J., Marier, N., and Maurice, O. (2018). Direct time-domain TAN model of 3D multilayer hybrid PCB: Experimental validation. IEEE Access, 6, 60645–60654.

Chapter 8

Coupling between EM field and multilayer PCB with MKME

Zhifei Xu[1]*, Yang Liu*[1]*, Blaise Ravelo*[1]*, Jonathan Gantet*[2]*, Nicolas Marier*[2]*, and Olivier Maurice*[3]

Abstract

This chapter is focused on modelling of radiated electromagnetic compatibility (EMC) coupling onto the multilayer printed circuit board (PCB). Kron's method integrates the electromagnetic (EM) emission, Taylor's and field-to-interconnect coupling models. The equivalent graph of the field-to-interconnect coupling is established. The modelling methodology consists in defining the primitive subnetwork elements. These primitive elements are represented by vias, interconnect lines and pads. Kron's graph equivalent to the EMC problem is elaborated. Finally, the coupling voltages are calculated via the tensorial equation translated from the graph. The radiated EMC Kron's model is validated with a four-layer PCB from 0.4 to 1.4 GHz by two scenarios of EM radiation. As proof of concept, a prototype of four-layer PCB was designed, fabricated, tested and simulated in full wave with a commercial three-dimensional EM tool. For the first case, the multilayer PCB was illuminated by plane wave emission propagating in different directions. The numerical computation from Kron's formalism was compared with simulation and measurement. The other case is the field-to-interconnect coupling between a microstrip I-line PCB, as an EM field emitter, and the multilayer PCB, as a receiver, in 1-m distance. For both cases, the simulated and calculated voltage couplings onto the multilayer PCB are in good agreement.

Keywords: Kron's method, modelling, methodology, multilayer PCB, EMC coupling

8.1 Introduction to R-EMC analytical modelling

In addition to the emission, analytical methods of electromagnetic compatibility (EMC) radiated coupling modelling were also proposed [1–5]. These coupling

[1]IRSEEM/ESIGELEC, Rouen, France
[2]AFSCET (Association Française de Science des Systèmes), Paris, France
[3]ArianeGroup, Paris, France

analytical methods are based on Taylor's [1], Agrawal's [2] and Rachidi's [3] model depending on the considered field illuminating the printed circuit board (PCB). Nevertheless, following with the forecast on the high-speed PCBs [6], a lot of effort still have to be made for predicting the EMC-radiated coupling effect onto the PCB trace structures. Most of the performed research work [7–12] about the PCB trace as electrical interconnect trees modelling was generally performed to the SI analysis without the EM radiation consideration.

Various approaches and simulation tools have been developed to predict the behaviour of multilayer PCBs with EM coupling from a holistic perspective. Despite the noise suppression technique [13], a simpler modelling method is needed for the multilayer interconnects as power delivery network [14,15]. Innovative design methodology and modelling tools are needed to predict the interconnect effects on the PCB performance as EMC. In this way, advancements and innovations on EMC engineering must be developed.

8.2 Bibliography of MKME formalism on EMC of PCB

MKME (Modified Kron's Method for EMC) formalism is an extension of Kron's method applied to EMC analysis. So far, a few research works [16–18] are available on the PCB coupled with EM illumination by using MKME. In [16], the case of EMC-radiated emission was studied. A simple microstrip line radiated EMC is modelled with MKME. The model was validated only with the far-field-radiated emission. In [17], the MKME was associated with the stochastic approach for assessing planar PCB coupling. Then, an analytical radiated EM coupling with a planar single layer PCB is proposed in [17] by considering Taylor's approach. The few research works [19,20] were published about the analytical modelling of 3D multilayer PCB with the MKME. Here we propose to apply MKME formalism on radiated EM coupling problems on multilayer PCB in this chapter, because it gives the compact solution and provides the conversion between mesh and moment spaces for solving the EM coupling problems.

8.3 Recall on MKME mesh space to moment space definition

We will discuss the mesh space to moment space in this chapter following with application examples. The moment space will be employed to solve the radiated EM coupling problems.

A flow circulating on a mesh of a surface denoted as S, generating a moment. The space of moment is denoted as *Mom* [21]. From this space, we will finally introduce the space of the networks considers the interactions of the far field between subnetworks, such as the coupling between two-point sources radiating.

In these different levels of space, there are precise physical interactions. For example, in the space of the nodes, we will be able to elaborate on the actions of the forces of Newton. The electric fields will act as an induced potential difference between two nodes at the ends of the branch in the space of the branches. The magnetic field will be introduced into the space of the meshes between the closed circulation of flow currents. Finally, all the networks will constitute the universe representing the global system. In other cases, this super-network can be introduced into a larger hyper-network [22].

Two elementary networks presented here containing one mesh in each network, J^1 and J^2, are the mesh currents on the surfaces S_1^1 and S_2^2. The two moments are presented as in Figure 8.1.

The space of moment can represent the remote interactions without electrical contact and link between electrical subnetworks. If we consider the coupling between two antennas' loops with a long-distance respect to the far-field condition, then each antenna can consider the other as a radiant source. As a consequence, the propagation function of the field can be transformed into coupling generated by mutual inductance in the space of the meshes. The magnetic field B_ω and the moment source m^q are connected by a function as follows:

$$B_\omega = \alpha_{\omega q} m^q \tag{8.1}$$

where α is a function, including the delayed potentials, attenuations, and is generally expressed in the frequency domain. The moment m^q has a relation with the mesh current J^ω flow by the following expression:

$$m^q = S_\omega^q J^\omega \tag{8.2}$$

The derivation in the time domain of the moment is marked as in mechanics by adding a point above the letters:

$$\dot{m}^q = S_\omega^q \dot{J}^\omega \tag{8.3}$$

Components of S_ω^q represent the surfaces of each mesh of the network, the matrix S is square and invertible, (8.2) represents the base change from mesh to moment.

A moment of emitter network produced a coupling mesh, through the moment, of the remote receiver network, an inserted source is represented in the following expression with the rotational field B_ω:

$$E_p = S_p^\omega \dot{B}_\omega \tag{8.4}$$

with

$$\dot{B}_\omega = \alpha_{\omega q} \dot{m}^q \tag{8.5}$$

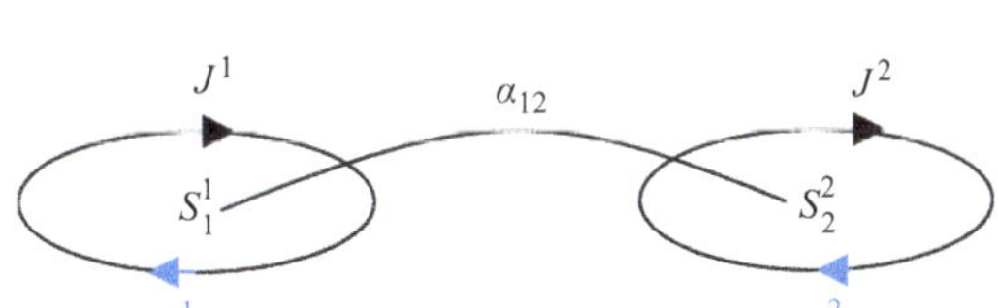

Figure 8.1 Mesh to moment space transformation [23]

Now, we can write the mesh current induced on one network by the mutual coupling from the excitation of the other network:

$$E_t = M_{tp}J^p \tag{8.6}$$

From (8.4) and (8.7), we can get

$$\begin{aligned} M_{tp}J^p &= S_p^{\omega}\alpha_{\omega q}m^q \\ &= S_p^{\omega}\alpha_{\omega q}S_p^q J^p \end{aligned} \tag{8.7}$$

Then,

$$M_{\omega q} = S_{\omega}^{\omega}\alpha_{\omega q}S_q^q \tag{8.8}$$

As a result, this last equation represents the base change from a characteristic function of the propagation of the field toward a function taking account the physical nature as the mutual coupling impedance between two meshes of two remote networks. Characterization of the system directly in the mesh space is relatively complicated. MKME proved that the base changes of branch and meshes allow an easier equation setting. Changes in topological spaces will allow the flexible manipulation of all the equations of a complex system.

8.4 Recall on field coupling with MKME formalism

8.4.1 Electric coupling

The subsection presents a case that the electric field coupled to a line that illustrates typically the interaction of the space of the branches [22].

The divergence of the electric field is expressed in the integral form using the theorem from the Green–Ostrogradski equation. The flow of the electric field vector through a closed surface is then equal to the integral of the divergence of this vector on the volume delimited by this surface. On the other hand, the integral expression of Gauss's theorem gives us the expression of the flux Φ. It is assumed that the flux area of the field is small enough for the field to be constant [22]. Deriving to show the current, we obtain, as a function of the flux Φ_e,

$$q = \Phi_e\varepsilon_0 \rightarrow i = j\omega\varepsilon_0\Phi_e \tag{8.9}$$

In the case of electrical coupling onto a line [22], there is a current source that finally reports the electrostatic influence of the field on the line. By inserting the previous obtained current source in parallel with a capacitor, based on the properties of Thevenin and Norton, a voltage generator in series with the capacitor can be placed:

$$V(j\omega) = \frac{1}{j\omega C}j\omega\varepsilon_0\Phi_e = \frac{h}{j\omega\varepsilon_0 S}j\omega\varepsilon_0\Phi_e = hE \tag{8.10}$$

The potential difference induced by the Coulomb force is finally obtained by the integration of the work of the field on the height h of the line. In the case of crosstalk coupling, the current source will be obtained in linear with the coupling capacitance

C_{12} and the voltage source from the emitter line with dimension L_0. V_{10} is the voltage source applied on the first line:

$$i(j\omega) = j\omega C_{12} V_{10} L_0 \tag{8.11}$$

Take the example of a monopole antenna, Z_{ant} is the equivalent impedance, excited by a source e_0 with internal impedance R_0. This antenna generates an electric field in its environment that is captured by a microstrip line element with length dz, shown in Figure 8.2.

The current source generated by the electric field on the element of the line is replaced by a mutual impedance Z_{ce} realizing the coupling between the two subnetworks as shown in Figure 8.3.

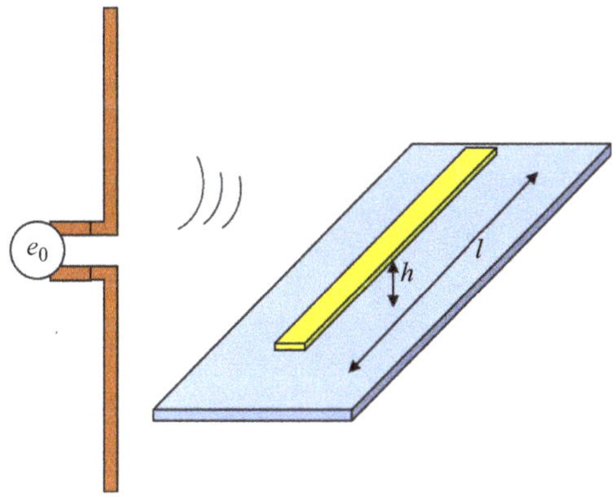

Figure 8.2 Description of normal electric field coupling

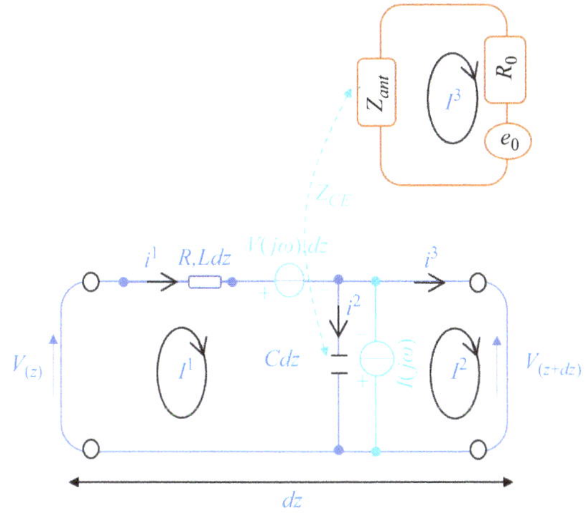

Figure 8.3 Equivalent topology of electric field coupling

Rather than representing the current source in a source tensor I^s, a mutual impedance characterizing the electrical coupling is coupled to the capacitor Cdz. This ability is parallel in the circuit. It is the reason that the electric coupling will interact in the space of the branches and not in the space of the meshes. To write the impedance tensors in mesh space, MKME connection matrix C allows to change space from branch to mesh [22].

$$\begin{cases} i^1 = I^1 + 0I^2 + 0I^3 \\ i^2 = I^1 - 1I^2 + 0I^3 \\ i^3 = 0I^1 + 1I^2 + 0I^3 \\ i^4 = 0I^1 + 0I^2 + 1I^3 \end{cases} \Rightarrow C = \begin{bmatrix} 1 & 0 & 0 \\ 1 & -1 & 0 \\ 0 & 1 & 0 \\ 0 & 0 & 1 \end{bmatrix} \tag{8.12}$$

At first, let us write the impedance matrix z' in the space of the branch without involving the electrical coupling impedance matrix m_E:

$$z' = \begin{bmatrix} R + j\omega L & 0 & 0 & 0 \\ 0 & \frac{1}{j\omega C} & 0 & 0 \\ 0 & 0 & 0 & 0 \\ 0 & 0 & 0 & R_0 + Z_{ant} \end{bmatrix} \tag{8.13}$$

$$m_E = \begin{bmatrix} 0 & 0 & 0 & 0 \\ 0 & 0 & 0 & Z_{CE} \\ 0 & 0 & 0 & 0 \\ 0 & Z_{CE} & 0 & 0 \end{bmatrix} \tag{8.14}$$

The matrix impedance in the space of the branches with the couplings:

$$z = z' + m_E \tag{8.15}$$

The coupling in the space of the branch appears as the terms m_{24}, in other words, between branches 2 and 4. By applying matrix C and matrix z with MKME transformation, we can get automatically matrix impedance Z' in the mesh space:

$$Z' = \begin{bmatrix} R + j\omega L + \frac{1}{j\omega C} & -\frac{1}{j\omega C} & Z_{CE} \\ -\frac{1}{j\omega C} & \frac{1}{j\omega C} & -Z_{CE} \\ Z_{CE} & -Z_{CE} & R_0 + Z_{ant} \end{bmatrix} \tag{8.16}$$

Let us look at the positions of the coupling impedance elements Z_{CE}. The coupling happens between meshes 1 and 3 but also meshes 2 and 3. This is because the coupling is simultaneously associated with both mesh currents I^1 and I^2. For this reason, electrical coupling will be easy to present in the mesh space.

8.4.2 Magnetic coupling

To illustrate the coupling in magnetic field, we will associate with the topological mesh space, we choose the interaction between a magnetic loop and an LC cell of the line [22] presented later.

The magnetic field interacts with the surface of the element of the line producing a magnetic field flux ϕ with the following equation (Figure 8.4):

$$\phi = \iint_S \vec{B}\,\vec{dS} \tag{8.17}$$

By applying Stokes theorem (8.18),

$$\phi = \iint_S \vec{rot}\vec{A}\,\vec{dS} = \oint_{dl} \vec{A}\,\vec{dl} \tag{8.18}$$

Then we can obtain the expression of the equivalent coupling source e_0 in the frequency domain with a Laplace transform:

$$e_0(j\omega) = j\omega SB \tag{8.19}$$

The equivalent EMF in the case of crosstalk coupling between two lines with length L_0 can be written in the following form:

$$e_0 = j\omega L_{12} I^1 L_0 \tag{8.20}$$

By definition, the magnetic field generates a voltage source across the mesh in which the EMF is coupled. The total voltage is obtained by the summation of the voltage sources across the inductance. For the inductive effects due to the closed circulations of the currents, we therefore suggest working in the space of the meshes. Rather than modelling the coupling by an EMF source, we use mutual impedance Z_{CM} [22]. The mutual impedance will be carried on a cord connecting the two virtual nodes arranged at the centre of the meshes of the two subnetworks as shown in Figure 8.5. The coupling appears between these two meshes and not between branches. In the case of more complex networks, if we use branch space to represent the magnetic coupling, several meshes could have been excited. This representation therefore allows coupling exclusively between meshes. In the same way for electrical coupling, we write the impedance matrices in the space of the branch and then changed to the space of the

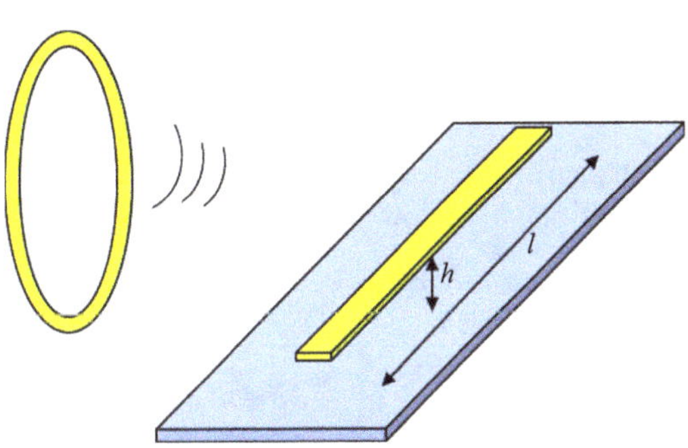

Figure 8.4 Description of normal magnetic field coupling [22]

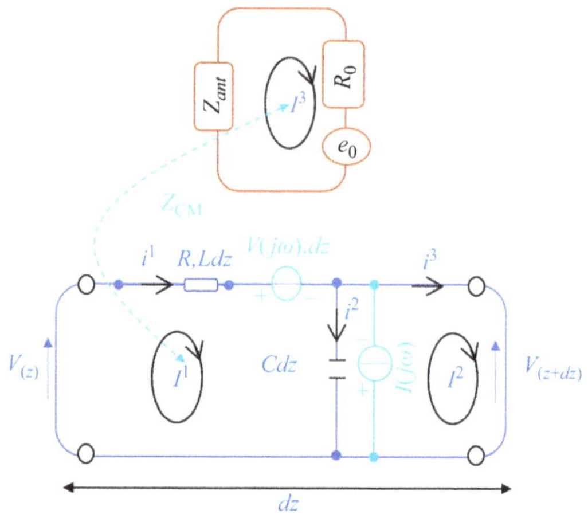

Figure 8.5 Equivalent topology of magnetic field coupling [22]

mesh by the connection matrix to make couplings between meshes. The connection matrix is the same as in the previous subsection [22]:

$$z' = \begin{bmatrix} R + j\omega L & 0 & 0 & 0 \\ 0 & \frac{1}{j\omega C} & 0 & 0 \\ 0 & 0 & 0 & 0 \\ 0 & 0 & 0 & R_0 + Z_{ant} \end{bmatrix} \tag{8.21}$$

Therefore, in the space of the mesh we obtain the following matrix that does not take into account the effect of magnetic coupling Z_M:

$$Z' = \begin{bmatrix} R + j\omega L + \frac{1}{j\omega C} & -\frac{1}{j\omega C} & 0 \\ -\frac{1}{j\omega C} & \frac{1}{j\omega C} & 0 \\ 0 & 0 & R_0 + Z_{ant} \end{bmatrix} \tag{8.22}$$

$$M_{CM} = \begin{bmatrix} 0 & 0 & Z_M \\ 0 & 0 & 0 \\ Z_M & 0 & 0 \end{bmatrix} \tag{8.23}$$

Finally, the impedance matrix of the two networks connected by a magnetic coupling impedance is written as follows:

$$Z = Z' + M_{CM} = \begin{bmatrix} R + j\omega L + \frac{1}{j\omega C} & -\frac{1}{j\omega C} & Z_M \\ -\frac{1}{j\omega C} & \frac{1}{j\omega C} & 0 \\ Z_M & 0 & R_0 + Z_{ant} \end{bmatrix} \tag{8.24}$$

In this way, the matrices representing many different physical interactions of the system complex can be added to the matrix impedance of the system [22]. The final impedance matrix obtained in the mesh space will then be representative of the global system.

8.5 MKME model for 3D multilayer PCB illuminated by EM plane wave

This section is focused on the MKME modelling methodology applied to the multilayer PCB aggressed by EM plane wave radiation. The graph topology and the abstracted analytical model of the PCB trace interacting with the EM wave are elaborated.

8.5.1 Formulation of the problem

Figure 8.6(a) depicts the scenario of the EMC problem under study. It illustrates the perspective 3D view of the multilayer PCB under radiated EM aggression. The considered POC is a four-layer PCB with relative permittivity ε_r and total thickness h. It is composed of traces and the ground plane. The arbitrarily shaped interconnects are placed on Layers 1, 2, 4 (Table 8.1).

To set the analytical description, let denote $j\omega$ the complex angular frequency, c the vacuum light celerity and the wave number $k = \omega/c$. The PCB structure is illuminated by the uniform EM propagating in different directions in the reference

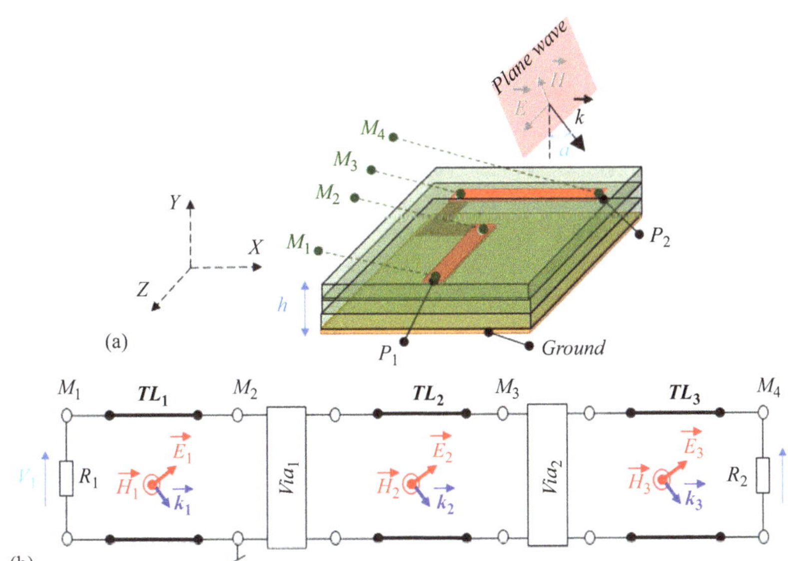

Figure 8.6 (a) Structure of 3D multilayer PCB trace illuminated by EM plane wave with angle a respect to Z axis and (b) the electrical equivalent circuit [24]

Table 8.1 Distribution of the interconnect shapes

"I"-shape	"L"-shape	"I"-shape line
$TL_1 = TL_{M1M2}$	$TL_2 = TL_{M2M3}$	$TL_3 = TL_{M3M4}$

system (O, e_x, e_y, e_z). By denoting that a is the oblique angle compared to the Z-axis and defined by electric field, magnetic field and wave vector, respectively,

$$\vec{E}(j\omega) = E(j\omega)\vec{e}_z \tag{8.25}$$

$$\vec{H}(j\omega) = H(j\omega)[\sin(a)\vec{e}_x - \cos(a)\vec{e}_y] \tag{8.26}$$

$$\vec{k}(j\omega) = k[\cos(a)\vec{e}_x - \sin(a)\vec{e}_y] \tag{8.27}$$

Figure 8.6(b) depicts the equivalent schematic of the EM field illuminated PCB shown in Figure 8.6(a). It is composed of elementary TLs $TL_{1,2,3}$, and two vias Via_1 and Via_2.

The PCB is terminated by resistive loads R_1 and R_2 placed at nodes M_1 and M_4, respectively. The calculation methods of coupling voltages $V_1 = V_{M1}$ and $V_2 = V_{M2}$ induced by the EM radiation are developed in the next sections [23].

8.5.2 MKME model establishment

The model of the multilayer PCB–plane wave interaction scenario can be accomplished quickly following the MKME general methodology. From each subsystem, the radiated EMC MKME model aims to determine current and voltage variables in the mesh or branch spaces. In the present study, the primitive elements of multilayer PCB system consist of interconnect lines, vias, pads and anti-pads, loaded by identical resistances $R_{1,2} = R_0 = 50\ \Omega$.

8.5.2.1 Graph topology establishment

The MKME graph represents the EM interaction between the interconnects and the local wave from the incident fields $\overrightarrow{E}_{incident}$ and $\overrightarrow{H}_{incident}$. The graph is built from Taylor's model described in Figure 8.7(a) and (b). It is implemented by considering the interconnects and via which are meshed as several Telegrapher's cells coupled to the EM field. It should be recalled that due to the GND plane reflection, the electric and magnetic fields in medium 1 (air) are approximated equal to

$$\begin{cases} E_{t-air}(j\omega) = 2E_{incident}(j\omega) \\ H_{n-air}(j\omega) = 2H_{incident}(j\omega) = 2E_{incident}(j\omega)/Z_{air} \end{cases} \tag{8.28}$$

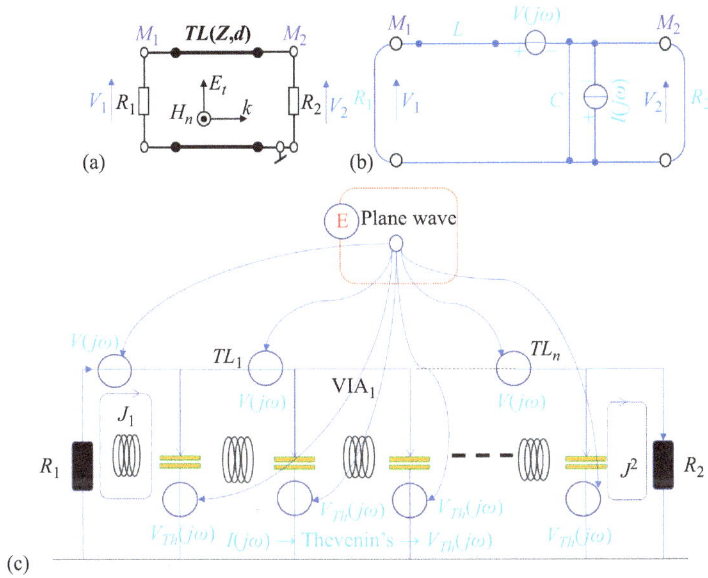

Figure 8.7 *(a) Plane wave injected to the transmission line, (b) Taylor's lossless cell and (c) MKME equivalent graph of the structure [23,24]*

with $Z_{air} = 120\pi\,\Omega$, where t and n represent the transversal and normal to the PCB. Acting as a multilayer structure, the electric and magnetic fields penetrating in medium 2 (FR4) depend on the substrate relative permittivity must be given by

$$\begin{cases} E_{t-FR4}(j\omega) = 2E_{incident}(j\omega)/\varepsilon_r \\ H_{n-FR4}(j\omega) = 2H_{incident}(j\omega) \end{cases} \tag{8.29}$$

The multilayer structure MKME graph is shown in Figure 8.7(c). It is composed of B branches and M meshes. The interconnect line losses are integrated in the MKME model by adding resistance elements. These cell characteristics are defined in function of the PCB microstrip and strip line parameters. The number of cells is defined in function of the desired frequency band to reach better accuracy. The electric and magnetic fields generate Taylor's current, and voltage sources are given as in the following equations [17,18]. The current sources, in parallel with capacitance in Taylor's model, have been transformed to voltage source in series with the capacitance in the topology in Figure 8.7(c) based on Thevenin's theorem expressed as $V_{th}(j\omega)$.

$$V(j\omega) = j\omega\mu_0 H(dl^n)hdl \tag{8.30}$$

$$I(j\omega) = j\omega c E(dl^n)hdl \tag{8.31}$$

where dl is the per-unit length, μ_0 is the vacuum permeability and $n = 1, 2, 3$.

8.5.2.2 Graph to tensorial object

The physical laws governing the system under investigation are explored from relations between the branch and mesh currents I^b and J^m, respectively, based on the graph of Figure 8.7(c). The MKME analysis is performed with the Einstein tensorial notation between the current contravariables:

$$Z_{ab} = \begin{bmatrix} Z_{b1} & \cdots & 0 & \cdots & 0 \\ \vdots & \ddots & \vdots & \iddots & \vdots \\ 0 & \cdots & Z_{b10} & \cdots & 0 \\ \vdots & \iddots & \vdots & \ddots & \vdots \\ 0 & \cdots & 0 & \cdots & Z_{bn} \end{bmatrix} \tag{8.32}$$

where $a, b = 1, 2, \ldots, B$ and $m, n = 1, 2, \ldots, M$. The vias, overall combined and interconnect source covariables in branch space induced by the EM field are expressed, respectively, as

$$e_{via(m)} = \begin{bmatrix} 0 & 0 & 0 & 0 \end{bmatrix} \tag{8.33}$$

$$e_a = \begin{bmatrix} e_{TL(1)} & e_{via(1)} & e_{TL(2)} & e_{via(2)} & e_{TL(3)} \end{bmatrix} \tag{8.34}$$

$$e_{TL(n)} = \begin{bmatrix} H(dl^1, TL_n) & 0 \\ 0 & E(dl^1, TL_n) \\ \vdots & \vdots \\ H(dl^n, TL_n) & 0 \\ 0 & E(dl^n, TL_n) \end{bmatrix} \begin{bmatrix} j\omega\mu_0 hdl \\ h \end{bmatrix} \tag{8.35}$$

The branch and mesh connectivity matrix yielded from relation:

$$C_n^b = \begin{bmatrix} 1 & 0 & 0 & \cdots & \cdots & 0 & 0 \\ \vdots & \vdots & \vdots & \vdots & \vdots & \vdots & \vdots \\ 0 & 0 & 1 & -1 & 0 & \cdots & 0 \\ 0 & 0 & 0 & 1 & 0 & \cdots & 0 \\ \vdots & \vdots & \vdots & \vdots & \vdots & \vdots & \vdots \\ 0 & 0 & 0 & 0 & 0 & \cdots & \cdots \end{bmatrix} \tag{8.36}$$

The sources covariable voltage and impedance in mesh space can be obtained.

8.5.3 Validation results

The developed MKME model is validated with a 3D multilayer PCB designed in a CST MWS® environment. The results are obtained in frequency domain by considering the EM plane wave illumination. Two cases of an EMC scenario modelled by the MKME are presented and validated by both simulations and measurement. The frequency-domain computations are performed from 0.4 to 1.4 GHz with 401 samplings.

Table 8.2 PCB board parameters

PCB layout parameters					
Board width	**Board length**	**Board height**	**Substrate permittivity**	**Loss tangent**	**Metal thickness**
103 mm	103 mm	1.6 mm	4.5	0.012	0.035 mm
Via and interconnect parameters					
Via diameter		**Pad diameter**	**Interconnect width**	**Length/Each**	
1 mm		1.2 mm	0.5 mm	20/40 mm	

8.5.3.1 Description of the POC

The POC PCB acts as a four-layer PCB with metallization in copper (Cu) implemented on an FR4-epoxy dielectric substrate. This PCB is supposed to be placed in the EM environment submerged by uniform plane wave EM radiation. The conductors have identical rectangular section with physical width and thickness. The PCB conductor is constituted by three interconnect lines: two via and a ground plane on the bottom layer. The PCB layout, via and TL parameters are presented in Table 8.2. Giving these parameters, the coupling voltages across terminal ports 1 and 2 are calculated with the MKME. The numerical calculations with the Python program are discussed in the next sections [24].

8.5.3.2 Case 1: frequency-independent field with different angles

In this case, the PCB is assumed to be radiated by the uniform harmonic EM plane wave with field strength approximated, defined based on the transverse electromagnetic (TEM) cell characteristics used in the experiments. As described in Figure 8.6(a), this field illuminates the PCB in different propagation directions. The associated wave vector is defined by the incidence angle a varied from 0° to 60°. In the present paragraph, the plane wave illumination is defined by the incident angles $a = 0°$, 30°, 45°, 60°. The computed coupling voltage spectrums, $|V_1 MKME(j\omega)|$, are compared with the simulations from 0.4 to 1.4 GHz, because the maximum frequency of the TEM cell we used is 2 GHz. The CST MWS full wave simulations were performed with transient solver and automatic meshing. The MKME calculation results and CST simulations are plotted in Figure 8.8 with different incident degrees. The coupling voltages present an average level of around 65 dBμV. Acting as a linear passive circuit, the coupling voltages are proportional to the illuminating EM field intensity. As expected, the computed voltages from MKME are well correlated with the simulations. The coupling voltage across port 1 presents a zero transmission between 1 and 1.2 GHz. Despite the correlation between the coupling, there are notable deviations

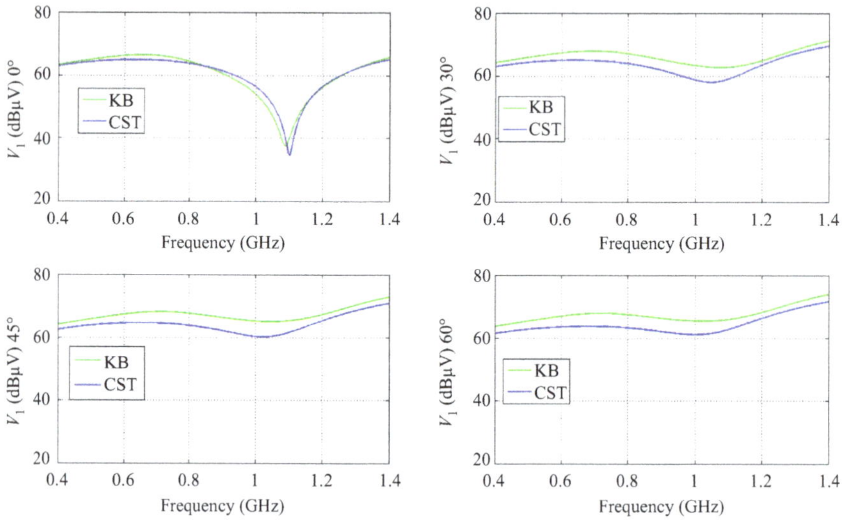

Figure 8.8 Spectrums of V_1 from the MKME model and CST MWS from the PCB aggressed by the plane wave

Table 8.3 Absolute maximum differences between the calculated and simulated voltages

a	0°	30°	45°	60°
ΔV_1 (dBμV)	4.5	1.06	1.6	1.75
ΔV_2 (dBμV)	9.4	6.8	7.4	7.6

in the considered frequency band. The maximal absolute differences between the calculated and simulated coupling voltages defined by

$$\Delta V_{1,2} = \left| V_{1,2_{MKME}} - V_{1,2_{CST}} \right|_{\max} \tag{8.37}$$

The results are summarized in Table 8.3 for the different values of angle a.

8.5.3.3 Case 2: illuminated field square waveform

In this case, the illuminating EM plane wave is supposed as a periodical square waveform signal with intensity $E_m = 10\,\text{kV/m}$ and time-period equal to $T_0 = 10\,\text{ns}$. During the simulation and computation, this plane wave is assigned in the frequency domain. Based on the signal-processing theory, the wave source is defined as a pulse function spectrum.

By sweeping the operation frequency from 0.4 to 1.4 GHz, V_1 and V_2 were computed. Figure 8.9 plots the comparisons of coupled voltages V_1 and V_2 calculated from the MKME model (solid line) and simulated (dotted line) from CST MWS. It

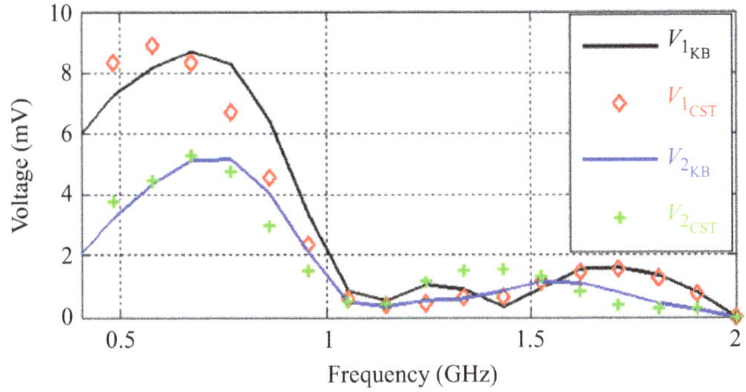

Figure 8.9 Comparison between V_1 calculated from MKME model and CST MWS simulations with broadband square wave pulse spectrum

can be underlined that a good agreement between the calculated and simulated results is found. In correlation with the radiated field signal, the coupling effects present a higher magnitude in the frequency band under 1 GHz. They present a maximum value at about 0.7 GHz. The maximal deviations ΔV_1 and ΔV_2 are lower than 2 mV.

8.5.3.4 Comments on the calculated and simulated results

The main deviations between the calculated and simulated coupling voltages onto the multilayer PCB are mainly due to the following reasons:

1. the 3D meshing numerical inaccuracies of the structure especially around the metal conductor;
2. the calculated effective permittivity of the substrate that may change the frequency and
3. the inaccuracies related to Taylor's cell approximations.

8.5.3.5 Experimental results

The four-layer PCB photographed in Figure 8.11(a) was fabricated. As depicted in Figure 8.11(b), to validate the proposed MKME model, S-parameter measurements were performed with TEM cell FCC-TEM-JM3 and VNA E5071C from Agilent Technologies® operating from 100 kHz to 8 GHz. The TEM cell basic structure is presented in Figure 8.10.

The transversal and normal fields are approximated by

$$E_t = \frac{V_{septum}}{\varepsilon_r h} \tag{8.38}$$

$$H_n = \frac{V_{septum}}{\eta h} \tag{8.39}$$

where V_{septum} is the input voltage amplitude, h is the separation between the board and the septum and η is the air impedance.

As illustrated by Figure 8.11(c), the measured ("meas.") coupling voltages are in good correlation with measurements up to 1 GHz. The spikes appeared above 1 GHz

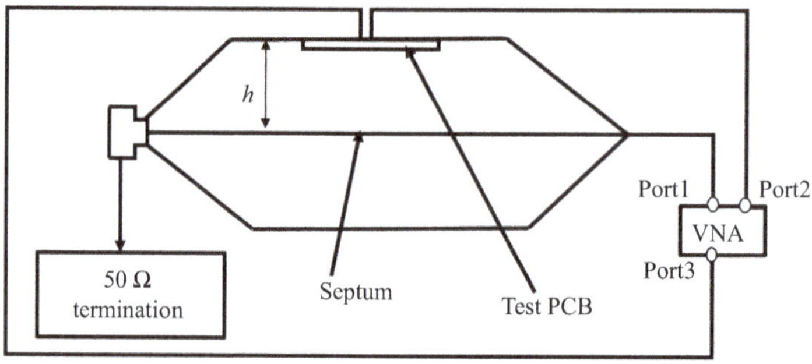

Figure 8.10 Basic structure of the TEM cell [23]

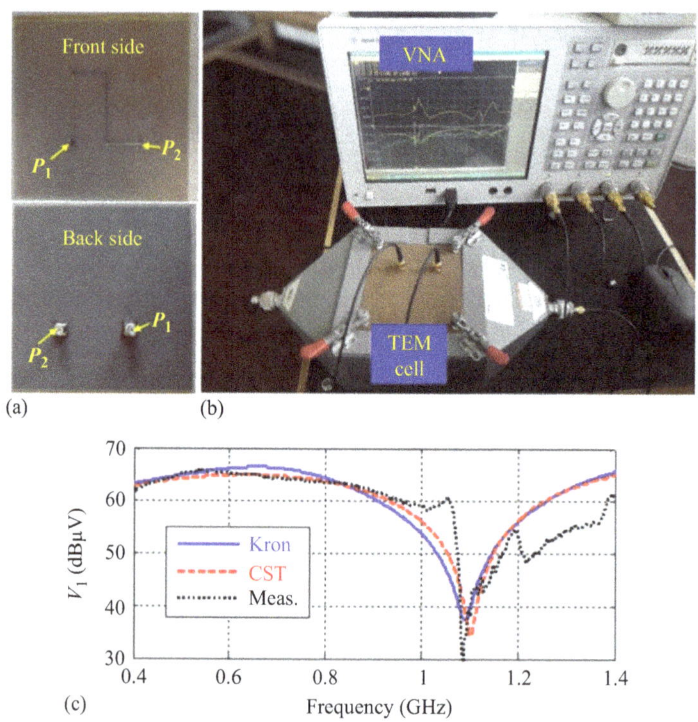

Figure 8.11 Photograph of (a) the PCB prototype (physical size: 103 mm × 103 mm × 1.6 mm) and (b) the TEM cell-based experimental setup. (c) Comparison between calculated from MKME model, CST MWS and measurement with broadband plane wave excitation [24]

in the measurement results are principally due to different effects. For the performed experimental setup, the most significant EM effects are as follows:

1. the variation of the EM field penetrating into the multilayer structures,
2. the discrepancies of the substrate relative permittivity and tangential loss over the frequency,

3. the imperfections of the PCB fabrication,
4. the TEM cell influences that may induce self-resonances in the considered frequency band and
5. also, the lack of consideration of realistic perturbations from the TEM in the computed model.

And in this case, our work is different from the work of Op'land from ESEO Lab [18]; they have done only the coupling to a microstrip line; however, we have done the waves interacting with multilayer PCB with angles. This is an improved study of MKME for EM coupling to multilayer structures.

8.6 Conclusion

As reported in this chapter, the MKME enables one to perform efficiently the R-EMC analysis of multilayer PCB. The definition of MKME mesh and moment spaces has been presented. The general basic field coupling with MKME has been developed. Then, an original radiated EMC model of 3D multilayer PCB radiated by EM field is investigated. The MKME modelling methodology is described. It was shown analytically that the model enables one to group all interactions into an assembly of subnetwork implemented in the same tensor space. The MKME graph combines the branch and mesh subnetworks and integrates the electrical, EM and physical parameters of the problem. The interaction between the EM field and the multilayer conductor interconnect lines is expressed by the distributed current and voltage sources defined from Taylor's model. The POC PCB was designed, simulated, fabricated and measured. The MKME model computations were run with the Python program and simulated with CST MWS simulations confirm the validity of the proposed unfamiliar model. As a result, the calculated, simulated and tested coupling voltages across the four-layer PCB prototype from 0.4 to 1.4 GHz are in good correlation.

References

[1] Taylor C, Satterwhite R, and Harrison C. The response of a terminated two-wire transmission line excited by a nonuniform electromagnetic field. IEEE Transactions on Antennas and Propagation. 1965;13(6):987–989.

[2] Agrawal A, Price H, and Gurbaxani S. Transient response of multiconductor transmission lines excited by a nonuniform electromagnetic field. IEEE Transactions on Electromagnetic Compatibility. 1980;EMC-22(2):119–129.

[3] Rachidi F. Formulation of the field-to-transmission line coupling equations in terms of magnetic excitation field. IEEE Transactions on Electromagnetic Compatibility. 1993;35(3):404–407.

[4] Ying W, Zheng Z, and Ooi BL. Electromagnetic coupling on airborne structures and systems using NEC. In: 2008 Asia-Pacific Symposium on Electromagnetic Compatibility and 19th International Zurich Symposium on Electromagnetic Compatibility. Zurich: IEEE; 2008.

[5] Leseigneur C, Baudry D, Ravelo B, *et al.* Near-field coupling model between PCB and grounded transmission line based on plane wave spectrum. The European Physical Journal Applied Physics. 2013;64(1):11001.

[6] Fan J, Ye X, Kim J, *et al.* Signal integrity design for high-speed digital circuits: progress and directions. IEEE Transactions on Electromagnetic Compatibility. 2010;52(2):392–400.

[7] Kahng AB, and Muddu S. An analytical delay model for RLC interconnects. IEEE Transactions on Computer-Aided Design of Integrated Circuits and Systems. 1997;16(12):1507–1514.

[8] Averill RM, Barkley KG, Bowen MA, *et al.* Chip integration methodology for the IBM S/390 G5 and G6 custom microprocessors. IBM Journal of Research and Development. 1999;43(5.6):681–706.

[9] Ruehli AE, and Cangellaris AC. Progress in the methodologies for the electrical modeling of interconnects and electronic packages. Proceedings of the IEEE. 2001;89:740–771.

[10] Eudes T, and Ravelo B. Analysis of multi-gigabits signal integrity through clock H-tree. International Journal of Circuit Theory and Applications. 2012;41(5):535–549.

[11] Ruan A, Yang J, Wan L, *et al.* Insight into a generic interconnect resource model for Xilinx Virtex and Spartan series FPGAs. IEEE Transactions on Circuits and Systems II: Express Briefs. 2013;60(11):801–805.

[12] Khan ZA. A novel transmission line structure for high-speed high-density copper interconnects. IEEE Transactions on Components, Packaging and Manufacturing Technology. 2016;6(7):1077–1086.

[13] Eriksson K, Gunnarsson SE, Nilsson PA, *et al.* Suppression of parasitic substrate modes in multilayer integrated circuits. IEEE Transactions on Electromagnetic Compatibility. 2015;57(3):591–594.

[14] Shringarpure K, Pan S, Kim J, *et al.* Formulation and network model reduction for analysis of the power distribution network in a production-level multilayered printed circuit board. IEEE Transactions on Electromagnetic Compatibility. 2016;58(3):849–858.

[15] Wen S, Zhang J, and Lu Y. Modeling and quantification for electromagnetic radiation of power-bus structure with multilayer printed circuit board. IEEE Transactions on Components, Packaging and Manufacturing Technology. 2016;6(1):79–86.

[16] Rogard E, Azanowsky B, and Ney MM. Comparison of radiation modeling techniques up to 10 GHz—application on a microstrip PCB trace. IEEE Transactions on Electromagnetic Compatibility. 2010;52(2):479–486.

[17] Kasmi C, Maurice O, Gradoni G, *et al.* Stochastic Kron's model inspired from the Random Coupling Model. In: 2015 IEEE International Symposium on Electromagnetic Compatibility (EMC). Dresden: IEEE; 2015.

[18] Land SO, Perdriau R, Ramdani M, *et al.* Kron simulation of field-to-line coupling using a meshed and a modified Taylor cell. In: 2013 9th International Workshop on Electromagnetic Compatibility of Integrated Circuits (EMC Compo). Nara: IEEE; 2013.

[19] Xu Z, Liu Y, Ravelo B, *et al.* Multilayer power delivery network modeling with modified Kron's method (MKM). In: 2017 International Symposium on Electromagnetic Compatibility – EMC EUROPE. Angers: IEEE; 2017.
[20] Xu Z, Liu Y, Ravelo B, *et al.* Modified Kron's TAN modeling of 3D multilayer PCB. In: 2017 11th International Workshop on the Electromagnetic Compatibility of Integrated Circuits (EMCCompo). Saint Petersburg: IEEE; 2017.
[21] Maurice O. La compatibilité électromagnétique des systèmes complexes. Publications HS, editor; 2007.
[22] Leman S. Contribution à la résolution de problèmes de compatibilité electromagnétique par le formalisme des circuits electriques de KRON. Univ. Lille; 2009.
[23] Xu Z. Tensorial analysis of multilayer printed circuit boards, PhD thesis; 2019.
[24] Xu Z, Ravelo B, Maurice O, *et al.* Radiated EMC Kron's Model of 3-D multilayer PCB aggressed by broadband disturbance. IEEE Transactions on Electromagnetic Compatibility. 2019:1–9.

Chapter 9

Conducted emissions (CEs) EMC TAN modelling

Olivier Maurice[1], Zhifei Xu[2], Yang Liu[2], and Blaise Ravelo[2]

Abstract

The electromagnetic compatibility (EMC) conducted emission (CE) of printed circuit boards (PCBs) constitutes the most attractive research topic of EMC research engineers due to the fascinating challenging aspect related to the electronic component and PCB design complexity. The present chapter proposes an attempted EMC CE theory of the conducted EMC PCB emission. In this case, the PCB is assumed as a hybrid system comprising

- Lumped devices: frequency dependent $R(f)$, $L(f)$ and $C(f)$ components;
- Active components: as integrated circuits, including the die and packaging parameters that may behave as a nonlinear device;
- Passive elements: vias, pads and anti-pads, and interconnect TLs. The EMC CE model of the PCB system will be developed by considering some different standards perturbation signals.

The originality of the present EMC CE theory will be the elaboration of the transfer impedance matrices relating the contravariables represented by the active component internal activities as current tensor sources and the covariables voltage tensors. The EMC theorization will provide the way to establish the twice covariables transfer-impedance tensors. To highlight the feasibility of the EMC CE tensorial analysis of networks (TAN) model illustrative, examples of PCB used in automotive and PC will be treated. Discussion will be made about the strength and the weakness of the developed model in function of the PCB complexity and also the bandwidth of the EMC noises. Then, the accuracy, advantages and limits of the EMC CE TAN model will be made at the end of the chapter.

Keywords: TAN approach, Kron's method, Kron–Branin's model, frequency domain analysis, modelling methodology, PCB analysis, SI analysis, PI analysis, EMC analysis

[1]IRSEEM/ESIGELEC, Rouen, France
[2]ArianeGroup, Paris, France

9.1 The ICEM model and the EMC problem

ICEM means *Integrated Circuit (IC) Emission Model*. We have already evoked this model in Chapter 2. It is a generic approach that can be used for any IC. In this model, the noise source is represented by a current generator with its own impedance. This source is followed by an interface network linking the chip with the printed circuit board (PCB), then by the tracks on the PCB and its connections with other PCB in the equipment or some connectors and their harnesses. The problem can be decomposed in parts from the noise source in one device to the electromagnetic compatibility (EMC) equipment test following a standard setup. Figure 9.1 shows a typical situation in the case of an EMC-conducted emission (CE) test.

In this test we recognize the PCB with its components, the connections to the equipment connectors, the harnesses going from the equipment to the load which is represented by an LISN (*line impedance stabilization network*). This filter is here to simulate the power supply network of the system. It has the advantage to be perfectly well-known and allow making robust computations. The harness is 1–3 m long, depending on the standards. It is often located 5 cm up to a conductive ground that represents the system grounding. The noise voltage u_m is measured across the LISN resistance or it can be also the current at the input of the LISN. Other measurements can be realized on other signals than the power supply, but the principles remain the same.

Our problem is to be able to predict the noise amplitude u_m coming from noisy source existing on one PCB, inside the equipment. To solve this problem we will try

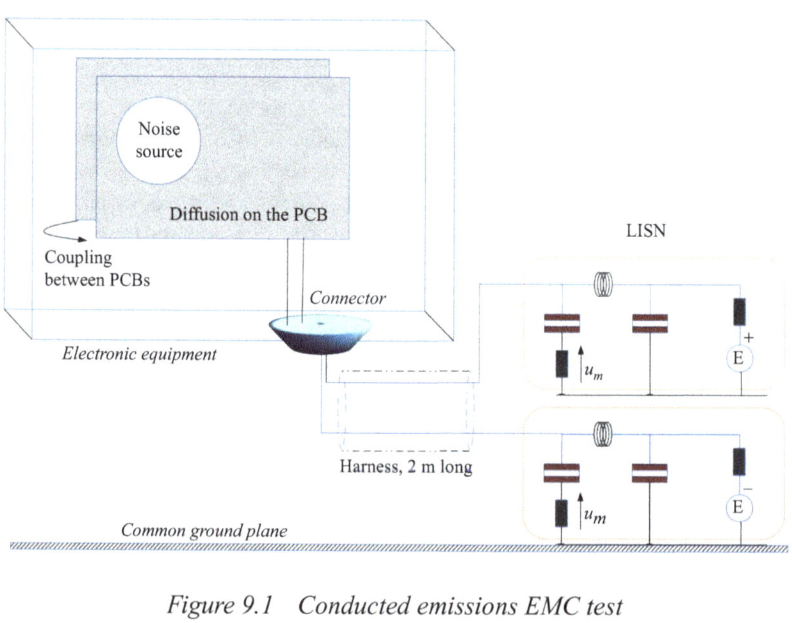

Figure 9.1 Conducted emissions EMC test

to list. The various models available and necessary to construct the whole model of the system under test.

9.2 Noise source

We often better know the current associated with some IC activity than any variation of voltage. This is logical as regulators are used to guarantee a constant value of voltage on the power supplies. Anyway, we will study the noise sources defined in current or in voltage.

9.2.1 Current noise source

Virtual meshes are used to incorporate current source in the mesh space. It consists of a mesh of voltage U unknown and with a known current J. If we inject this source on an impedance as shown in Figure 9.2, we can define two meshes: one with the current source J imposed and one with an unknown mesh current k. A first mesh (the single and original mesh of this circuit) is composed of impedances z_1 and z_2 with mesh current k. This mesh is not equipped with any electromotive force (EM force). The equation in the mesh space associated with this mesh is

$$0 = (z_1 + z_2)\,k \text{ [V]} \tag{9.1}$$

Now we inject a source current J on impedance z_1. A second mesh appears associated with this new source. It is made of the voltage U, and the impedance z_1. So, the equation associated with this mesh is

$$U = z_1 J \text{ [V]} \tag{9.2}$$

One impedance is shared between the meshes k and J: z_1, both currents having the same direction. The two meshes coupled have for coupled equations:

$$\begin{cases} 0 = (z_1 + z_2)\,k + z_1 J \\ U = z_1 k + z_1 J \end{cases} \tag{9.3}$$

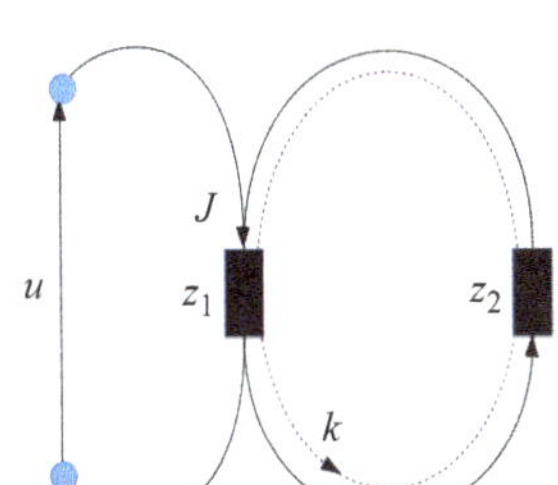

Figure 9.2 IA modellization

Finally, this system has for current vector $k^i = [k, J]$, for voltage covector $e_j = [0, U]$ and for impedance operator:

$$\zeta = \begin{bmatrix} z_1 + z_2 & z_1 \\ z_1 & z_1 \end{bmatrix} \; [\Omega] \tag{9.4}$$

In a first step we can solve the unknown current: $k = -z_1/(z_1 + z_2)J$. Then we can solve the unknown voltage $U = z_1(k + J)$.

Through this technique we can introduce IA models in tensorial analysis of networks (TAN) formalism and in the tensorial equation $e_i = \zeta_{ij}k^j$.

9.2.2 Thermal noise source

The noise can come from IC activity. In that case, the spectrum characteristic of the noise depends on this activity. Or it can come from other physical process, like the thermal noise. In that last case, we must remember the principle properties of this kind of noise.

Thermal noise is first defined in power spectrum density (PSD) $\bar{p}$. This PSD is obtained by integrating the power over all the frequency band:

$$\bar{p} = \frac{1}{f_2 - f_1} \int_{f_1}^{f_2} dP(f) \; [\mathrm{W}]\,[\mathrm{Hz}]^{-1} \tag{9.5}$$

or by making the numerical addition of the power measured at each frequency over the frequency band of analysis δf (often called *resolution bandwidth*).

$$\bar{p} = \sum_{f_1}^{f_2} \frac{P(f_i)}{(f_i + \delta f/2) - (f_i - \delta f/2)} \; [\mathrm{W}]\,[\mathrm{Hz}]^{-1} \tag{9.6}$$

This PSD must be equal to kT, where k is Boltzmann's constant (1.38×10^{-23} $[\mathrm{m}]^2\,[\mathrm{kg}]\,[\mathrm{s}]^{-2}\,[\mathrm{K}]^{-1}$) and T the temperature in Kelvin. So if we want to find an equivalent EM force to this PSD, we want to say that a matched generator of *emf* (EM force) e and resistance R developing the power $e^2/(4R)$ across R must be equal to $\bar{p}$ times the frequency band Δf. Or in other words:

$$e = \sqrt{4R\bar{p}\Delta f} \; [\mathrm{V}] \tag{9.7}$$

the same reasoning applied to the current leads to

$$J = \sqrt{\frac{4}{R}\bar{p}\Delta f} \; [\mathrm{A}] \tag{9.8}$$

Sometimes, normalized emf and current are used for writing

$$\bar{e} = \frac{e}{\sqrt{\Delta f}} = \sqrt{4R\bar{p}} \; [\mathrm{V}][\sqrt{f}]^{-1} \tag{9.9}$$

$$\bar{J} = \frac{J}{\sqrt{\Delta f}} = \sqrt{\frac{4}{R}\bar{p}}\ [\mathrm{A}][\sqrt{f}]^{-1} \tag{9.10}$$

As an example of thermal noise source, we have seen that in the ICEM model, the noise source is the IC activity and the impedance is the bulk capacitor. If the average digit duration is τ, the transition pulses coming from the intermediate state of the NAND doors makes pulses about ten times less than the digit duration, so around $\tau/10$. For digits of 10 ns, the calls of current inside the chip are around 1 ns. If the chip consumption is 200 W, for a 5 V logic, the peak current is around 40 A. It means that the noise coming from the NAND doors change of states is equivalent to pulses of 40 A peak and 1 ns duration. The period being 10 ns, it leads to a spectrum with one ray every 100 MHz until 1 GHz. $P(f_i) = 20\ [\mathrm{W}], \bar{p} = 200\ [\mathrm{nW}]\,[\mathrm{Hz}]^{-1}$ and $\bar{J} = 3.6\ [\mathrm{mA}]\,[\sqrt{\mathrm{Hz}}]^{-1}$ as $R = 5/40\ [\Omega]$.

9.3 IC package

Before reaching the PCB level, the noise source must be analysed with the package network that connect the chip to the PCB, inside the component. We can consider first order network and second order network, these networks being the bridges between one chip access and the PCB. The set of all the networks and their couplings constitute the component package.

9.3.1 First or second order access network (AN)

A first order access network (AN) is an *RLC* circuit. *R* is the resistance of the wire, *L* and *C* constitute a resonator where *C* is the capacitance between the wire and the PCB reference and *L* the inductance associated with the loop of the wire. Similar to the previous analysis, we obtain the metric of this first architecture.

$$\zeta = \begin{bmatrix} z_1 + R + \frac{1}{Cp} + Lp & z_1 \\ z_1 & z_1 \end{bmatrix}\ [\Omega] \tag{9.11}$$

z_1 is the impedance of the noise source (it can be a capacitor for ICEM or another operator).

For a second order circuit we have

$$\zeta = \begin{bmatrix} R_2 + \frac{1}{C_2p} + \frac{1}{C_1p} + L_2p & \frac{1}{C_1p} & 0 \\ \frac{1}{C_1p} & z_1 + R_1 + \frac{1}{C_1p} + L_1p & z_1 \\ 0 & z_1 & z_1 \end{bmatrix}\ [\Omega] \tag{9.12}$$

This tensor is obtained for the current direction shown in Figure 9.3.

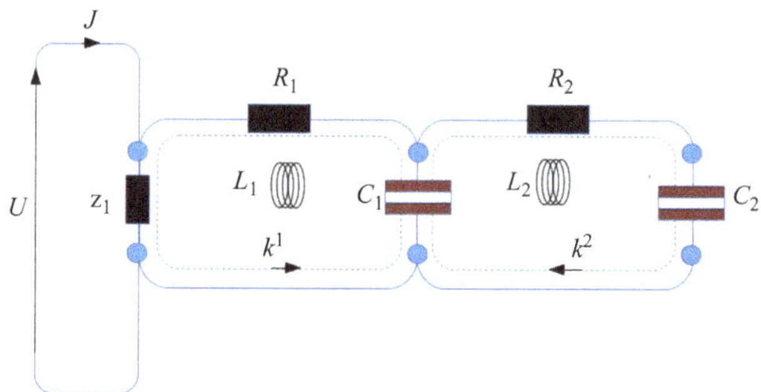

Figure 9.3 Package of second order

9.3.2 N order AN

Once the first order is built, we can see that the boundaries for any added cell RLC are defined by the capacitors on each side. Cell N will share with the previous cell the capacitor C_{N-1}, etc. If more, all the capacitors, resistances and inductances from the AN are the same, the impedance operator becomes (keeping the same current convention)

$$\zeta = \begin{bmatrix} R + \frac{2}{Cp} + Lp & \frac{1}{Cp} & 0 & 0 & \dots \\ \frac{1}{Cp} & R + \frac{2}{Cp} + Lp & \frac{1}{Cp} & \dots & \dots \\ \dots & \dots & \dots & \dots & \dots \\ \dots & \dots & \frac{1}{Cp} & z_1 + R_1 + \frac{1}{Cp} + Lp & z_1 \\ \dots & \dots & 0 & z_1 & z_1 \end{bmatrix} [\Omega] \tag{9.13}$$

Starting from the first current source operator ζ_0 (9.11), we add as diagonal element as many cells $R + (2/Cp) + Lp$ as the order is higher. If ζ_α are the impedances of these cells, we have

$$\zeta_0 \oplus_\alpha \zeta_\alpha \ [\Omega] \tag{9.14}$$

where $\oplus$ is the operator direct summation. It enables to add new diagonal elements the operands pointed out by its arguments. So if A is a matrix or a tensor and B another one, $A \oplus B$ is a super matrix or new tensor made of the two objects A and B as diagonal components. We accept $\oplus_a z_a = z_1 \oplus z_2 \oplus \cdots \oplus z_N$. But it remains to reach ζ (9.13) to add the coupling terms $1/(Cp)$ through another tensor ξ. Finally a package of any order is represented by the operation:

$$\zeta_N = \zeta_0 \oplus_\alpha \zeta_\alpha + \xi \ [\Omega] \tag{9.15}$$

We may wonder what is the best order to model a chip access. In fact for a AN of length X (distance between the chip port and the component boundary) and a maximum working frequency F, the needed order O is given by (*int* is the integer number function):

$$O = int\left(10\frac{X}{\lambda}\right) + 1,\ \lambda = \frac{3.10^8}{F}\ [\text{m}] \tag{9.16}$$

9.3.3 Couplings between AN

Depending on the package structure, couplings may exist between the various access. There are today more than 16 kinds of packages. But for all of them, we can model the access ports independently of all other pins, then looking at the coupling between these accesses. It is clear that depending on the kind of package considered, the access models and the couplings would not be the same.

As the wires are very thin, the capacitive part of the coupling is often negligible. The couplings between AN can be in first approximation reduced to the mutual inductances between the loops. Anyway, we will study both kinds of coupling as in the case of ball grid array packages, this tendency may be in contradiction.

9.3.3.1 Mutual inductance couplings

Suppose that we have two operators ζ_1 and ζ_2 associated with two ANs. We do not care for the moment of their orders. These two operators appear as two matrices, and we can realize their direct summation:

$$\zeta = \zeta_1 \oplus \zeta_2\ [\Omega] \tag{9.17}$$

All the diagonal components of ζ_1 or ζ_2 are associated with cells of the corresponding AN. When two cells include each an inductance L_1 and L_2, we must add a mutual inductance M_{12} as extra-diagonal coupling component. We can create the magnetic coupling tensor μ that will contain all the mutual inductances. The rule is

$$\text{if } \exists L_i \,/\, L_i \in \zeta_{ii} \text{ AND } \exists L_j \,/\, L_j \in \zeta_{jj} \Rightarrow \mu_{ij} = -M_{ij}p\ [\Omega] \tag{9.18}$$

The mutual inductance M_{ij} can be estimated using Neumann's formula (already evoked in Chapter 2):

$$M_{ij} = \frac{\mu_0}{4\pi}\oint_{c_1}\oint_{c_2}\frac{d\mathbf{c}_1 \cdot d\mathbf{c}_2}{R_{12}}\ [\text{H}] \tag{9.19}$$

where c_1 and c_2 are the closed circulation over each AN loop and R_{12} the average distance between the two ANs. We will see how to measure these values as in general, we miss geometrical information to make the analytical calculation.

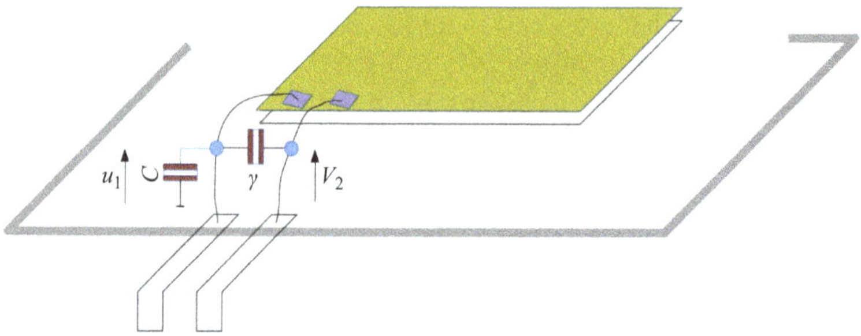

Figure 9.4 Capacitive coupling

9.3.3.2 Capacitive couplings

Knowing the small sizes of the wires inside the components, we can estimate the capacitive coupling simply by looking at the capacitor network between two neighbour accesses. Figure 9.4 shows the architecture taken into account.

This leads to

$$u_1 = \frac{\gamma}{\gamma + C} V_2 \ [\mathrm{V}] \tag{9.20}$$

If $V_2 = z_2 k^2$, this gives

$$u_1 = \frac{\gamma}{\gamma + C} z_2 k^2 \Rightarrow z_{12} \approx \frac{u_1}{k^2} = z_2 \frac{\gamma}{C} \ [\Omega] \tag{9.21}$$

with the assumption that $\gamma \ll C$.

As though for the mutual inductances, a coupling tensor Q can be created having uniquely extra-diagonal components. It is a band matrix as it concerns principally neighbour accesses. So the components concerned are $Q_{n,n+1}$ or $Q_{n-1,n}$.

9.4 Synthesis of the package impedance operator construction methodology

The first step of this synthesis is to construct the model associated by considering a kind of IC access. They can be first order RLC networks or higher orders, made from the coupling of elementary first order RLC (note that the order here refers to the number of dimensions in the meshes space and not to the differential equation order). Once this modelling performed, we call each AN ζ^{AN}.

The impedance operator of the set of AN impedances ζ^N is obtained making the direct summation of all the ζ^{AN} and with adding mutual inductance and capacitance couplings. This is realized adding tensors μ and Q. This can be written as

$$\zeta^N = \zeta^{AN} + \mu + Q \ [\Omega] \tag{9.22}$$

9.5 Computing the package model

Due to the wide variety of components, the values of the package model are measured. We can have some order of values using studies already available in publications (like [1]). We discuss now how these measurements can be carried out. The considered example is a generic one and the approach can be applied to any kind of package.

For illustrating the approach to conduct measurements, we consider a component with three power supply accesses. Each of them has two ports Vss and Vdd. The three power supplies are connected to the IA and its impedance. This typical architecture is presented in Figure 9.5.

Each access has its own equivalent schematic as already seen. If we use first order mesh dimension, they are RLC networks with one half capacitor on the pin side $C/2$ and one half capacitor on the IA side. This second half capacitor is in parallel with the impedance z. We can consider this term as included in z.

Our problem is to determine by measurements R, L, C, z and IA values. We make two major assumptions:

1. The package circuit is linear and independent of the signal amplitudes.
2. Dispersions between various components are of second order.

Another consideration can be used. EMC is the science of dispersion and uncertainties. Our purpose is not to guarantee some levels, as the natural dispersion between components is currently around 6 dB. Our purpose is to guarantee that the CEs of such a component cannot be higher than some given value or, better, that they belong to a given interval.

All the impedance measurements are realized with the components disconnected to any sources. Its power supply is not connected neither any other functional signal.

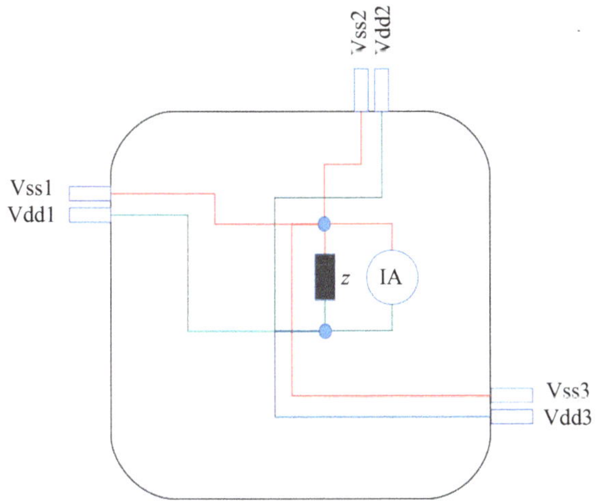

Figure 9.5 Power supply accesses on a component

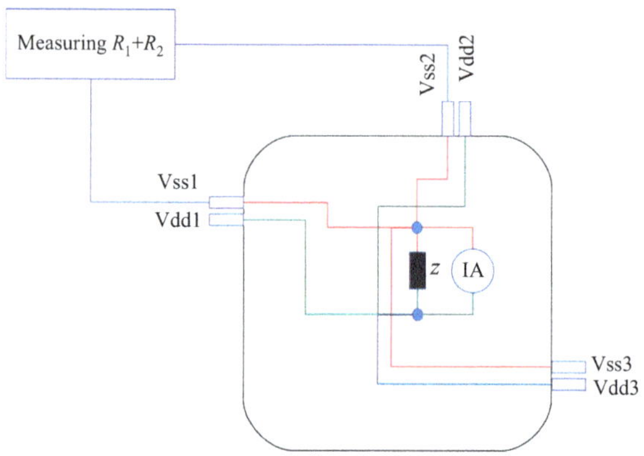

Figure 9.6 Measuring $R_1 + R_2$

That is why these measurements require a special PCB, made especially for making these measurements.

9.5.1 Measuring resistances

If we affect an index n for the components of the access *n*, we have to acquire three resistance values, R_1, R_2, R_3, for the wires and three RG_1, RG_2, RG_3 for the returns through the chip bulk. Using a milliohmmeter (better here than a network analyser) we can acquire three equations in each cases, connecting the milliohmmeter to two ports each time (see, e.g., Figure 9.6). If m_x are the measurements:

$$\begin{cases} R_1 + R_2 = m_1 \\ R_1 + R_3 = m_2 \\ R_2 + R_3 = m_3 \end{cases} \tag{9.23}$$

This gives us the DC values of the resistances. But it can be necessary to estimate the skin effect increasing their values in frequency. In practice it remains very difficult to measure this effect. For answering of the EMC needs, the simplest consists in taking a cut-off frequency around 1 kHz and to create a curve increasing in $\sqrt{f}$ from this frequency. Figure 9.7 shows the kind of estimation that can be made.

9.5.2 Measuring inductances

Following the same approach than for the resistances, we short-circuit some ports in order to measure the inductance of accesses associated with the complete way travelled in each case. As the wavelength can be considered very long compare to the chip dimensions, this permits to consider that we measure directly the inductance of the connected power supply access (see Figure 9.8 for illustration).

The measurements are realized by using a network analyser. The component must be mounted on a special board realized in order to make these measurements. The

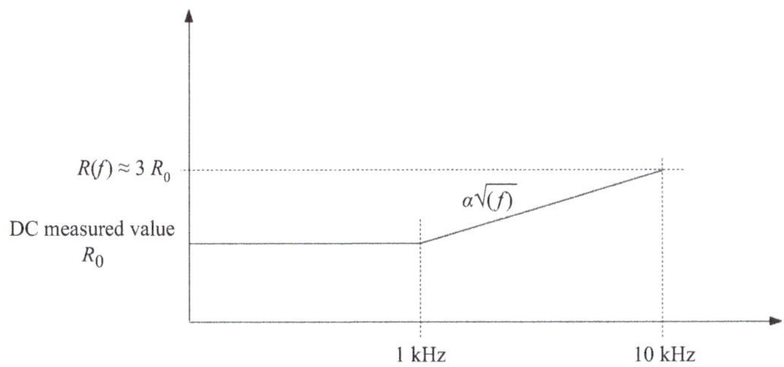

Figure 9.7 Skin effect estimation

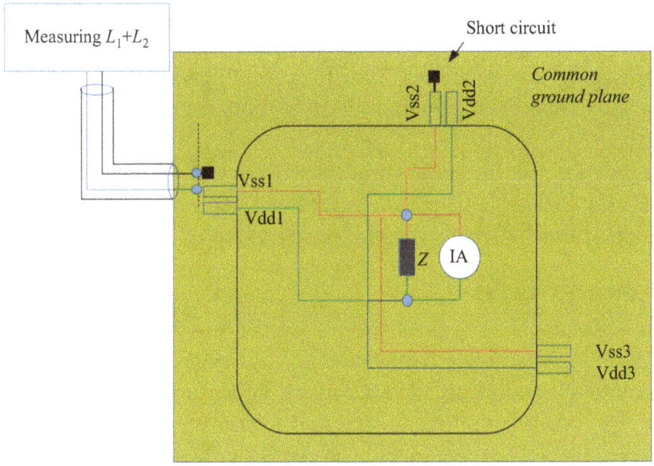

Figure 9.8 Measuring L_1

tracks making the interfaces between the SMA connectors of the network analyser and the component must be as shorter as possible, and with perfect PCB layout including a complete layer for the ground, similar to hyperfrequency PCBs.

All the inductances of the AN can be obtained using the same technique. Note also that the short-circuit must be very short (around one millimetre) in order not to add new inductance to the one measured.

The principle of these measurements is to exploit special frequency domain where lumped components of the model can be determined as easier as possible. If we short-circuit a line, the equivalent schematic is an RL circuit. To determine L, we must use a frequency for which $L\omega \gg R(\omega)$. This gives the low limit for the frequency. But it exists also a higher limit. If we reach the first mode of the circuit—its resonance—its equivalent circuit is no more an RL one. So we must respect also $\omega \ll \frac{1}{\sqrt{LC}}$.

There are two difficulties to respect these objectives. First, the resistance increases by skin effect and this rejects higher the low frequency limit. Second, the presence of z and the fact that z is unknown complicates the estimation of the higher limit.

9.5.2.1 Low frequency limit determination

We suppose that the coefficient α has been determined during the resistance measurement on one access. Our condition may be written as:

$$L\omega = 10\left(\alpha\sqrt{f}\right) \Rightarrow f_b = \left(\frac{10\alpha}{2\pi L}\right)^2 \quad [\Omega] \tag{9.24}$$

The paradox is that we do not know L before the measurements. But we can estimate its value, which is around 1 nH. And so, f_b is around 100 MHz. Let us remember this order of value.

9.5.2.2 High frequency limit determination

As we do not know the value of C, we can start from the chip dimension x to estimate the resonance using the equation $f_0 \approx v/(2x)$. This leads to resonance frequencies around some gigahertz. Taking the ratio $1/10$ we obtain a high frequency limit around $f_h = 300$ MHz.

We see that the measurement frequency band is a quite narrow one, included between 100 and 300 MHz. But as it covers at least one frequency decade, this is sufficient to determine correctly the inductance value for one access.

9.5.2.3 *L* measurement

The synthesis of our previous discussions is a curve where we can see the narrow region available to measure the inductance L. It is very important to understand that this region cannot be easily extended because the measurement facilities will always give a value for L. But if the frequency band of measurement is not in accordance with the facility performance and in accordance with the assumptions of the post analysis, the L value extract from these measurements would not be correct. Figure 9.9 illustrates the behaviour of the inductance impedance.

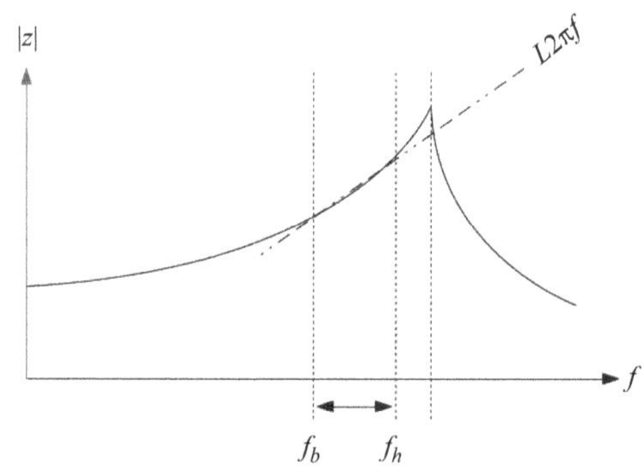

Figure 9.9 Measuring inductances

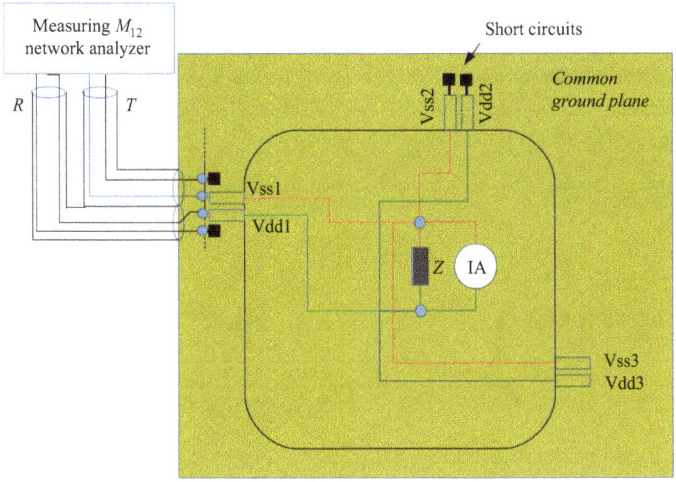

Figure 9.10 Measuring mutual inductances

9.5.3 Measuring mutual inductances

The mutual inductance coupling between two neighbour accesses can be written as $M = K\sqrt{L_1 L_2}$. Our objective is to acquire the coefficient K. We know that the mutual inductance must be inferior to each inductance. The measurement being quite difficult, this result is important to estimate the measurement validity.

Figure 9.10 shows the way to acquire the mutual inductance between two neighbour accesses. We can limit the mutual inductance measurements to these cases. Mutual inductances of accesses which are not neighbours being of second order.

The frequency limits of the measurements are determined following the same approach. Knowing the chip dimensions and its height allows to estimate the mutual inductances through Neumann's formula. This gives an information more to estimate the pertinence of the measurements.

Third, we know that coupling two neighbour loops leads to a coefficient of coupling K quite weak, around 0.1. This gives a third estimation of the mutual inductance value.

When we realize the experimental setup given in Figure 9.10, we see that the measurement on the transmit port T is influenced by the part of signal R propagating through the impedance z.

9.5.4 Measuring capacitance

Using open-circuit condition for the common mode lines of the accesses (rather than the short-circuit for the inductance and at the same location), we can measure the values of line capacitance. We emphasize that the dielectric can present losses. But if we make the measurements sufficiently low in frequency, we can hope to accept these losses as negligible. Under this assumption, three measurements give us the three capacitance values as for the resistances.

The paradox here is that the low frequency limit is determined compared to the conductance g. And we had made the assumption that the dielectric losses were negligible. Writing the dielectric:

$$\varepsilon = \varepsilon_C - j\varepsilon_g \ [\mathrm{F}]\,[\mathrm{m}]^{-1} \tag{9.25}$$

The capacitance expression in a generic approach becomes (for $\alpha = 0$ and $p = j\omega$)

$$\varepsilon \frac{S}{h} p = \left(\varepsilon_C - j\varepsilon_g\right) \frac{S}{h} p = Cp + g\omega \ [\mathrm{S}] \tag{9.26}$$

and so, in impedance

$$Z = \frac{1}{Cp + g\omega} = \frac{g\omega}{g^2\omega^2 + C^2\omega^2} - j\omega \frac{C}{g^2\omega^2 + C^2\omega^2} \ [\Omega] \tag{9.27}$$

As we measure the impedance, we want to verify that

$$\omega \frac{C}{g^2\omega^2 + C^2\omega^2} \ll \frac{g\omega}{g^2\omega^2 + C^2\omega^2} \Rightarrow C \ll g \tag{9.28}$$

As we must respect here a relation of inferiority, and knowing the typical values of conductance (until 1 GΩ), we can take this inequality as true. It means that typically for the capacitance measurement, there are no low frequency limit well defined. In practice and due to the network analyser performances, we start the measurements in general around 10 kHz.

The high frequency limit follows the same criteria as for the inductance measurements. In general, we search for a −20 dB Decade^{-1} slope in the impedance measurement curve that is the good trace to acquire the capacitance value.

We may wonder also what is the influence of the other capacitances on one access capacitance measurement? The differential capacitance (or mutual capacitance) between lines are smaller than the capacitance of each line. Anyway, it is in series with the neighbour line capacitance. So, to the intrinsic capacitance of an access is added a parasitic capacitance given by

$$C_{added\ to\ C_1} = \frac{\gamma_{12} C_2}{\gamma_{12} + C_2} \ [\mathrm{F}] \tag{9.29}$$

We must accept the assumption that this added term is negligible compared to C_1. This makes easier the AN components acquisition, and in any cases, this approach does not allow to predict the noise level generated by the equipment, taking into account all the uncertainties. With this postulate we can make one measurement giving $C_1 + C_2 + C_3$ (we cannot isolate each capacitor from the other). From this measurement we can extract the capacitance values using the knowledge of the resistance values. Knowing that the structure remains homogeneous all over the chip, we can consider that the wires are all of same material. Taking the smaller of all, for example R_1, we can compute the ratio of length between the accesses by calculating

$$\alpha_2 = \tfrac{R_2}{R_1} \quad \alpha_3 = \tfrac{R_3}{R_1} \tag{9.30}$$

These ratio are maintained for the capacitances and

$$C_1 + C_2 + C_3 = C_1\,(1 + \alpha_2 + \alpha_3) \;\; [\mathrm{F}] \tag{9.31}$$

Knowing α_2 and α_3 we can deduce C_1 then C_2 and C_3.

For capacitances, the curves have the profiles shown in Figure 9.11.

9.5.5 Measuring mutual capacitance

The mutual capacitance is acquired using a similar technique as the mutual inductances, but with open-circuit conditions. Frequency limits also follow the same idea.

To understand the assumptions and the approximations made, we represent the measurement conditions of Figure 9.12.

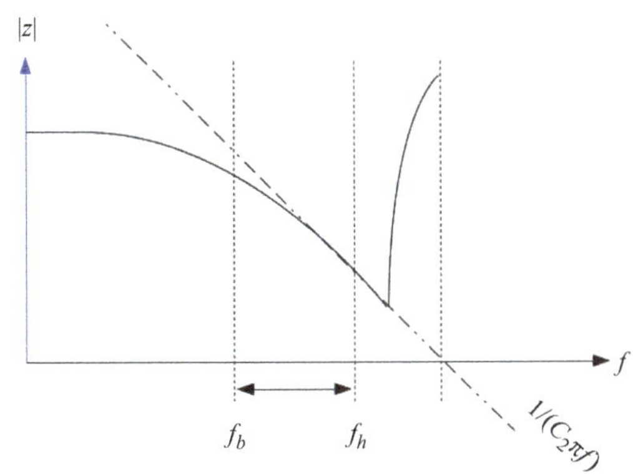

Figure 9.11 Illustrative curve of capacitance measurement

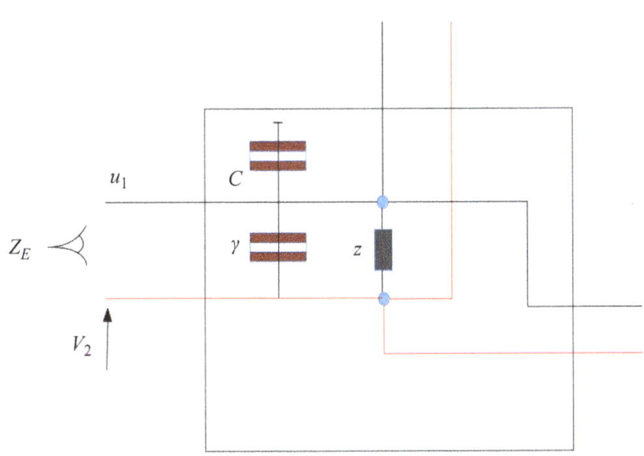

Figure 9.12 Measuring mutual capacitances

Knowing the capacitances C_i, we can make two measurements: one $u_1(V_2)$ and the other Z_E, the input impedance in differential between two accesses.

We have seen that the first measurement can be approximated by

$$u_1 = \frac{\gamma}{C} V_2 \text{ [V]} \tag{9.32}$$

If we use the index i for all the Vss and i' for all the VDD, the capacitor C comes from all the accesses capacitors in parallel. So

$$C = \sum_i C_i \text{ [F]} \tag{9.33}$$

We use here the same approximation as before. For the impedance in differential we can write

$$Z_E = p^{-1} \left[\sum_i \left\{ \gamma_{ii'} + \frac{C_i}{2} \right\} + z \right]^{-1} \quad [\Omega] \tag{9.34}$$

Noting

$$\gamma = \sum_i \gamma_{ii'} \text{ [F]} \tag{9.35}$$

We have

$$\begin{cases} u_1 = \frac{1}{C}(\gamma + z)\, V_2 \\ 1 = \left(\gamma + z + \frac{C}{2}\right) Z_E \end{cases} \Rightarrow \begin{cases} \gamma + z = C u_1 / V_2 \\ \gamma + z = Z_E^{-1}\,(1 - C/2Z_E) \end{cases} \tag{9.36}$$

which gives us approximations of γ and z. Note that using the same technique of ratios as for the capacitances, we obtain

$$\gamma = \gamma_{11'}\,(1 + \alpha_2 + \alpha_3) \tag{9.37}$$

the knowledge of $\gamma_{11'}$ gives us the values of $\gamma_{22'}$ and $\gamma_{33'}$.

All the parameters being complexes, the measurements must be done using a network analyser and exploiting both real and imaginary parts of the results.

9.6 Acquiring the IA and complete component model for conducted emissions

To make this measurement, the component must be power supplied.

At this step we have acquired all the components of the AN under some restrictive approximations. It remains to determine the noise source IA. The whole AN can be represented by an operator ζ_{AN}. When measuring the noise source IA, the only source of energy in this network is the IA source itself. As there are no *emf*, the problem can be written for given load conditions on all the accesses:

$$0 = \zeta_{AN\,xy} k^y \text{ [V]} \tag{9.38}$$

One of the k^y is the IA unknown source and some of the impedances include in ζ_{AN} are the loads on the accesses. Let R_m be one of these loads. We measure on R_m the output current $k^{(b)}$: the index is written between parentheses to indicate that it is a single value. (b) does not explore all the dimensions of the problem but points out only one abstract value. For example, if the measurement is made on port 3, $(b) = 3$. This measured current can be associated with the source covector T_a:

$$k^{(b)} = Y^{(b)a} T_a \text{ [A]} \tag{9.39}$$

where Y is the inverse of the operator ζ_{AN}. As there was no source except the IA J, we can deduce the structure of T:

$$T_a = [0, \ldots, 0, \pm zJ, 0, \ldots, 0] \text{ [V]} \tag{9.40}$$

All the components of T are zero except the one associated with J. The result is a set of equalities of the form:

$$Y^{(m)(a)} (\pm zJ)_{(a)} = k^{(m)} \text{ [A]} \tag{9.41}$$

where (m) is pointing out the measured port and (a) the source number. As we have estimated z from previous measurements, we obtain

$$J = \zeta_{AN,(am)} k^{(m)} \text{ [A]} \tag{9.42}$$

This measurement and post-treatment can be reproduced as many times as there are ports connected to the IA. It is interesting to exploit all these possibilities to extract an average value of J. By knowing the numerous approximations we have made, making more than one time, the acquisition of J gives an indicator on the impact of these approximations. If the various results were too much dispersive, probably that the construction of the AN should be reviewed.

9.7 Coupling between blocks in the chip

If the chip includes various separate blocks, coupling of shared bulks can appear as a resistance linking these blocks. This resistance can be acquired using milliohmmeter as for the accesses resistances. This interblocks component can be estimated only once the resistances are known as often it appears like a series component with the resistances of each block.

9.8 Conducted emissions of power electronics

Today there are four major kinds of electronics:

- analogical, low power electronics;
- numerical electronics, including big ICs;
- power electronics; and
- optical electronics.

Looking at this list, it is clear that big ICs and power electronics are the main EMC and noise sources. Having explored the ICs' modeling, let us begin now with the analysis of power electronic EMC CEs.

Power electronics can be summarized in three major parts: an input filter, a power chopper and an output filter with a load. In general, we can consider that the load must respect some requirements, knowing the output filter, in order to be compliant with the standards of CEs. That is why we consider the set: output filter plus load.

There are two elements not so easy to model in this architecture: the power chopper itself and the common mode filter if it is included in the PCB system.

9.8.1 Power chopper

The power chopper is a set of electronic commutator, often implemented with MOS-FET transistors that allow to drive the current coming from the power supply (where the energy is stocked) to the load in a desired way.

These commutators have their own electronic schematic used when functional studies are conducted. But these circuits can ask for long-time computations, not always converging and not obviously adapted for the EMC needs. That is why we develop [2] a specific model, very flexible and which can be used for any kind of power chopper—and giving needed information for EMC: i.e., rise time, spectrum and nonlinear behaviours.

The model key comes from the use of domain functions that can simulate any kind of commutators. We now detail these functions and their use.

There are various functions that can be used to define domains. What we call domain is a part of space where depending on some parameters, a law is valid. The basic idea is that a same object follows different laws depending on some parameters. A simple example is a resistance. If we take for parameter the temperature, the resistance increases abnormally if the temperature increases also beyond to a given threshold. We can model this saying that the resistance follows various laws depending on the temperature.

We can notice that there is a domain for which the resistance is in a transition state where it becomes to follow a new law, while it continue to respect a first law for part. It is half in a normal regime, and half in an abnormal regime. So our domain functions must be chosen in order to authorize simultaneously different states.

After several tries, sigmoid functions seems to be easier to use and set.

Sigmoid functions $\mathscr{D}$ are defined by

$$\mathscr{D} = \left[1 + ce^{-a(q-q_0)}\right]^{-1} \tag{9.43}$$

where q_0 is the threshold of state change, c the function value at threshold and a the change speed. To define a domain on an interval, we can consider the analytical function:

$$\mathscr{D}_{[q_0,q_1]} = \left[1 + ce^{-a(q-q_0)}\right]^{-1} - \left[1 + ce^{-a(q-q_1)}\right]^{-1} \tag{9.44}$$

And if this interval is associated with the parameter q (which is evidently the case here), we write

$$\overset{q}{\mathscr{D}}_{[q_0,q_1]} = \left[1 + ce^{-a(q-q_0)}\right]^{-1} - \left[1 + ce^{-a(q-q_1)}\right]^{-1} \tag{9.45}$$

Often a real physical behaviour can be very well approximated using simplest functions defined on domains. Let us take the example of a diode. The voltage/current relation follows a physical law expressed by

$$V = \frac{KT}{q} \ln\left(1 + \frac{i}{i_0}\right) \ [\mathrm{V}] \tag{9.46}$$

In practice, when the current is positive and the voltage across the diode superior to 0.6 [V], the diode behaves like a very weak resistance R_D with an *emf* of 0.6 [V] in series. When the current is negative, the diode behaves like a very high resistance R_n implying a weak negative current near to i_0. We exclude in this representation the avalanche phenomena.

Finally the diode becomes associated to the model:

$$z_d = R_0 + \overset{V}{\mathscr{D}}_{[0,+\infty]}(R_D - R_0) + \overset{V}{\mathscr{D}}_{[-\infty,0]}(R_n - R_0) \ [\Omega] \tag{9.47}$$

V being computed across the diode and the source covector is modified including

$$T = 0.6 \ [\mathrm{V}] \tag{9.48}$$

R_0 is the classical default impedance of the diode. It also permits to robustify the numeric computation of the model if somewhere the parameters i or V may be badly defined.

To implement numerically this case, we choose a parameter giving the voltage across the diode. If the diode is driven by a generator e of self-impedance R_G, the parameter is $u - e - R_G i - 0,6$. If u is positive, the diode is on, if not, the diode is off. The Python program indicated in Annexe 9.A gives an example of such a computation.

The curves obtained with this program are shown in Figure 9.13.

Looking at the example of the diode, we understand that it is quite easy to model a commutator with these domain functions. Each MOSFET or other field effect transistor used as commutator will be modelled by this kind of function. This avoids long-time processing if using exact electronic formulations and gives the same electromagnetic behaviour for EMC. Why deprive yourself of such benefits? Moreover, it gives a better way to understand what happens in the circuit. At the same time, it does not prevent from using sometimes classical models for electronics. Classical gain function β may be used, for example, for NPN transistors.

Let us study now a branch of a power chopper.

It generally comprises two commutators, each of them being driven by a dedicated circuit that we call control command circuit (ccc). These circuits are themselves driven by a microprocessor (microcontroller) that optimizes a command law. This command law has for objective to deliver the optimal power supply perfectly matched to the associated loads. It is often possible to integrate implicitly the ccc in order to simply consider the control signal sent to each commutator. In the EMC area, this is the best way to work, avoiding by there to model more complex circuits. We can take a look to a single branch, including two commutators with a capacitor as load and a battery for energy source. Figure 9.14 represents this circuit.

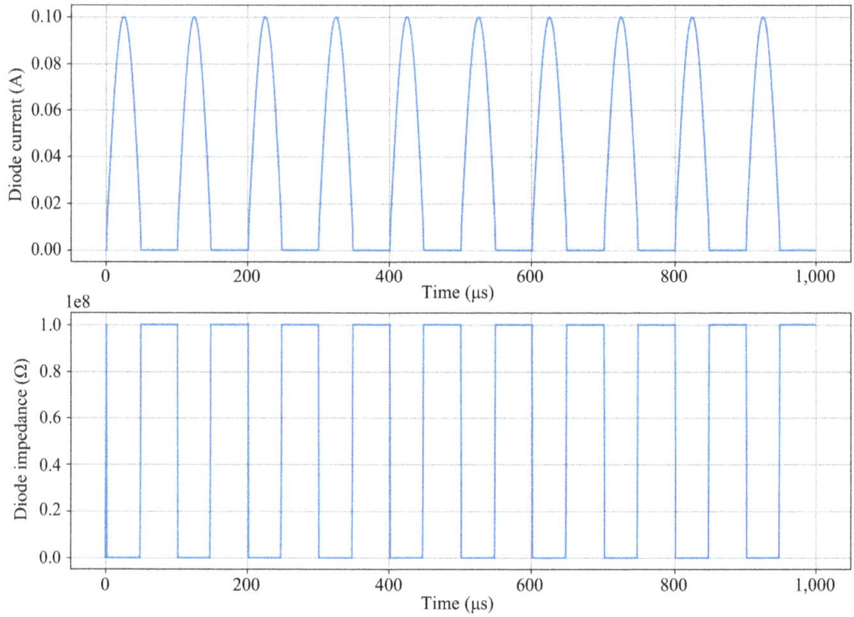

Figure 9.13 Diode current

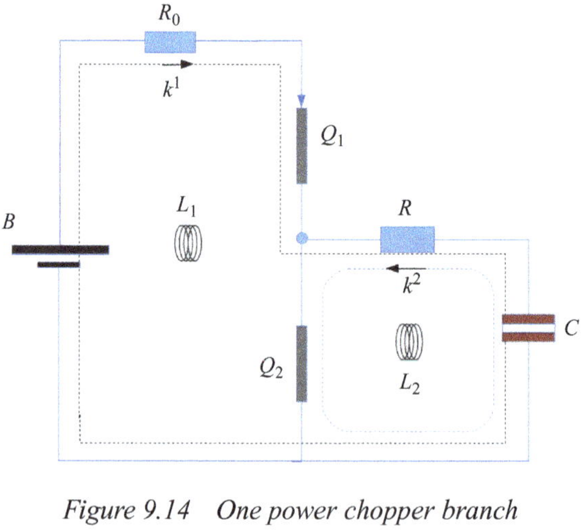

Figure 9.14 One power chopper branch

The first mesh of this circuit has an impedance:

$$z_{11} = R_0 + Q_1 + R + \frac{1}{Cp} + L_1 p + L_2 p \ [\Omega] \tag{9.49}$$

In this impedance operator, Q_1 is the operator of the first commutator, driven by some implicit ccc. For the second mesh we have

$$z_{22} = R + \frac{1}{Cp} + L_2 p + Q_2 \ [\Omega] \tag{9.50}$$

Between both meshes, the coupling impedance is $-(R + 1/(Cp))$.

So the complete impedance operator of the circuit is given by

$$\zeta = \begin{bmatrix} R_0 + Q_1 + R + \frac{1}{Cp} + L_1 p + L_2 p & -\left(R + \frac{1}{Cp}\right) \\ -\left(R + \frac{1}{Cp}\right) & R + \frac{1}{Cp} + L_2 p + Q_2 \end{bmatrix} [\Omega] \tag{9.51}$$

We have then two possibilities:

1. to directly model the time changing of impedance of the commutators and
2. to model their ccc using a signal and associating to the commutator its controlled impedance.

The first strategy uses a macromodel higher than the second one which is generally implemented with more detailed circuits. Depending on the objectives of the simulation, one strategy can be better than the other. In the next paragraphs, both techniques are investigated.

9.8.1.1 Direct commutation modelling

In that case we represent the commutator directly a Laplace's function impedance translating their command. For an increasing impedance reaching the value z_m with a rise time of τ seconds, the Laplace's function is

$$Q_1(p) = z_m \left(\frac{1 - e^{-\tau p}}{\tau p^2} \right) \ [\Omega] \tag{9.52}$$

By taking into account the pulse variation, the previous impedance can be rewritten as follows:

$$Q_1(p) = z_m \left(\frac{1 - e^{-\tau p}}{\tau p^2} \right) \left[1 - e^{-Tp} \right] \ [\Omega] \tag{9.53}$$

T being the pulse duration ($T > \tau$). As we want Q_1 to have a low value of impedance rather than a high one during the pulse, we finally use

$$Q_1(p) = \frac{z_m}{p} - \left\{ z_m \left(\frac{1 - e^{-\tau p}}{\tau p^2} \right) \left[1 - e^{-Tp} \right] \right\} \ [\Omega] \tag{9.54}$$

Q_2 follows the same law but delayed of one pulse. This leads to

$$Q_2(p) = \left[\frac{z_m}{p} - \left\{z_m\left(\frac{1-e^{-\tau p}}{\tau p^2}\right)\left[1-e^{-Tp}\right]\right\}\right]e^{-(T+\tau)p}\ [\Omega] \tag{9.55}$$

We have the advantage of this approach for being able to study the problem directly in the harmonic domain. If we look for example to the current k^2 conducted to the load, this current can couple with other line in the harness it uses to reach the load.

We can compute the answer of this structure with two inductance L_2 assumption: one for 1 μH value, which is the natural inductance of the line, and one with an added inductance increasing the whole inductance to 100 μH. Figure 9.15 shows both results obtained with the program described in Annexe 9.B.

On both curves we have plotted the standard limit authorized for the noise in red, knowing that the functional need is of 100 dBμA at 10 kHz. It is clear that the

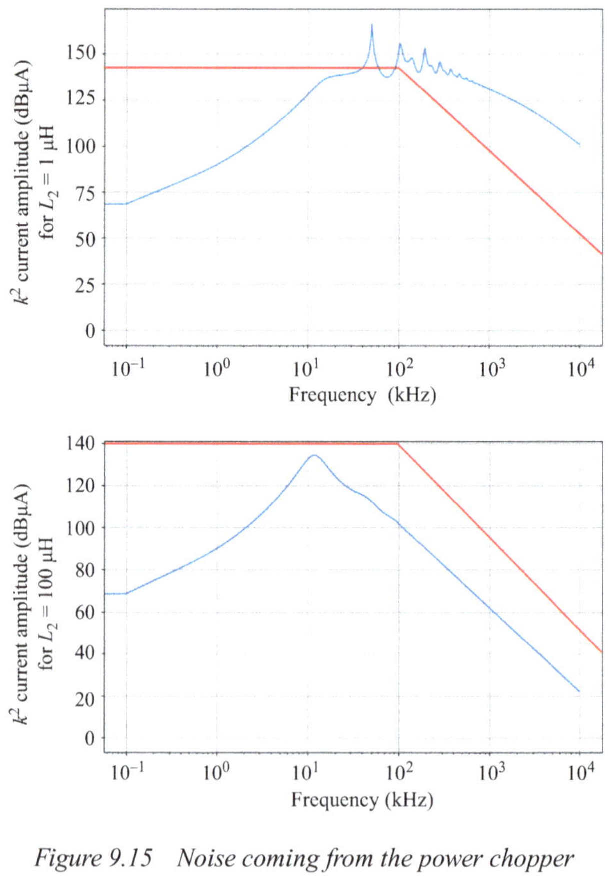

Figure 9.15 Noise coming from the power chopper

second solution permits to reach the requirements whereas the first structure gives an inacceptable noise.

9.8.1.2 Commutation through ccc

In that case we define a junction whose conductivity depends on the voltage applied on its gate input. This function $G(u)$ is defined by

$$G(u) = R_{on} + R_b \left(1 - \left[1 + ce^{-a(u-u_t)}\right]^{-1}\right) \ [\Omega] \qquad (9.56)$$

where R_b is the resistance value of the junction when the gate voltage is inferior to a threshold u_t. R_{on} is its value when the gate voltage is over the threshold u_t.

Having defined the impedance operator of one commutator, the whole working of the power chopper depends now entirely of the control signals applied on their gates. We can study our previous circuit by using domain function to fix the commutator impedance and a command law supposed to come from another circuit. Figure 9.16 shows the result we obtain for similar parameters as the last solution in the previous modelling. As the spectrum changes depending on time, the EMC analyser integrates the amplitude measured to deliver the root mean square (rms) value of the signal. This value is defined by (N is the number of sample and f the time signal amplitude)

$$rms = \sqrt{\frac{1}{N}\int_t dtf^2(t)} \qquad (9.57)$$

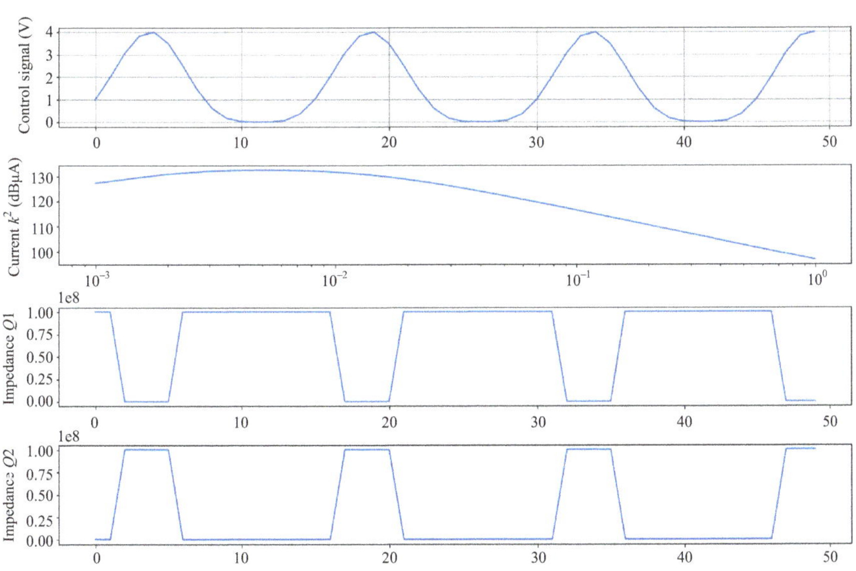

Figure 9.16 Spectrum measured by the EMC analyser

That is what we compute in program Annexe 9.C, which gives the results presented in Figure 9.16.

9.8.2 The generic power chopper

A generic power chopper (GPC) can be imagined, which is able to give all the results obtained with any N branches power chopper, under the unique condition of multiplying its instantiation. This GPC has two branches. And if you want to model a three-branch power chopper, you will need two GPCs. Three basic structures of this power chopper are shown in Figure 9.17.

We can see that by driving the four commutators Q_1–Q_4 we can create any kind of signal form in the load Z_L. This load can be a pole of a N poles machine, and so N GPC can reproduce the working of a N branches power chopper. The principle consists in the reproduction of the current of the N branches power chopper in all loads by using N GPC, each of them being driven in order to reach this equality between the currents.

The GPC can be studied in a different ways that usual and shows the multiple possibilities of the TAN to model commutators.

The GPC circuit can be seen through its branches. Figure 9.18 shows its circuit with the branch numbers in circles and the directions of the branch currents shown with red arrows.

Following the classical approach, we connect branches to meshes through the connectivity matrix Λ. Then, as detailed in Chapter 2, we elaborate the impedance operator in the mesh space using the transformation formula:

$$\zeta_{\alpha\beta} = \Lambda_{\alpha}^{a} z_{ab} \Lambda_{\beta}^{b} \ [\Omega] \tag{9.58}$$

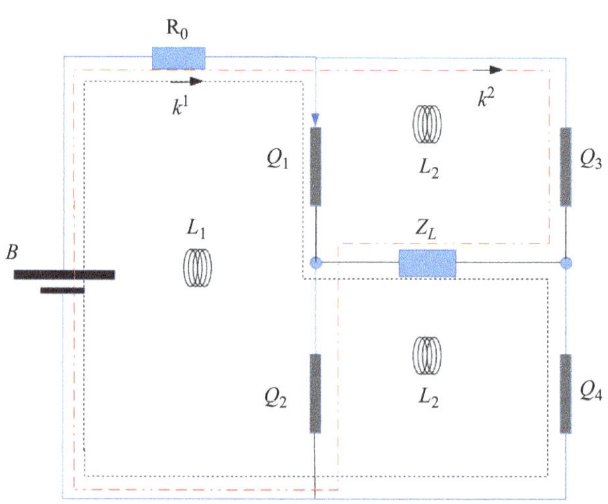

Figure 9.17 Generic power chopper

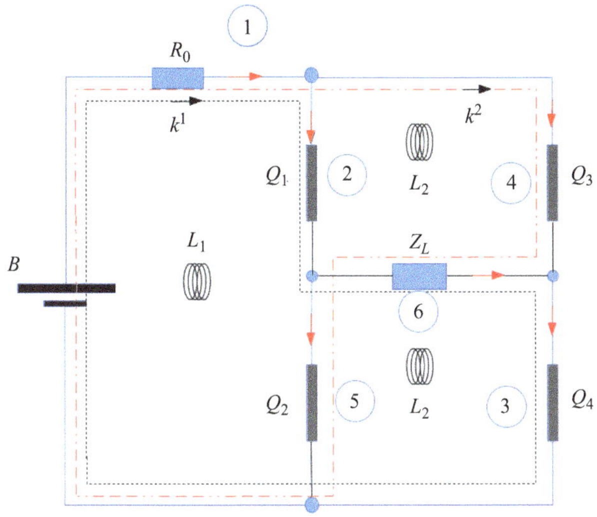

Figure 9.18 Generic power chopper through branches

Following Figure 9.18 this connectivity matrix is:

$$\Lambda = \begin{bmatrix} 1 & 1 \\ 1 & 0 \\ 1 & 0 \\ 0 & 1 \\ 0 & 1 \\ 1 & -1 \end{bmatrix} \tag{9.59}$$

With this connectivity, we obtain the impedance operator ζ defined by

$$\zeta = \begin{bmatrix} R_0 + Q_1 + Z_L + Q_4 + (L_1 + L_2)p & R_0 + L_1 p - z_L \\ R_0 + L_1 p - z_L & R_0 + Q_3 - Z_L + Q_2 + (L_1 + L_2)p \end{bmatrix} [\Omega] \tag{9.60}$$

The commutators Q_1–Q_4 are driven by control signals. The commutator transition effect on the one branch of the power chopper influences the variation of the desired current in the load.

Now what happens if branches 4 and 5 do not belong to a circuit made of a single mesh? It means that the connectivity matrix becomes

$$\Lambda = \begin{bmatrix} 1 \\ 1 \\ 1 \\ 0 \\ 0 \\ 1 \end{bmatrix} \tag{9.61}$$

The impedance matrix in the branch space is defined by

$$z = \begin{bmatrix} R_0 & 0 & 0 & 0 & 0 & 0 \\ 0 & Q_1 & 0 & 0 & 0 & 0 \\ 0 & 0 & Q_4 & 0 & 0 & 0 \\ 0 & 0 & 0 & Q_3 & 0 & 0 \\ 0 & 0 & 0 & 0 & Q_2 & 0 \\ 0 & 0 & 0 & 0 & 0 & z_L \end{bmatrix} [\Omega] \tag{9.62}$$

and so

$$\Lambda^T z = \begin{bmatrix} R_0 & Q_1 & Q_4 & 0 & 0 & z_L \end{bmatrix} \tag{9.63}$$

giving

$$\Lambda^T z \Lambda = \begin{bmatrix} R_0 + Q_1 + Q_4 + z_L \end{bmatrix} [\Omega] \tag{9.64}$$

We can consider also that branches 4 and 5 belong to mesh 1, but that the branches 2 and 3 do not belong to this mesh. Our objects become

$$\Lambda = \begin{bmatrix} 1 \\ 0 \\ 0 \\ 1 \\ 1 \\ -1 \end{bmatrix} \tag{9.65}$$

In that case,

$$\Lambda^T z \Lambda = \begin{bmatrix} R_0 + Q_3 + Q_2 - z_L \end{bmatrix} [\Omega] \tag{9.66}$$

We see that with the exception of the inductances that are added after the branch to mesh transformation, we obtain the impedances ζ_{11} and ζ_{22} of the global problem. But the coupling terms are not taken into account if we solve the problem by taking each impedance separately. However, as the impedance of the commutators in off state present a high value, we can neglect in first approximation the effect of the couplings. The reported emf is equal to $(R_0 + L_1 p - z_L)k^o$, k^o being the current of the commutators in off state. As $k^o = 0$, this emf is also null.

So if the commutations are sufficiently separated, the working of the circuit can be studied looking at each couple of activated commutators separately. We exploit here the fact that this separability is acceptable.

Another solution consists of memorizing the final currents and voltage at the end of one state, and use them as initial conditions for the next state.

Under the simplest case of separable states we can model the alternance of the commutators using a variable connectivity that changes with time. If $\overset{t}{\mathscr{D}}_1$ is the time domain where the current k^1 is activated and if $\overset{t}{\mathscr{D}}_2$ is the time domain where the current k^2 is activated, the connectivity can be

$$\Lambda(t) = \begin{bmatrix} 1 \\ \overset{t}{\mathscr{D}}_1 \\ \overset{t}{\mathscr{D}}_1 \\ \overset{t}{\mathscr{D}}_2 \\ \overset{t}{\mathscr{D}}_2 \\ -1 \end{bmatrix} \tag{9.67}$$

To solve the problem starting from the branch space, we proceed by establishing the connectivity, transforming of the impedance operator made of all components except the inductances, and then we add the inductances $\mathscr{L}_{\beta\alpha}$. Synthesizing, we do

1. $i^a = \Lambda^a_\alpha(t)k^\alpha$ [A];
2. $e_b = z_{ba}i^a + U_b \Rightarrow e_b = z_{ba}\Lambda^a_\alpha(t)k^\alpha + U_b$ [V];
3. $\breve{e}_\beta = \Lambda^b_\beta e_b + \mathscr{L}_{\beta\alpha}pk^\alpha$ [V]

$\breve{e}$ is the covector of the *emf* in the mesh space. Due to the fact that we operate the transformation independently of the inductances, there are no time derivative of the connectivity Λ. This remains simple while the need of this kind of time derivative would have added complex terms associated with the time derivation of the connectivity.

9.9 Other nonlinear noise sources

Many other kinds of nonlinear electronics may be studied. But the two previous cases analysed give all the needed techniques to solve all these other cases.

We have seen some nonlinear operator of impedance, the use of domain function or time-dependent connectivity. We have also seen that Laplace's transform can be used respecting the fixed impedance condition but on finite time domains. Taking care of the initial conditions at each start of domain, the Laplace's transform condition can be used even for impedances depending on time. We must at least verify that they remain constant on each domain duration.

Royer's oscillator was studied through the domains technique, as some spark gap protections, etc.

9.10 From the component to the PCB connectors

At this step, we have determined the whole equivalent model of a chip, including the package through the AN model and the noise sources existing on the chip. On all chip accesses, tracks wear the signals from the chip to various other components or to some PCB connectors. In any cases, these structures can create coupling of noise from one track to another. The main propose of the present paragraph is to study the noise diffusion. The properties of couplings between tracks can be used in the case of transmissions between components as in the case of transmissions from a component to a connector.

The noise can be generated from various components simultaneously. Depending on the customer requirements, the noise can be measured at least on power supply wires or more on all signals. Even in the case where the noise is acquired on the power supplies only, the measured noise can come from coupling between a noisy signal and the power supply access and this on the PCB. In fact that is often the case because the power supply networks is generally well filtered in order to guarantee a good working of the components.

So our problem here is to determine the good PCB layout to avoid this uncontrolled noise diffusion.

The various components on the PCB can be seen as separated groups of meshes CG_i. Between these groups there are tracks used to exchange signals. We have seen in Chapter 2 that these links can be modelled as cords b_{ij} between meshes. So we can study the whole PCB as an impedance operator with the groups as diagonal matrix elements and the links as extra-diagonal interactions.

To this basic structure comes to be added parasitic interactions π_{ij} like cross-talks between tracks, etc. The objective of any PCB layout strategy is to guarantee the good transmission of signals through the elements b_{ij} and to limit the parasitic interactions π_{ij} at the minimum, ideally to zero.

Optimizing the PCB layout means to obtain the impedance operator as diagonal as possible.

Starting from a purely diagonal matrix representing only the set of components on the PCB, we can add the various links and couplings. Basically there are no relations between the location of a component in the ζ matrix and its location on the PCB.

A graph of relations can be made establishing the links between the components.

We take, for example, the circuit which presents the graph given in Figure 9.19. It is a typical ccc including a microcontroller, a sensor and an actuator. Other peripheral circuits like a regulator are present on this electronic. On the graph presented in Figure 9.19, we connect the components in relation. This graph shows the links existing between the CG_i.

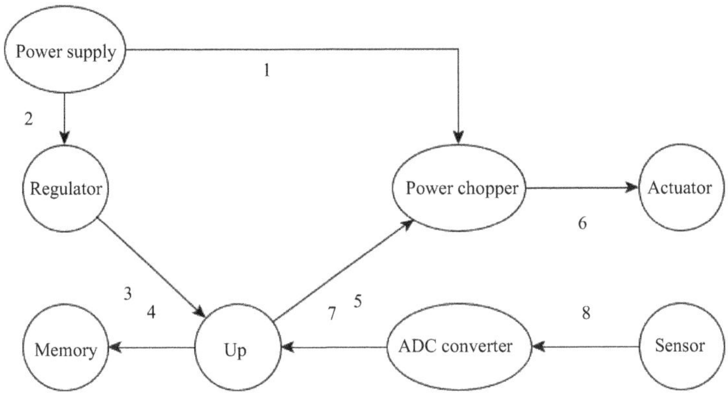

Figure 9.19 Graph of relations between devices

We can browse this graph following main groups of links between components. By organizing the component locations on the PCB following this graph and its main way, we can obtain a matrix principally diagonal. In our case, link 5 (as link 4) must be added to the previous matrix and makes appear extra-diagonal components. We can approximate this matrix organization (we start from the actuator—*a* means actuator, *w* power chopper, *s* power supply, *r* regulator, *u* microprocessor, *m* memory, *A* ADC converter and *n* sensor):

$$\zeta = \begin{bmatrix} a & \zeta_{aw} & 0 & 0 & 0 & 0 & 0 & 0 \\ 0 & w & \zeta_{ws} & 0 & \zeta_{wu} & 0 & 0 & 0 \\ 0 & 0 & s & 0 & 0 & 0 & 0 & 0 \\ 0 & 0 & \zeta_{rs} & r & 0 & 0 & 0 & 0 \\ 0 & 0 & 0 & \zeta_{ur} & u & 0 & \zeta_{uA} & 0 \\ 0 & 0 & 0 & 0 & \zeta_{mu} & m & 0 & 0 \\ 0 & 0 & 0 & 0 & 0 & 0 & A & \zeta_{An} \\ 0 & 0 & 0 & 0 & 0 & 0 & 0 & n \end{bmatrix} [\Omega] \tag{9.68}$$

The elements ζ_{xy} are the operators associated with the links between the components. We see that the matrix is a band matrix. Only the term ζ_{wu} is out of this band. The band width except for ζ_{wu} is of one component.

Now we must translate the neighbourhood between the components when they are implemented on the PCB. Figure 9.20 shows a possible implementation in the case of a specific requirement: here the use of only one connector for all the signals.

The ellipses with the dot lines indicate possible parasitic interactions. These interactions add new terms in our impedance operator ζ. If we consider the chosen locations for the components shown in Figure 9.20, and noting $\pi_{\alpha\beta}$ the parasitic interactions, we obtain the operator (we must complete the previous matrix with the

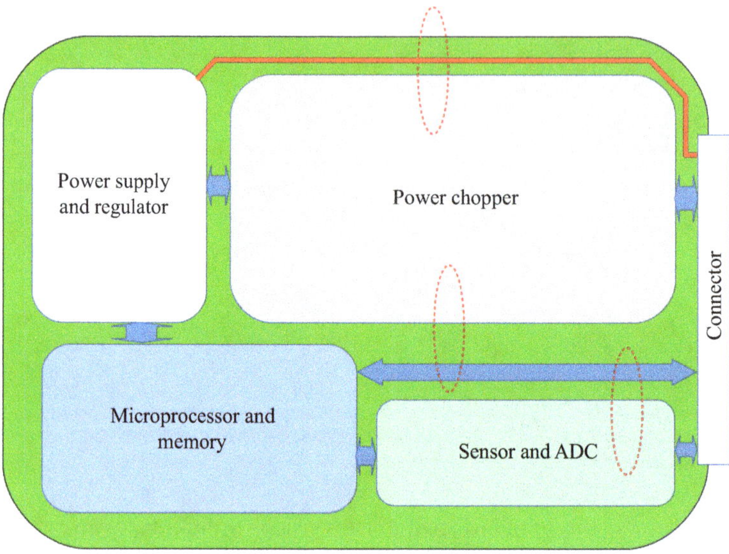

Figure 9.20 Possible implementation

four accesses Fx to the connector that can be seen as four loads and add the links to these accesses and the couplings between these links. Likewise we note x_i, the coupling terms, coming from the line going to the connector):

$$\zeta = \begin{bmatrix} a & \zeta_{aw} & 0 & 0 & 0 & 0 & 0 & 0 & 0 & 0 & 0 & 0 \\ 0 & w & \zeta_{ws} & 0 & \zeta_{wu} & 0 & 0 & 0 & 0 & x_{wFw} & 0 & 0 \\ 0 & 0 & s & 0 & 0 & 0 & 0 & 0 & x_{sFs} & 0 & 0 & 0 \\ 0 & 0 & \zeta_{rs} & r & 0 & 0 & 0 & 0 & 0 & 0 & 0 & 0 \\ 0 & 0 & 0 & \zeta_{ur} & u & 0 & \zeta_{uA} & 0 & 0 & 0 & 0 & 0 \\ 0 & 0 & 0 & 0 & \zeta_{mu} & m & 0 & 0 & 0 & 0 & 0 & 0 \\ 0 & 0 & 0 & 0 & 0 & 0 & A & \zeta_{An} & 0 & 0 & 0 & 0 \\ 0 & 0 & x_{sFs} & 0 & 0 & 0 & 0 & n & Fs & -\pi_{sw} & 0 & x_{nFn} \\ 0 & x_{wFw} & 0 & 0 & 0 & 0 & 0 & 0 & -\pi_{ws} & Fw & -\pi_{wu} & 0 \\ 0 & 0 & 0 & 0 & 0 & 0 & 0 & 0 & 0 & -\pi_{uw} & Fu & -\pi_{un} \\ 0 & 0 & 0 & 0 & 0 & 0 & 0 & x_{nFn} & 0 & 0 & -\pi_{nu} & Fn \end{bmatrix} [\Omega] \tag{9.69}$$

For this PCB architecture, the "EMC layout" objective is to reduce the number of parasitic interactions π. How could we proceed?

The various kinds of signals existing can be categorized in families:

- DC power supplies;
- dynamic power signals;
- digital signals;
- analogical signals.

These signals can be seen as *EMC families* and, basically, EMC methodologies give rules to manage these signals in their neighbourhoods. Intuitively, high-level power signals must be separated from low-level signals. But this kind of constraint appears often like a very hard requirement. Reducing electronic volumes, decreasing weights, etc., come in contradiction with this requirement.

Separating signals is the most robust solution: it is a prevention solution that avoids the sensitive signals to be disturbed by noisy ones. This segregation approach implies sufficient dimensions. When this space is not available, the only other solution is to create this electromagnetic distance (EM distance) through added protections. A fundamental concept managing EMC for PCB is the prevention–protection (PP) diagram.

In our case, the subject is the noise diffusion. Anyway, the physical phenomenon uses the same impedance operator. Looking for conducted noises transmitted outside the PCB, our criterion is the noise level measured on each pin of the connector. Once this criterion defined, we can locate each level computed under one architecture assumption on the PP diagram.

9.10.1 PP diagram

A PP diagram is shown in Figure 9.21.

Our research and development target is to obtain a solution inside the safe zone, for a standard requirement level of noise A dBμA.

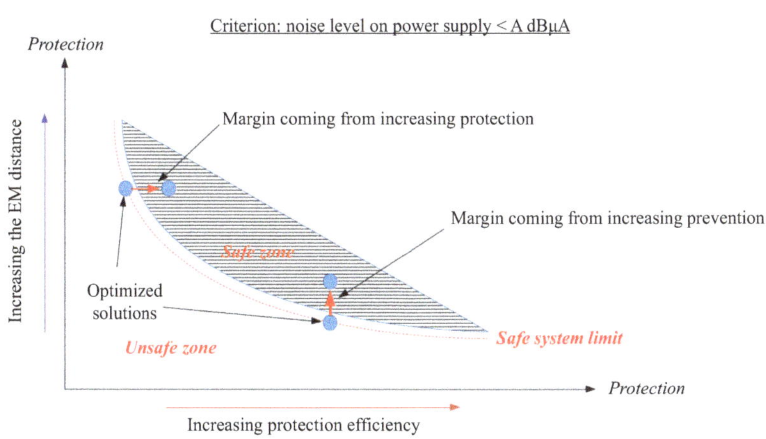

Figure 9.21 Prevention–protection diagram

As an example we will consider the noise level conducted on the power supply wires. The authorized level for this noise is defined in EMC standards and constitutes the EMC-CE requirements. So we are first interested in computing the noises transmitted on the power supply pin of the output connector. Now, we may wonder on the organization of this PP diagram?

9.10.2 Interaction matrix and architecture decision

The interaction matrix is the simplest way to organize the EMC work, in emissions as in immunity. In this matrix, all signals appear in row and column. The row titles point out the receivers. The column titles point out the emitters. When an emitter can communicate its noise to some receiver, the corresponding cell is painted in red. Another solution consists in noting in the cell the maximum noise level transmitted to the receiver. In all cases, this indicates that some work must be done in order to reduce this noise diffusion. For our example and focusing on the power supply, we obtain the interaction matrix shown in Figure 9.22.

We have to study the noise transmission from the power chopper to the power supply for the architecture presented in Figure 9.20. We have previously computed the power chopper noise. It remains to compute the coupling between the tracks of the power chopper and the power supply.

Classical tracks on PCB are microstrips. Microstrips are particular lines and were deeply studied (see, e.g., [3]). For a microstrip of width W and height h over the ground plane, the characteristic impedance and the effective relative permittivity are (for $W/h < 1$)

$$z_c = \frac{\eta}{2\pi\sqrt{\varepsilon_{re}}} \ln\left(\frac{8h}{W} + 0.25\frac{W}{h}\right) \; [\Omega] \tag{9.70}$$

with

$$\varepsilon_{re} = \left[\frac{\varepsilon_r + 1}{2} + \frac{\varepsilon_r - 1}{2}\right]\left[\left(1 + 12\frac{h}{W}\right)^{-1/2} + 0.041\left(1 - \frac{W}{h}\right)^2\right] \; [\mathrm{F}]\,[\mathrm{m}]^{-1} \tag{9.71}$$

	Signals			
	w	s	up	n
w		■ (red)		
s				
up				
n				

Figure 9.22 Interaction matrix

where ε_r is the dielectric permittivity of the PCB insulating. Like for line, this characteristic impedance is associated with the TEM mode. But as we work on emissions, the signals use microstrips for functional signals. In fact, they use microstrip in TEM mode.

Knowing the characteristic impedance gives us an estimation of the microstrip capacitance per meter. But to estimate the coupling between two microstrip, we must know the capacitance between two neighbour microstrips. The mutual inductance can be calculated using Neumann's formula.

Between the two tracks of the microstrips, there are two capacitors: one for which the field lines go through the air C_{ta}, and one for which the field lines go through the dielectric C_{td}. These two capacitors are in parallel and their addition constitute the coupling capacitance between the two microstrips. We have [3], keeping the same geometrical assumption:

$$\begin{cases} C_{ta} = \frac{\varepsilon_0}{\pi} \ln\left(2\frac{1+\sqrt{k'}}{1-\sqrt{k'}}\right) \overset{k}{\mathscr{D}}_{[0,0.5]} + \pi\varepsilon_0 \left[\ln\left(2\frac{1+\sqrt{k'}}{1-\sqrt{k'}}\right)\right]^{-1} \overset{k}{\mathscr{D}}_{[0.5,1]} \\ C_{td} = \frac{\varepsilon_0\varepsilon_r}{\pi} \ln\left[\coth\left(\frac{\pi S}{4h}\right)\right] + 0.65 C_f \left[\frac{0.02h}{S}\sqrt{\varepsilon_r} + \left(1 - \frac{1}{\varepsilon_r^2}\right)\right] \end{cases} \tag{9.72}$$

where S is the distance between the two microstrips, $k = S/(S+2W)$ and $k' = \sqrt{1-k^2}$. C_f is the capacitor between the top of the track and the common ground. The propagation capacitance is located between the down of the track and the common ground:

$$C_f = \frac{1}{2}\left(\frac{\sqrt{\varepsilon_{re}}}{c z_c} - \varepsilon_0\varepsilon_r\frac{W}{h}\right) \tag{9.73}$$

We have already seen in Chapter 2 as how to compute the capacitive cross-talk.

The difficulty comes from the fact that various capacitances are involved in microstrips. Figure 9.23 shows these different capacitors.

In fact, the capacitive coupling coefficient α is given by

$$\alpha = u_2\left[1 + \frac{C_p + C_f + C_f'}{C_{ta} + C_{td}}\right]^{-1} \tag{9.74}$$

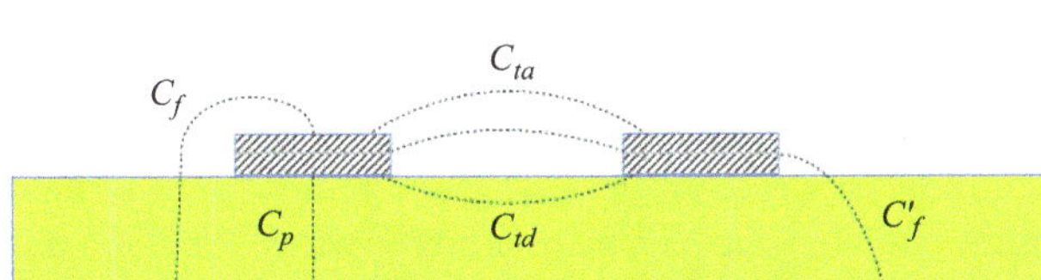

Figure 9.23 Microstrip capacitances

u_2 being the noise level on the neighbour microstrip. This formula is an approximation but sufficient to evaluate the risk of the noise diffusion. Finally, the impedance operator of the microstrip couple looks like

$$\zeta_{pcb} = \begin{bmatrix} z_0 + \frac{2}{C_1 p} & -\frac{2}{C_1 p} & 0 & 0 \\ -\frac{2}{C_1 p} & \frac{1}{C_1 p} + L_1 p + R_1 & -Mp & -\alpha z_c \\ 0 & -Mp & \frac{1}{C_2 p} + L_2 p + R_2 & -\frac{2}{C_2 p} \\ 0 & -\alpha z_c & -\frac{2}{C_2 p} & \frac{2}{C_2 p} + z_L \end{bmatrix} [\Omega] \quad (9.75)$$

where z_0 is the source impedance.

It is the link with the previous matrix created for the noise source inside the component. The outputs of the PCB are represented by the two open-circuit outputs $C_1/2$ and $C_2/2$ (the line structures retained here are pi structures C–L–C).

Other numberings, other current signs, etc., can be used to write this problem. Anyway, the final impedance operator should have a similar form. As seen in Chapter 2, if the wavelength is shorter than the microstrip length, Brain's models should be used. But as seen also in the same chapter, Branin's model has the same dimension as the RLC low frequency model (dimension 2×2). The extra-diagonal elements may be different, but the whole matrix appearance should remain the same.

The α and M numbers are components of the global π_{ws} coupling function. By taking some particular tracks to evaluate the noise transmitted to output (the work cannot be done for all the signals. It is better to limit the studies to one example of each kind of couplings), we verify if one architecture is compliant with the requirements of CEs. The last work to do consists of connecting the equipment to the harness of EMC test with its LISN as load. That is the purpose of the next paragraph. But before looking at this step, we must begin by discussing on the box influence in the first time. Then, in the second time, we should investigate on the interconnection branches between the components to the microstrip network.

We may wonder if the couplings through vias should be studied or not? Vias are of very small dimensions. As a matter of fact, it is quite unlikely that two vias may be neighbours, in more the fact that their small dimensions create small coupling capacitances. In general, the couplings coming from the microstrips connected to the vias are much more higher and of first order compared to the vias.

9.10.3 Box influence

The resonance of PCB box cannot exist if the propagation signal wavelength is widely larger than the box physical dimensions. The fact that the box constitutes a closed environment changes only the characteristic impedances of wires and the distribution of capacitors and inductances between the various conductors existing in this volume.

While if the frequency reaches the first mode of resonance of the fulfilled box, it can influence the coupling between microstrips or wires. We study this second case using the simplest model as possible. More complex models are available but out of the scope of this book [4–6], and the model we will study can be used in a generic approach.

A cavity can be considered for one eigenmode as a resonator. This resonator is an RLC circuit. The capacitive component can be determined in electrostatique mode. By knowing the eigenmode frequency, the inductance component can be deduced. It remains to evaluate the losses to determine the resistive component.

The resonances of a cavity is completely defined by its empty volume. The form of this volume determine the modes of resonances.

The draw of the field lines follows some golden rules. In static, any conductor wears an unique charge: positive or negative. In dynamic, the field lines cannot cross each other.

Once the field functions are estimated, their replacement in d'Alembert's equation gives the dispersion. And the dispersion contains the eigenmodes. In one dimension, d'Alembert's equation is for the propagation of a field u:

$$\frac{\partial^2 u}{\partial x^2} - \frac{1}{v^2}\frac{\partial^2 u}{\partial t^2} = 0 \tag{9.76}$$

We can see in Figure 9.24 a cavity fulfilled with a PCB. We choose three axes to describe the field propagation: two axes rectangular and one curvilinear. A plan that wears the field lines on the (r, z) surface moves following the axe z. If we can associate a field function to each direction, we can hope to find the modes.

In this example, a first mode is possible if the electric field obey to the law: $\mathbf{E} = E_0\mathbf{u}_r$, the space base being $\mathbf{u}_z, \mathbf{u}_r, \mathbf{u}_\theta$. If the frequency increases, we can have

$$\mathbf{E} = E_0\mathbf{u}_r \cos\left(n\pi\frac{\theta}{\Theta}\right) e^{j(\omega t - kz)} \; [\mathrm{V}]\,[\mathrm{m}]^{-1} \tag{9.77}$$

If $n = 0$, the mode is called *trans-electromagnetic*.

In a space having N dimensions, d'Alembert's equation becomes

$$\sum_i v_i^2 \frac{\partial^2 \mathbf{u}}{\partial x_i^2} - \frac{\partial^2 \mathbf{u}}{\partial t^2} = 0 \tag{9.78}$$

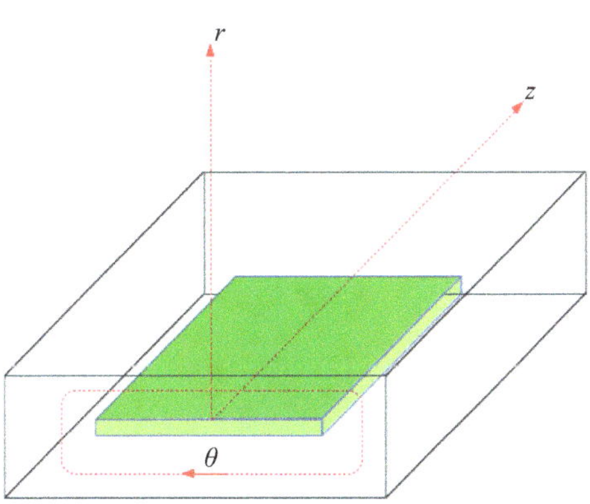

Figure 9.24 Fulfilled cavity modes

Replacing (9.77) in (9.78) we obtain

$$-v_z^2 k^2 \mathbf{E} - v_\theta^2 \left(2\frac{n\pi}{\Theta}\right)^2 \mathbf{E} + \omega^2 \mathbf{E} = 0 \tag{9.79}$$

This leads to (if $v_z = v_\theta$)

$$k = \sqrt{\frac{\omega^2}{v_z^2} - \left(2\frac{n\pi}{\Theta}\right)^2} \; [\mathrm{m}]^{-1} \tag{9.80}$$

This is the dispersion of the waveguide without any limit conditions. If we short-circuit these extremities we obtain for the field:

$$\mathbf{E} = E_0 \cos\left(2n\pi\frac{\theta}{\Theta}\right) \sin\left(m\pi\frac{z}{Z}\right) \mathbf{u}_r \; [\mathrm{V}]\,[\mathrm{m}]^{-1} \tag{9.81}$$

and the dispersion becomes

$$f_{mn} = \frac{v_z}{2\pi}\sqrt{\left(2\frac{n\pi}{\Theta}\right)^2 + \left(\frac{m\pi}{Z}\right)^2} \; [\mathrm{Hz}] \tag{9.82}$$

The first possible mode is TE_{10} with the convention $TE_{mn} \rightarrow TE_{z\theta}$. If we consider the mode TE_{11}, knowing the length Z and Θ, we can calculate the frequency f_{11}. Knowing the frequency we must determine the equivalent capacitance for this first mode in order to construct the equivalent resonator.

The determination of the capacitance comes from the electromagnetic energy stored in the volume. We can write the equivalence:

$$\frac{1}{C}CV^2 = \varepsilon_0 \int_z dz \int_\theta d\theta \int_r dr \mathbf{E} \cdot \mathbf{E} \; [\mathrm{J}] \tag{9.83}$$

With $\mathbf{gradV} = -\mathbf{E}$ this leads to:

$$C = \varepsilon \frac{Z\Theta}{R} \; [\mathrm{F}] \tag{9.84}$$

Having an estimation for C and knowing f_{11} we can calculate the inductance L_{11} by

$$L_{11} = \frac{1}{4\pi^2 f_{11}^2 C} \; [\mathrm{H}] \tag{9.85}$$

If the capacitance can be supposed as a fixed parameter depending on the frequency, the inductance depends if fact on the frequency in order to follow the mode frequency changing.

We can estimate the quality coefficient Q by

$$Q_{11} = \frac{3V}{2S}\sqrt{\pi f_{11} \mu \sigma} \tag{9.86}$$

where V is the empty volume of the cavity and S the surface of all conductive walls. Knowing the quality coefficient, we can obtain the losses:

$$R_{11} = \frac{L_{11}\omega_{11}}{Q_{11}} \; [\Omega] \tag{9.87}$$

At this level we have the three components of our resonator.

We have construct this resonator to evaluate the coupling between microstrips or tracks using the cavity. Each object is coupled with the cavity fields, and this can be the cause of a noise diffusion.

We know each graph associated with each object: microstrip or cavity. We just lack the cords representing the electromagnetic interactions between them.

The interaction between a microstrip and the cavity is equal to the interaction between the cavity and the microstrip. This allows to choose the simplest calculation of the interaction.

Knowing the field distribution, we can compute the emf induced in a microstrip. We can symbolize the relation between the emf and the magnetic field by $e = \alpha \cdot B$. As seen previously, we have the relation in energy:

$$\frac{1}{2}L_{11}i^2 = \oint_v dv\frac{B_{11}^2}{\mu} \text{ [J]} \tag{9.88}$$

We can also symbolize the magnetic field integration by $\beta \cdot B^2$, giving:

$$B = \sqrt{\beta^{-1}\left(\frac{\mu}{2}L_{11}\right)}i \text{ [T]} \tag{9.89}$$

i being the current in the resonator. Finally

$$\frac{e}{i} = \zeta_{mc} = \alpha \cdot \sqrt{\beta^{-1}\left(\frac{\mu}{2}L_{11}\right)} \text{ [}\Omega\text{]} \tag{9.90}$$

By computing (9.88) this relation replaces β by a geometrical factor.

In general we can make links between the fields and the potentials through the equivalences in energies. This is the way to transpose the relations in field, in impedances into the graph of the studied system.

Figure 9.25 shows the graph of the coupling previously studied. This shows the process of coupling of far microstrips not cross-talked by neighbourhood.

It is clear that we would not study all possible couplings in the equipment. Usually we study the particular case of two microstrips existing on the PCB and we change

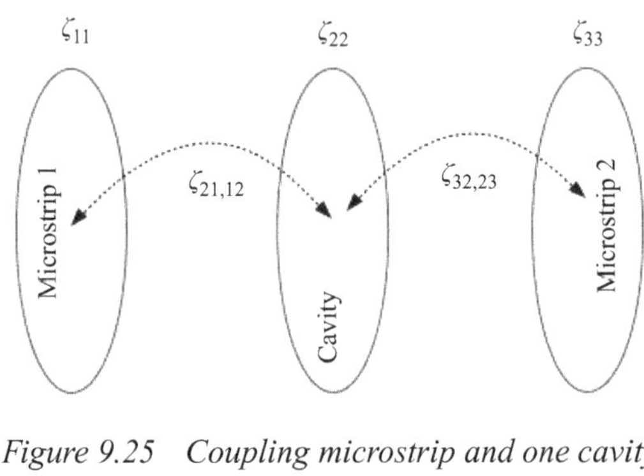

Figure 9.25 Coupling microstrip and one cavity

the parameters (distances, properties, loads, etc.) to detect if one situation can lead to unacceptable noise diffusion. After that, we can see if some microstrip design is similar to this identified riskable situation.

9.10.4 Connecting the component to the microstrip network

Once the operator is determined for the component on one side and the PCB on the other side, it remains to connect both objects. It introduces the concept of frontier. The fact to connect two ports means to share a common impedance. The connection is realized in two steps:

1. determining the shared impedance;
2. determining the coupling cord.

We can resume one output of a component under the form of a single mesh with its own source. The output branch of this graph is a capacitance C_o. On the other side, we have a microstrip modelled using a telegraph's model of first order: *RLCG*. Figure 9.26 illustrates this operation.

The output impedance of the component PDN (passive description network) is the capacitor C_2, while the input of the line is the capacitor C_3 (this capacitor has for value the half of the line capacitance per meter multiplied by its length). When we connect both extremities, the equivalent capacitor becomes

$$C_{23} = C_2 + C_3 \ [\text{F}] \tag{9.91}$$

So, the cord function is known: it is a capacitance impedance given by

$$\zeta_{23} = \zeta_{32} = -\frac{1}{C_{23}p} \ [\Omega] \tag{9.92}$$

The structure of the impedance operator $\zeta_{connect}$ is made of the direct summation of ζ_{PDN} the operator of the component output and of ζ_L the line operator.

$$\zeta_{connect} = \zeta_{PDN} \oplus \zeta_L \ [\Omega] \tag{9.93}$$

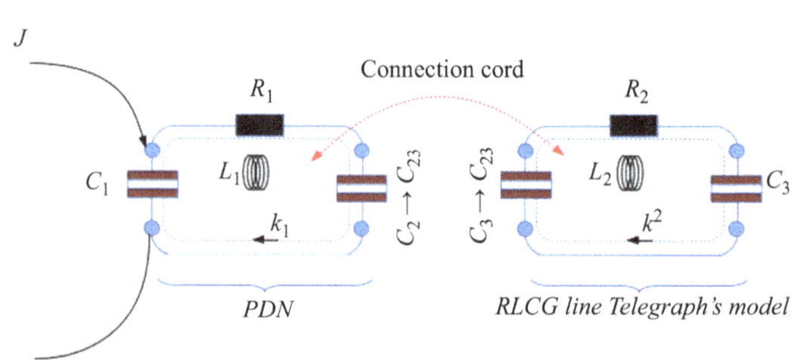

Figure 9.26 Connecting a component to one microstrip

To translate the connection, it remains just to replace C_2 and C_3 by C_{23} and to add the cord ζ_{23}:

$$\zeta_{connect} = \begin{bmatrix} \zeta_{PDN}\{C_2 \rightarrow C_{23}\} & -\frac{1}{C_{23p}} \\ -\frac{1}{C_{23p}} & \zeta_L\{C_3 \rightarrow C_{23}\} \end{bmatrix} [\Omega] \tag{9.94}$$

9.10.5 Multilayers PCB

As in the cavity couplings case, we never compute the whole problem on one step. This may create problems of very large dimensions, difficult to control and imply long time computations in the case of numerical applications. But to separate the problem in smaller ones, it suggests the PCB designer to respect some constraints. The ground planes management makes part of these constraints, and we will now study how they act and how to model them.

For the other layers, they are included between two ground layers. Microstrips are also waveguides but with different characteristic impedances. The question is the following: what is the coupling between two microstrips that belong to two different layers?

The first layer of the considered PCB is an open layer where microstrips behave like waveguides. Figure 9.27 shows the situation considered. One microstrip that wears some power signals is located on the second layer, framed by two ground layers. We wonder what can be the coupling with a microstrip located on the top layer?

We can take into account three kinds of couplings, numbered on the Figure 9.27 from 1 to 3. First coupling comes from the shared ground plane between the two microstrips. We call it generally the common impedance coupling. The second coupling comes from the transmission through holes in the ground layer of magnetic field. This is characterized using a special function called transfer impedance. This function can be used too for the shielding modelling. The third coupling comes from interactions between layers going through the PCB borders and passing eventually through couplings with the equipment cavity.

9.10.5.1 Common impedance and transfer impedance coupling

The transfer impedance coupling has two parts: the first one is the resistive part translating the voltage transmitted by a current circulation on a conductive structure. One inductive part translating the penetration of the magnetic field through apertures. The resistive part is often called “common impedance coupling” when speaking in

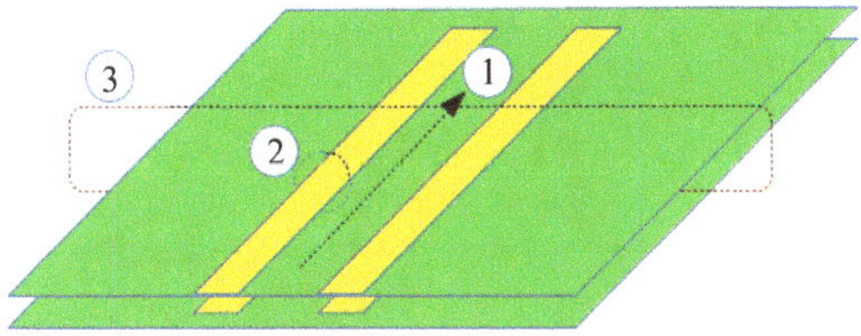

Figure 9.27 Coupling between microstrips that belong to different layers

general of this kind of coupling. Looking at Figure 9.28 we see that if a structure is shared between two microstrip, and if a current flow on it is coming from a signal transmitted by the first microstrip, this current develops a voltage seen by the second signal.

In DC and low frequencies (below the skin effect), the resistance is computed knowing the material resistivity and dimensions (length x and section S) using classical relation $\sigma^{-1}x/S$. Beyond the skin effect, the magnetic field cannot be transmitted through the conductive plate and the coupling decreases following the relation:

$$R_T = \frac{1}{\sigma}\frac{x}{S}e^{-\sqrt{\pi f \mu \sigma w}} \; [\Omega] \tag{9.95}$$

where w is the plate thickness.

The inductance coupling comes from the magnetic field transmission through openings. Let us take a look at Figure 9.29.

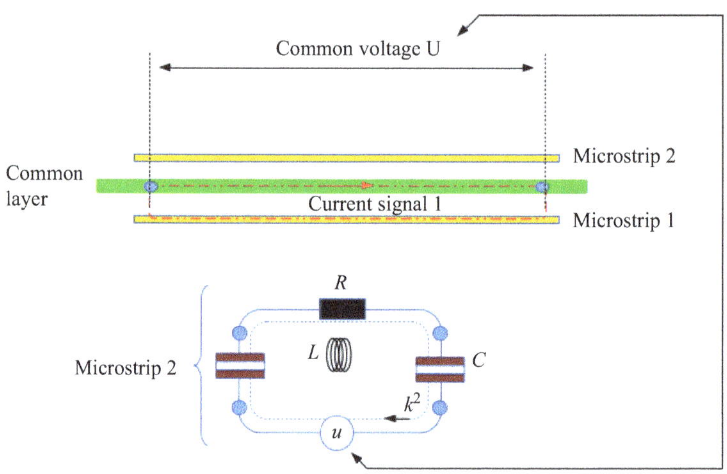

Figure 9.28 Coupling through shared impedance

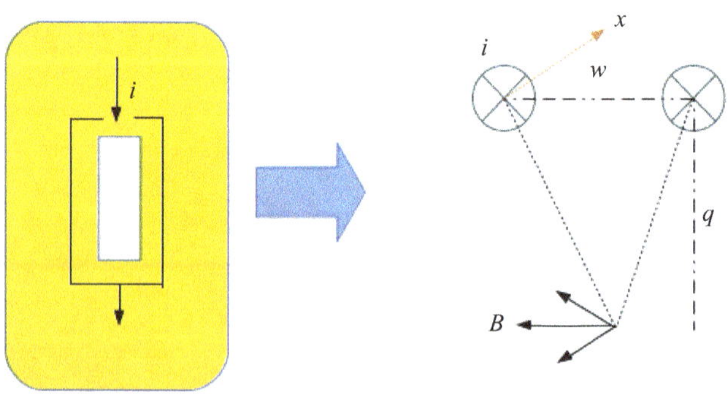

Figure 9.29 Coupling through transfer inductance

We can compute the magnetic field emitted by the current flowing on the opening borders. We find

$$B = \frac{\mu_0 iqx}{2\pi\left[q^2 + (w/2)^2\right]^{3/2}} \text{ [T]} \tag{9.96}$$

For a section s_0 of reception, we obtain the transfer function e/i:

$$\frac{e}{i} = -s_0 p \frac{\mu_0 qx}{2\pi\left[q^2 + (w/2)^2\right]^{3/2}} \Rightarrow L_T = s_0 \frac{\mu_0 qx}{2\pi\left[q^2 + (w/2)^2\right]^{3/2}} \text{ [H]} \tag{9.97}$$

The total transfer impedance that is the second coupling function between the microstrip is finally obtained by

$$Z_T = R_T + L_T p \text{ [}\Omega\text{]} \tag{9.98}$$

9.10.5.2 Couplings through PCB borders

In high frequency, the energy is principally concentrated under the track of the microstrip. This statement is always true even if the line is matched. For this reason in general on a PCB, microstrips radiate weakly. But the fact to have the two ground plates open on their borders create an open cavity (TGPOC) that can transmit the field to outside. We retrieve a situation similar to the situation of the coupling between microstrips through the equipment cavity. But this time the problem is of dimension 4. Figure 9.30 shows the interactions involved in this problem. The problem can be resumed to find both coupling TGPOC to cavity and microstrip to TGPOC. As we have said before, it is often simplest to search the coupling of the cavity fields to the microstrip rather than the inverse. This effect can also be verified even if the source is the microstrip. Anyway, the coupling functions are symmetric.

We may represent the TGPOC by a resonator as previously. But we can also compute the direct link including the two cavities in series.

If we imagine a magnetic field distribution between the two plates given by

$$B_y = B_{0y} \sin\left(m\pi \frac{x}{X}\right) \sin\left(n\pi \frac{y}{Y}\right) \text{ [T]} \tag{9.99}$$

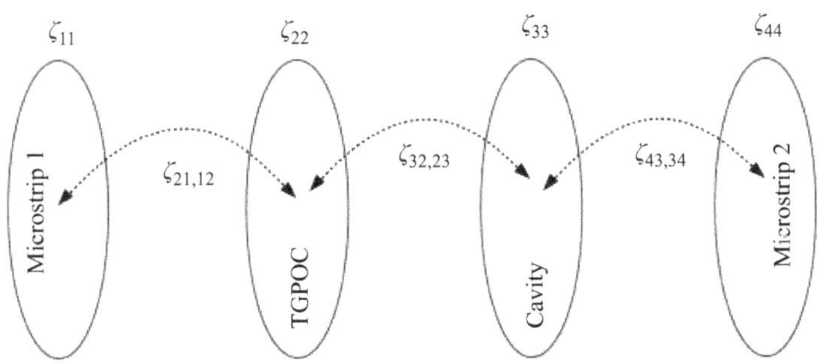

Figure 9.30 Coupling by TGPOC

z being the direction perpendicular to the plates. If the microstrip is oriented in the x direction, the EM force induced in the microstrip is

$$e = -hp \int_x dx B_{0y} \sin\left(m\pi \frac{x}{X}\right) \sin\left(n\pi \frac{y}{Y}\right) \text{ [V]} \tag{9.100}$$

where h is the height between the two plates.

Sometimes it seems complicated to compute some interaction functions. But we must always remember that we know how to write the relations between potentials and fields through energies, and we know the relations between the fields: they are Maxwell's equations. For the moment we have a link between the emf in the microstrip and the magnetic field in the TGPOC. The limit condition of the TGPOC is in electric field. Maxwell's equations tell us that

$$\frac{1}{c^2} p E_z = -\frac{\partial B_y}{\partial x} \Rightarrow E_z = hc^2 B_{0y} \sin\left(m\pi \frac{x}{X}\right) \sin\left(n\pi \frac{y}{Y}\right) \text{ [V] [m]}^{-1} \tag{9.101}$$

If we identify this field with the field developed in the cavity which is linked with the resonator current by

$$\frac{1}{Cp} i = h' E_z \text{ [V]} \tag{9.102}$$

h' being the height of the cavity.

Integrating (9.100) we obtain

$$\begin{cases} i = -h' C p B_{0y} \frac{c^2}{p} \left(\frac{m\pi}{X}\right) \cos\left(m\pi \frac{x}{X}\right) \sin\left(n\pi \frac{y}{Y}\right) \\ e = hpB_{0y} \left(\frac{X}{m\pi}\right) \cos\left(m\pi \frac{x}{X}\right) \sin\left(n\pi \frac{y}{Y}\right) \end{cases} \tag{9.103}$$

and finally

$$\frac{e}{i} = -\frac{hp}{h'c^2 C} \left(\frac{X}{m\pi}\right)^2 \text{ [}\Omega\text{]} \tag{9.104}$$

Merging the two cavities in one reduces the problem to a dimension 3.

9.10.6 Locating the solution on the PP diagram and conclusion on the EMC risk

For explaining the global approach of EMC risk determination, we take a simple example. This example consider a complete chain, from the component to the LISN filter of the EMC test. But in order to illustrate the technique without walnuting the reader in a long and complex case, we create a simple case with reduced models. The circuit presented in Figure 9.31 details this model.

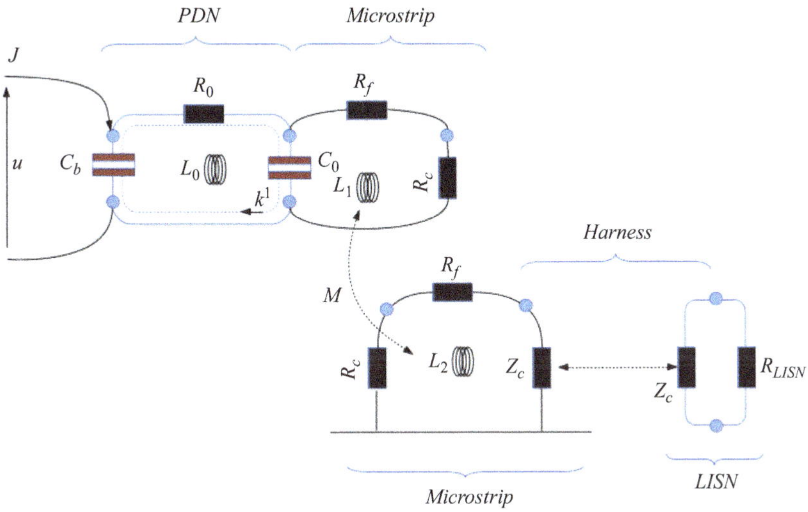

Figure 9.31 Complete chain for illustration

With practice we can easily establish the impedance operator of this system. First we write the diagonal elements:

$$\zeta = \begin{bmatrix} \frac{1}{C_b} & 0 & 0 & 0 & 0 \\ 0 & \frac{1}{C_b} + R_0 + L_0 p + \frac{1}{C_0 p} & 0 & 0 & 0 \\ 0 & 0 & \frac{1}{C_0 p} + R_f + R_c & 0 & 0 \\ 0 & 0 & 0 & R_c + R_f + z_c + L_2 p & 0 \\ 0 & 0 & 0 & 0 & z_c + R_{LISN} \end{bmatrix} [\Omega] \tag{9.105}$$

Then we add the coupling functions:

$$\zeta = \begin{bmatrix} \frac{1}{C_b} & -\frac{1}{C_b} & 0 & 0 & 0 \\ -\frac{1}{C_b} & \frac{1}{C_b} + R_0 + L_0 p + \frac{1}{C_0 p} & -\frac{1}{C_0 p} & 0 & 0 \\ 0 & -\frac{1}{C_0 p} & \frac{1}{C_0 p} + R_f + R_c & -Mp & 0 \\ 0 & 0 & -Mp & R_c + R_f + z_c + L_2 p & 0 \\ 0 & 0 & 0 & 2z_c e^{-\iota p} & z_c + R_{LISN} \end{bmatrix} [\Omega] \tag{9.106}$$

The harness is supposed to be matched. So we can translate the propagation function by a reported generator of emf $2z_c i$ and self-impedance z_c. We do not need in this case the transmission from the LISN to the microstrip. The interaction function is not symmetric.

J is a known current source and u an unknown voltage. The source covector is

$$T = \begin{bmatrix} u & 0 & 0 & 0 & 0 \end{bmatrix} \ [\mathrm{V}] \tag{9.107}$$

and the problem is $T_\alpha = \zeta_{\alpha\beta} k^\beta$. k is the mesh currents vector:

$$k = \begin{bmatrix} J \\ k^1 \\ k^2 \\ k^3 \\ k^4 \end{bmatrix} \ [\mathrm{A}] \tag{9.108}$$

Computing $k^\beta = y^{\beta\alpha} T_\alpha$ with $y = \zeta^{-1}$ excluding the first equation of the system, we obtain all the mesh currents with a source given by $1/(C_b p)J$. Then knowing the k^β, $\beta \in k$ without j we can solve the first equation and obtain u. What interests us is the fourth current k^4. The measured voltage in the EMC CE test is $z_c k^4$. We can compute the level obtained and search for the distance between the two microstrips for various cases of PCB layout that changes M and the measured level. This gives us a set of points where the level is equal to the compliant limit. We can report these points on the PP diagram (Figure 9.32).

Once a first choice is made, based on the compliance with the limit, we must take margin and consider uncertainties. Both can be taken into account in the margin. How

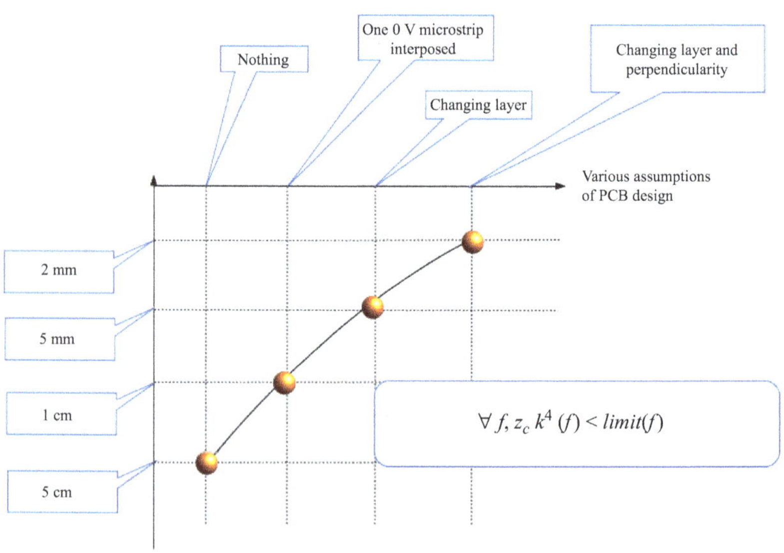

Figure 9.32 First location on the PP diagram

to conduct this analyse? First we can define the probability to go over the standard limit. We use a sigmoid function centred on the standard limit n_s:

$$P(l) = [1 + \exp(-\sigma(n - n_s))]^{-1} \tag{9.109}$$

It means that if the field level emitted by the electronic under test is equal to the authorized limit, we have one chance over two to be compliant, which is quite realistic! On the contrary, if the level is 20 dB under the limit (ten times lower), the probability of being not compliant is near to zero (10^{-5} for $\sigma = 10^6$). If the level is twice the limit, the probability of being noncompliant is 73%. So we see that this function translates quite well the risk depending on the distance of the level to the standard limit.

Now we define the probability that the electronic emits such a level. It means for us that the product $z_c k^4$ goes over the limit level n_s at each frequency. It can depend on the electronic activity, on the impedance values uncertainties, on the test bench dispersions, etc. For the solution currently chosen, the level $_e$ is equal to n_s. We can create a distribution defined by

$$P(e) = 1 - [1 + \exp(-\sigma(n - n_s))]^{-1} \tag{9.110}$$

The probability of having the emitter level superior to the threshold and the measurement compliance accepted is then given by the probabilities product. This product is computed and the corresponding results are plotted in Figure 9.33.

The product curve in yellow gives a risk of 0.25. This risk is quite high! The safety job may give us a target of 3×10^{-2}. It means that we must reduce the risk of

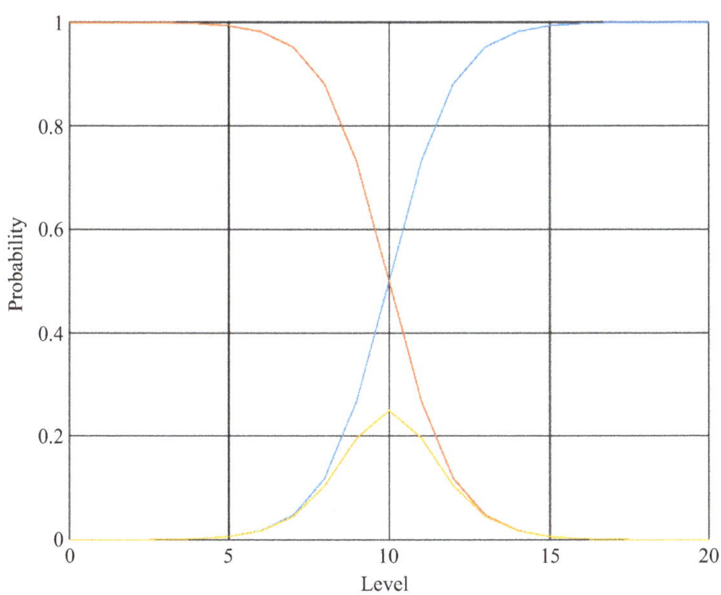

Figure 9.33 Probabilities

a factor 8. To reach this objective we can increase the distance to 5 mm and use both another layer and the perpendicularity of the microstrips. This gives us (we imagine) 20 dB of margin, and in this case, we cover the ×8 objective of safety. Figure 9.34 shows the operation realized.

The probability obtained is shown in Figure 9.35, where the probability product *has been multiplied by 1,000*. So the risk becomes, with this new solution, around 2×10^{-4}, which respects largely the safety objective.

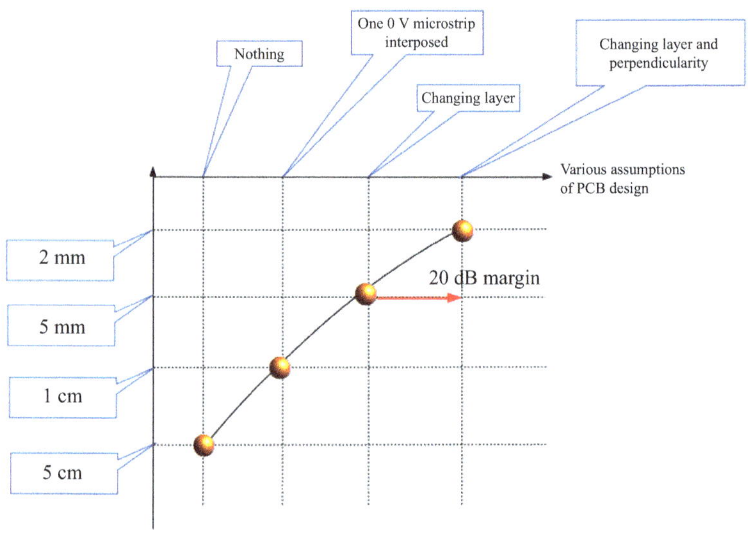

Figure 9.34 Taking a margin

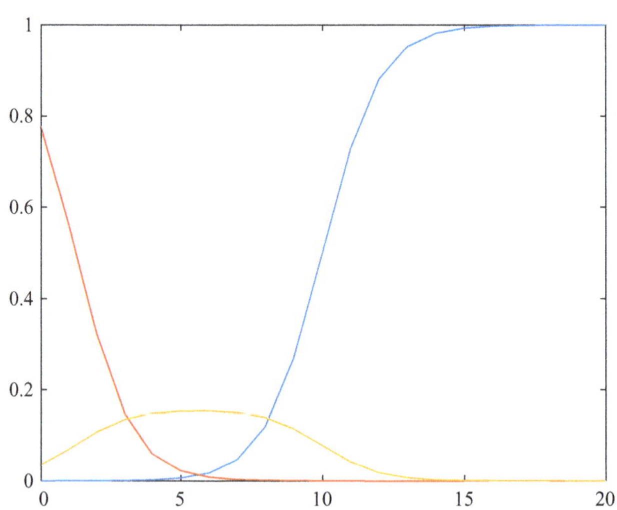

Figure 9.35 New margin—product × 1,000

It can be found usually that passing one step in EMC solutions gives a result far under the objective. This is because the effect of increasing the protections is not linear at all. Going further in the research, for other prevention distance, for example, this result may be optimized. The approach remains always the same and can be used by engineers in each kind of context.

9.11 Some indications on hyperfrequency modelling

Each time we have represented a line using telegraphist's model, we could have used Branin's model that has no frequency limitations. Its usage is a little more complex but has many more advantages if we want to work in a wide frequency band. As we treat here of the CEs, the frequency band is in general not very wide in the region where resonant frequencies are excited. That is why in many cases, models of first order are sufficient. If you need to cover wide frequency range, you will have to use Branin's model rather than telegraphist's one each time where we use these models. The same transformation must be applied to model the cavities, seen in this case as short-circuited waveguides.

Annexe 9.A

```
import numpy as np
import pylab as plt


def Dv(u):
    if u >= 0.:
        return 1.
    if u < 0.:
        return 0.


Ro = 1E9
Rn = 1E8
RD = 1E-3
RG = 50.
dt = 1E-6
T = 100E-6
N = 1000

k = np.zeros(N, dtype=float)
zd = np.zeros(N, dtype=float)
at = np.zeros(N, dtype=float)

V = 0.

for t in range(1,N):
    s = t * dt
```

```
    e = 5. * np.sin(2. * np.pi * s / T)
    zd[t] = Ro+Dv(e-RG*k[t]-0.6) * (RD-Ro) +
    (1.-Dv(e-RG*k[t]-0.6)) * (Rn-Ro)
    #
    To = Dv(e-RG*k[t]-0.6)*0.6
    k[t] = e / (RG + zd[t])
    V = zd * k[t]
    at[t] = s * 1E6

plt.subplot(2,1,1)
plt.plot(at, k)
plt.grid(True)
plt.xlabel(u'time [us]')
plt.ylabel(u'diode current [A]')
plt.subplot(2,1,2)
plt.plot(at, zd)
plt.grid(True)
plt.xlabel(u'time [us]')
plt.ylabel(u'diode impedance [ohm]')
plt.show()
```

Annexe 9.B

```
import numpy as np
import pylab as plt



Ro = 1.
L1 = 0.1E-6
C = 10E-6
L2 = 100E-6
R = 10.
zm = 1E6
B = 50.

fo = 1E2
N = 100000

k = np.zeros((N,2),dtype = complex)
ax = np.zeros(N,dtype = float)
res = np.zeros(N,dtype = float)

tau = 1E-6
T = 10E-6
Ts = [[B],[0.]]
for f in range(1,N):
    p = 1J*2.*np.pi*f*fo
    Q1 =zm/p-(zm*(1.-np.exp(-tau*p))/(tau*p**2.)*(1.-np.exp(-T*p)))
```

```
    Q2 =Q1*np.exp(-(T+tau)*p)
    zeta = [[Ro+R+1./(C*p)+L1*p+L2*p+Q1,-(R+1./(C*p))],
            [-(R+1./(C*p)),R+1./(C*p)+L2*p+Q2]]
    y = np.linalg.inv(zeta)
    k[f,:] = np.transpose(np.dot(y,Ts))
    res[f] = 20.*np.log10(abs(k[f,1])/1E-6)
    ax[f] = f*fo*1E-3

plt.plot(ax,res)
plt.grid(True)
plt.xlabel(u'Frequency [kHz]')
plt.ylabel(u'k2 current amplitude [dBuA] for L2 = 100uH')
plt.xscale('log')
plt.show()
```

Annexe 9.C

```
import numpy as np
import pylab as plt

Ro = 1.
L1 = 0.1E-6
C = 10E-6
L2 = 100E-6
R = 10.
zm = 1E6
B = 50.
Ron: float = 1E-3
Rb = 1E8

# Time definition of the control signal
N = 50
dt = 1E-6
u = np.zeros(N,dtype = float)
at = np.zeros(N,dtype = float)
for t in range(N):
    s = dt*t
    u[t] = (1.+1.*np.sin(2.*np.pi*s/15E-6))**2.
    at[t] = s*1E6

plt.subplot(4,1,1)
plt.plot(at,u)
plt.grid(True)
plt.xlabel(u'time [us]')
plt.ylabel(u'Control signal [V]')

M = 1000
k = np.zeros((M,2),dtype = complex)
axx = np.zeros(M,dtype = float)
```

```
res = np.zeros(M,dtype = float)
UC = np.zeros(M,dtype = float)
espioQ1 = np.zeros(N,dtype = float)#
espioQ2 = np.zeros(N,dtype = float)#

fo = 1E3
a = 100.
c = 0.5

for t in range(N):
    s = dt * t
    Q1 = Ron + Rb * (1. - 1. /
    (1. + c * np.exp(-a * (u[t] - 2.8))))
    Q2 = Rb - Q1 + Ron
    espioQ1[t] = Q1
    espioQ2[t] = Q2
    if Q1<1.:
        Ts = [[B], [0.]]
    if Q1>1E7:
        Ts=[[B],[B]]
    for f in range(1,M):
        p = 1J*2.*np.pi*f*fo
        zeta = [[Ro + R + 1. / (C * p) +
        L1 * p + L2 * p + Q1, -(R + 1. / (C * p))],
                [-(R + 1. / (C * p)),
                R + 1. / (C * p) + L2 * p + Q2]]
        y = np.linalg.pinv(zeta)
        k[f, :] = np.transpose(np.dot(y, Ts))
        res[f] = res[f]+abs(k[f, 1])**2.
        axx[f] = (f * fo) * 1E-6

plt.subplot(4,1,2)
plt.plot(axx,20.*np.log10(np.sqrt(res/N)/1E-6))
plt.xlabel(u'Frequency [MHz]')
plt.ylabel(u'Current K2 [dBuA]')
plt.xscale('log')
plt.subplot(4,1,3)
plt.plot(at,espioQ1)
plt.ylabel(u'impedance Q1')
plt.subplot(4,1,4)
plt.plot(at,espioQ2)
plt.ylabel(u'impedance Q2')

plt.show()
```

References

[1] Serpaud, S., Levant, J. L., Poiré, Y., Meyer, M., and Tran, S. (2009). ICEM-CE extraction methodology. In EMC Compo Proceedings, INSA Toulouse, France, (pp. 17–19).

[2] Durand, P., Boussandel, F., and Maurice, O. (2012). Power chopper modelling using the Kron's method. In Proceeding of the IASTED International Conference Engineering and Applied Science (EAS 2012). Colombo, Sri Lanka.

[3] Bahl, I. J. (2003). Lumped elements for RF and microwave circuits. Boston, MA: Artech House.

[4] Kasmi, C., Maurice, O., Gradoni, G., Antonsen, T., Ott, E., and Anlage, S. (2015). Stochastic Kron's model inspired from the random coupling model. In 2015 IEEE International Symposium on Electromagnetic Compatibility (EMC) (pp. 935–940). Dresden: IEEE.

[5] Leman, S., Demoulin, B., Maurice, O., Cauterman, M., and Hoffmann, P. (2009). Use of the circuit approach to solve large EMC problems. Comptes Rendus Physique, 10(1), 70–82.

[6] Maurice, O. (2012). Développement modal du couplage par ondes guidées sous un formalisme de Branin. https://hal.archives-ouvertes.fr/file/index/docid/684240/filename/braninHarmonique.pdf

Chapter 10
PCB-conducted susceptibility (CS) EMC TAN modelling

Olivier Maurice[1], Blaise Ravelo[2], and Zhifei Xu[2]

Abstract

The complementary aspect of the electromagnetic compatibility (EMC) emission is the susceptibility issue. The present chapter will focus on the EMC conducted susceptibility (CS) of hybrid printed circuit board (PCB) system with specifications described in the previous chapter. The tensorial description of the PCB susceptible components as mathematical sensitive functions integrating subdomain aspect will be originally introduced. The subdomain functions act as sigmoidal mathematical functions depending on the specification of the EMC perturbations. After the analytical description of the susceptibility functions, the workflow of the tensorial analysis of networks (TAN) modelling methodology indicating the routine algorithm of the EMC analysis will be provided. Then, random risk analyses table, including the objective functions indicating the EMC severity and the damage severity, will be addressed. To validate the EMC CS TAN model, illustration example of PCB system with the prediction of EMC severity quantification with classes A, B and C will be discussed. Then, the accuracy, advantages and limits of the EMC CS TAN model will be presented at the end of the chapter.

Keywords: TAN approach, Kron's method, Kron–Branin's model, frequency-domain analysis, modelling methodology, PCB analysis, SI analysis, PI analysis, EMC analysis

10.1 Disturbing mechanisms

Considering a whole system, the conducted immunity starts from the electromagnetic field coupling with harnesses. This coupling is then transformed into power transmitted to the input of components existing on the printed circuit board (PCB).

[1]ArianeGroup, Paris, France
[2]IRSEEM/ESIGELEC, Rouen, France

The first interaction is located between the electromagnetic field and some wires included in harnesses.

Once the energy is transported to the component boundary, this energy is transformed to noise changing the component working. Our first job consists of searching the model for the field-to-line coupling.

10.2 Field-to-line coupling

The electromotive force (emf) induced by the field on the line propagates its extremities. It appears as generators added on the two meshes representing the line under Branin's model.

Let us remember the basics of field-to-wire coupling. We can consider first the magnetic field coupling on a loop circuit.

10.2.1 Magnetic field coupling

There are various expressions of the field to loop coupling depending on the field involved. If we speak of the incident field, or the refracted one, the relations are not always the same. We consider a simple loop like a closed circulation, including any possible impedances. This loop belongs to a single plan. The incident magnetic field B_i is homogeneous over the whole loop surface S. The magnetic field vector makes an angle θ with the loop plan. The *emf* or e associated with the magnetic flux in the loop is defined by

$$e = -p\mathbf{S} \cdot \mathbf{B}_i \ \ [\mathrm{V}] \tag{10.1}$$

This *emf* has for dual the current J flowing in the loop, crossing all the impedances (i.e. all the media) existing on this travel. If ζ is the summation of all these impedances, the total magnetic field $(B_i + B_r)$ must respect the relation:

$$e + p\mathbf{S} \cdot \mathbf{B}_r = \zeta J \ \ [\mathrm{V}] \tag{10.2}$$

The refracted magnetic field B_r is defined by the electromagnetic inertia:

$$p\mathbf{S} \cdot \mathbf{B}_r = LpJ \ \ [\mathrm{V}] \tag{10.3}$$

and

$$e - LpJ = \zeta J \ \ [\mathrm{V}] \tag{10.4}$$

The closed loop circuit that embraces the magnetic flux can be composed of any kind of impedance operator. It can include capacitors, diodes, etc. What is important to recall is that in our expression, the magnetic field involved is the incident one.

The closed circulation can be in particular a small part of a line as shown in Figure 10.1. If we apply the previous relation to the line section identified by the abscissa z and $z + dz$ we can write

$$e = -phdzB_y \cos(\theta) = LpJ + RJ \ \ [\mathrm{V}] \tag{10.5}$$

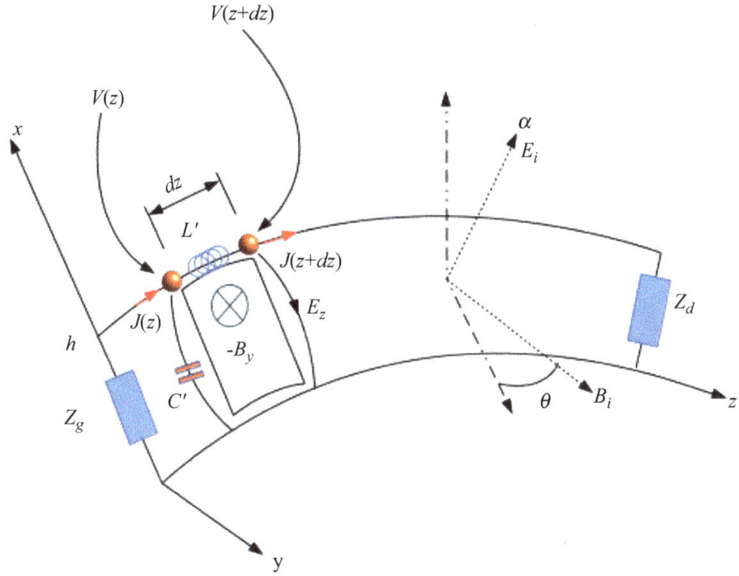

Figure 10.1 Field-to-line coupling

which gives

$$-p \int_x dx B_y \cos(\theta) = pL\frac{J}{dz} + \frac{dV}{dz} \tag{10.6}$$

Introducing the linear inductance per meter: $L = L'dz$:

$$-p \int_x d\mathbf{x} \cdot \mathbf{B}_l - pL'J + \frac{dV}{dz} \tag{10.7}$$

10.2.2 Electric field coupling

We have seen in Chapter 1 that the near electric field coupling—the coulombian one—is taken into account through the capacitance coupling. But looking at Faraday's law,

$$\oint_c d\mathbf{c} \cdot \mathbf{E} = -p\mathbf{S} \cdot \mathbf{B} \ \text{[V]} \tag{10.8}$$

Rather than covering the whole closed circulation, we can define a partial *emf* by integrating on a part of c. In this case it develops a voltage defined by

$$V = \int_x dx E_i \ \text{[V]} \tag{10.9}$$

This voltage induces loads and

$$qdz = C'dzV \Rightarrow J = -pC' \int_x dxE_x \cos(\alpha) \;\; [\mathrm{A}] \tag{10.10}$$

This voltage comes from the transverse far electric field effect.

Applied to the line, the difference of the currents between both abscissae is

$$J(z + dz) = J(z) - pC'dzV - pC' \int_x d\mathbf{x} \cdot \mathbf{E}_i \;\; [\mathrm{A}] \tag{10.11}$$

Finally,

$$\frac{dJ}{dz} + pC'V = -pC' \int_x d\mathbf{x} \cdot \mathbf{E}_i \tag{10.12}$$

When we compute the total *emf*, the transverse electric field total contribution is null. Effectively, as the current generators associated with the field have in the same direction, the total current added to the line is equal to zero. That is only on the line limits, where the coupling becomes asymmetric, that this contribution changes the *emf*.

We may have used the potential vector **A**. By replacing the magnetic field by

$$B_y = \frac{\partial A_z}{\partial x} - \frac{\partial A_x}{\partial z} \;\; [\mathrm{T}] \tag{10.13}$$

and replacing the transverse electric field by

$$E_x = -pA_x \;\; [\mathrm{V}]\,[\mathrm{m}]^{-1} \tag{10.14}$$

in (10.8) lead to the use of a single field. This shows that, often, discussions on electric versus magnetic field influence should be replaced by discussions on the influence of the time versus space derivations of the vector potential.

10.2.3 Conclusion on the field-to-line coupling fundamental processes

We can now remember that the objective is to calculate the emf induced on a line that follows any curvilinear abscissa z. Starting from the vector potential in Coulomb's gauge, we do not have to care anymore of the question of knowing about what is the electromagnetic field to be considered. Both are included in this field which is on the base of the generalized electromagnetic field definition $F_{\alpha\beta} = \partial_\alpha A_\beta - \partial_\beta A_\alpha$ (α is the declination and β, the elevation).

Let potential vector be linked with the current through a metric. Einstein had explained that contrary to what is often said, Maxwell's equations do not say us anything about loads. The fundamental link between loads and fields comes through the current density. The potential vector is defined by (u point out the basic vectors, u_R for the radius and $u_\perp$ for the transverse direction to the radius one)

$$\mathbf{A}_\alpha(b) = \mu_{\alpha\beta} \int_x dx \oint_{y,z} dydz \left[\frac{\mathbf{J}^\beta(y,z,a) \cdot d\mathbf{x}}{4\pi R_{ab}} \right] e^{-\frac{R_{ab}}{c}p} \left(\mathbf{u}_R \times \mathbf{u}_\perp \right) \; [\mathrm{W}][\mathrm{m}]^{-1} \tag{10.15}$$

The metric allowing one to go from the current to the field is the magnetic permeability μ. It is a tensor twice covariant translating the properties of the medium and its interaction with the field.

R_{ab} is the distance between the point where the current is observed, a, and the point where the field is observed, b. $\mathbf{u}_\perp$ is the vector perpendicular to the current direction and to the line of sight.

Once the field is associated with a source, it can create an *emf* as seen before, or a magnetomotive force through another object which has the dimensions of an inverse of a metric: ε.

$$J^\alpha\,(y,z) = p^2 \varepsilon^{\alpha\beta} \mathbf{S}_{yz} \cdot \mathbf{A}_{(\beta,x)} \; [\mathrm{A}] \tag{10.16}$$

We understand here that to go from the covector $\mathbf{A}$ to the vector J^α we need the inverse of a metric.

It can be conjectured from formulas (10.15) and (10.16) that the EMC analysis of the structure introduced in Figure 10.1 involves two kinds of spaces:

- the geometrical space-time referenced with the quadruplet, (x,y,z,ct), and the considered network space which is based on the current projection J^α.

We can describe all the branches of this last space in the space–time. So we speak of driving the network in the time–space. But it is easier to keep both spaces, knowing that one is described over the other, for identifying both the field-to-branch couplings and the branch location and number. Figure 10.2 shows this process.

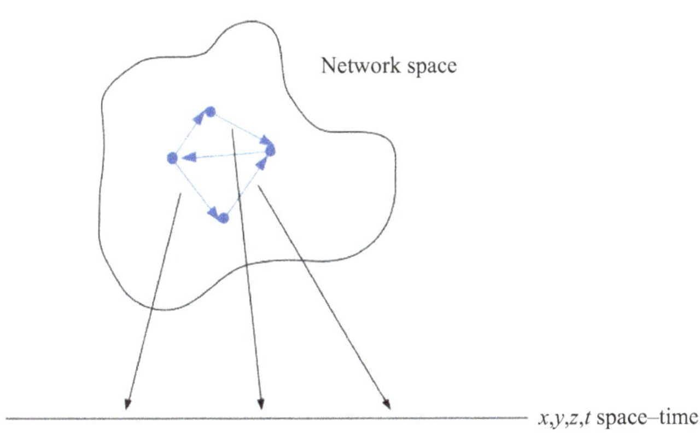

Figure 10.2 Relations between networks and space–time

10.3 Coupling to shielded cables

The shields for cables are nowadays well modellized. The transfer impedance function is the best characterization for these properties, even if its measurement remains difficult. And contrary to what is often said, the transfer impedance can be used at any frequency and the link with the shield efficiency was established a long time ago.

Transfer impedance of shielded cables is, at low frequencies and even in DC, a resistance. This resistance corresponds to the metallization resistance of the cable shield.

If the cable shield is a fulfilled conductive tube, the transfer impedance behaves like the transfer impedance of a shielded box (Figure 10.3).

At low frequencies, the voltage drop across the internal line is equal to the product of the shield resistance per the current flowing on the shield. If the cable shield does not have any apertures, when going over the skin effect cut-off frequency, no more total magnetic field is transmitted to the internal line and the level decreases

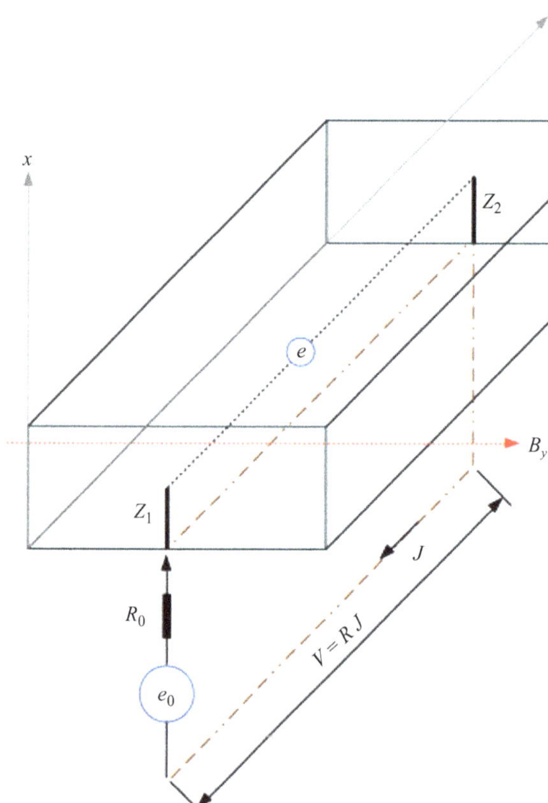

Figure 10.3 A rectangular shielded cable

until being null. So basically for a fulfilled cable shield, the transfer impedance seems like

$$z_T = R_T e^{-\sqrt{\pi f \mu \sigma \delta}} \ [\Omega] \tag{10.17}$$

where δ is the cable shield thickness. Now if there are any openers, they create a magnetic field inside the internal domain as studied in Section 9.10.5.1. In the case of shields made of braids, the skin effect drop does not exist and the transfer resistance remains constant.

For an aperture of length x, width w and a distance Y between the centred wire and the cable shield, the emf e induced is given by

$$e = -p\mu_0 \frac{ix}{2\pi} \left\{ \int_O^Y \frac{dy}{\left[y^2 + (w/2)^2\right]^{3/2}} - \int_Y^{2Y} \frac{dy}{\left[y^2 + (w/2)^2\right]^{3/2}} \right\} \ [\text{V}] \tag{10.18}$$

which gives a transfer inductance defined by

$$L_T = \mu_0 \frac{x}{2\pi} \left\{ \int_O^Y \frac{dy}{\left[y^2 + (w/2)^2\right]^{3/2}} - \int_Y^{2Y} \frac{dy}{\left[y^2 + (w/2)^2\right]^{3/2}} \right\} \ [\text{H}]\,[\text{m}] \tag{10.19}$$

In the case of a cable shield, it was shown [1] that the transfer inductance can be simply expressed by

$$L_T = \nu \frac{\alpha_m \mu_0}{\pi^2 D^2} \ [\text{H}]\,[\text{m}] \tag{10.20}$$

where ν is the number of holes, D is the cable shield diameter and α_m is given by

$$\alpha_m \approx \frac{d^3}{6} \tag{10.21}$$

where d is the hole diameter. At the end, the transfer impedance is defined by $z_T = R_T + L_T p$.

10.4 An example of a conducted source coming from an external field to harnesses coupling

We want to explain here simply using an illustration how the disturbance process is established from the external field to the component input. From the moment that the transmitted energy is conducted in the equipment, we can consider that the effect is considered as a conducted perturbation. The field coupling can come from an interaction between a harness and the field or through crosstalk coupling with another signal existing in the harness.

We consider here the first case that is the more often encountered. A shielded cable transmits a signal from a generator to the equipment under test. A load fixes the impedance on the generator side and in the equipment, the wire is connected to a load on a PCB through a shielded connector. This load is the input of a component and the whole system is exposed to a radiated field created by an antenna in the context of the electromagnetic compatibility (EMC) test.

Figure 10.4 shows the configuration considered to determine the conducted noise.

The cable shield is connected to the ground plane at each extremity. The external field injects a current on the cable shield. Through the transfer impedance of the cable shield, the noise is conducted until some input of component is available. First problem is to model the shielded cable and its interaction with the external field.

The EMC test in susceptibility is not so trivial to understand. Contrary to what may seem logical, the electric field is polarized perpendicular to the line axis and the magnetic field is near to be parallel to the line axis. It is parallel to this axis at the line centre in all cases. A way to compute the field to line coupling is to use the potential vector. It is near to be perfectly perpendicular to the EMC test ground plane. The variation of $\mathbf{A}_x$ depending on y defines the magnetic field in the z direction:

$$|\mathbf{B}_z| = \left|\frac{\partial \mathbf{A}_x}{\partial y}\right| \quad [\mathrm{T}] \tag{10.22}$$

and the emf e induced in the line is given by

$$|e| = \left|ph \int_0^Y dy \mathbf{B}_z\right| = \left|ph \int_O^Y dy \frac{\partial \mathbf{A}_x}{\partial y}\right| \quad [\mathrm{V}] \tag{10.23}$$

Due to the problem symmetry, we can write

$$|e| = 2\left|ph \int_O^{Y/2} dy \frac{\partial \mathbf{A}_x}{\partial y}\right| \quad [\mathrm{V}] \tag{10.24}$$

The emitting antenna is centred on the line centre. The potential vector is the maximum at the same point where the distance between the line and the antenna is the shortest one. After that, when looking at the extremities, the potential vector has decreased as the distance increases. For this reason, the derivative of the potential

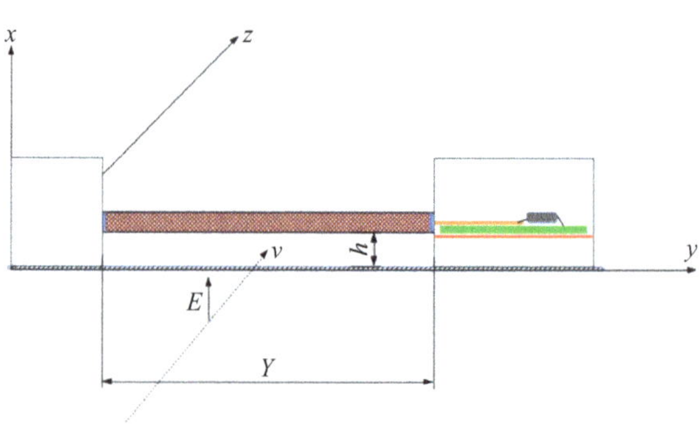

Figure 10.4 Radiated immunity EMC test

vector changes its sign on each side of the centre. So in low frequencies, the global emf induced tends to zero.

If the potential vector comes from a dipole, its expression is

$$\mathbf{A}_x(y) = \frac{\mu_0}{4\pi} \int_0^{X_a} dx \frac{i_0}{\sqrt{d^2 + y^2 + x^2}} \cos\left(n\pi \frac{x}{2X_a}\right) e^{-\frac{p}{c}\sqrt{d^2+y^2+x^2}} \ [\mathrm{W}]\,[\mathrm{m}]^{-1} \tag{10.25}$$

classical circuit consists of connecting the (+) input to zero and to link the output with the (−) input through a resistance RC. X_a is the dipole height and d the direct distance between the antenna and the line. If the dipole is of 1.5-m height, the first mode occurs for 100 MHz. So $n = 1$ for this frequency and 0 for below that. The amplitude i_0 is the current in an RLC resonator representing the antenna. By replacing (10.25) in (10.24) we can obtain the cord $e/i_0 = \beta$. We can establish the graph of our problem. Figure 10.5 shows this graph (voluntarily made without any source).

The impedance operator associated with this graph can be established knowing the impedance of the frontier between the external cable and the PCB microstrip. A simple way to fix this limit is to place a resistance of high value on the frontier. Knowing the limit of resistive load, R_L, we can write the mesh impedance as follows:

$$\zeta = \begin{bmatrix} L_1p + R_T & -Z_T & 0 \\ -Z_T & L_2p + R_T + Z_1 + R_L & -R_L \\ 0 & -R_L & L_3p + Z_2 + R_s \end{bmatrix} \ [\Omega] \tag{10.26}$$

The source covector T is defined by

$$T = \begin{bmatrix} e & 0 & 0 \end{bmatrix} \ [\mathrm{V}] \tag{10.27}$$

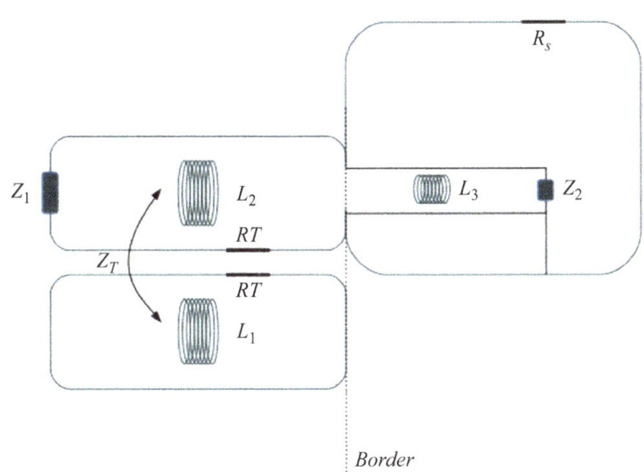

Figure 10.5 Radiated immunity EMC test graph

and to know the mesh currents, we solve

$$T_\alpha = \zeta_{\alpha\beta} J^\beta \ \text{[V]} \tag{10.28}$$

The level induced across the input impedance of the circuit Z_2 is given by $Z_2 J^3$, and the corresponding power:

$$P = \frac{1}{4}\left\{ Z_2 J^3 J^{*3} + Z_2^* J^{*3} J^3 \right\} \ \text{[W]} \tag{10.29}$$

Power and voltage are the two informations that are necessary to conclude on the disturbance risk.

We notice that the energy transmitted by the field-to-line coupling and the cable shield transfer impedance may have been transmitted by other processes. For example, it may be transmitted by a crosstalk with a power signal in a harness or again by couplings on the PCB of the equipment that communicates with the disturbed one.

Once we have evoked the process creating the noise conducted to the electronic, we can begin studying its effect.

10.5 In-band component disturbance risk

Any component can be seen as a filter, even nonlinear if needed. If we attack the circuit with a step function $\Gamma(p) = 1/p$, it answers with a filtered profile following

$$s(p) = \frac{1}{p(p+a)} \rightarrow \frac{1}{a}\left(1 - e^{-at}\right) \ \text{[V]} \tag{10.30}$$

If a signal is presented on the input of this circuit with the definition

$$e(p) = \frac{1}{p+b} \ \text{[V]} \tag{10.31}$$

the output is

$$s(t) = \frac{e^{-at}}{b-a}\left(1 - e^{(a-b)t}\right) \leftarrow \frac{1}{(p+a)(p+b)} \ \text{[V]} \tag{10.32}$$

If $b \gg a$ it gives

$$s(t) \approx \frac{e^{-at}}{b}\left(1 - e^{-bt}\right) \ \text{[V]} \tag{10.33}$$

As a is sufficiently small, this output is near to be the input signal. While in the case of $a \gg b$, the output approaches

$$s(t) \approx \frac{e^{-bt}}{a}\left(1 - e^{-at}\right) \ \text{[V]} \tag{10.34}$$

In the first case, the circuit transmits the signal presented to its input. But if the input signal is too fast, the output is the intrinsic circuit response. The in-band assumption is the case where $b \gg a$.

The in-band disturbance is also called jamming. Its study is realized as a classical signal-to-noise ratio study. If the impact of the noise can be relatively easily analysed, the question of the disturbance criterion remains difficult to define. Ideally, it should be linked with the electronic functionality. For example, in the case of a transmission bus, the criterion can be a bit error rate (BER).

It is not always simple to link the component sensibility to a common criterion not depending on the application. That is why if we know the relations between a simple hardware criterion and the functional disturbance, it is better to characterize the component versus this criterion. It will be simplest after this characterization in various different applications. That is why we will prefer to measure, for example, the probability of modifying bits rather than acquiring directly the BER. But sometimes it is too much difficult to realize this link between a susceptibility observable and the functional parameters. In that case the characterization is based on the functional disturbance.

For the basic hardware criterion, we can discuss canonical cases such as the digital circuits or analogue circuits.

10.5.1 Digital circuits

On any digital circuit, we have two thresholds: one low threshold V_{iL} and the other high threshold V_{iH}. Between these two levels, the circuit behaviour is probabilistic (deterministically unknown). We can use the same sigmoid functions as previously to describe this probabilistic input V_i:

$$V_i = \left[1 + e^{-(u-V_{iH})}\right]^{-1} \cdot H + \left(1 - \left[1 + e^{-(u-V_{iL})}\right]^{-1}\right) \cdot L \text{ [V]} \tag{10.35}$$

By defining both probabilities of a low input (L) and high input (H) through a probability, we define completely the digital circuit behaviour. Once an input digital circuit is modelled, it is out of question to model the whole digital circuit. The rest of the circuit is modelled using behavioural function. By this combination, it becomes possible to model a whole digital circuit but keeping an indispensable simplicity for making the modellization reachable. Some particular inputs are modellized, which represent all similar inputs.

One input susceptibility characterized gives the way to understand the whole circuit susceptibility. Sometimes it is necessary to consider simultaneous disturbances on more than one input in order to explain the functional disturbance. Anyway, the approach remains the same in all cases.

10.5.2 Analogue circuits—operational amplifiers

One major difference between analogue circuits is that they work in differential mode. In fact, the disturbance must be computed for the differential mode. We can take the example of an operational amplifier (OA). By definition of their working principle, their signal outputs s are defined by

$$s = A_0 \left(u_p - u_m\right) \text{ [V]} \tag{10.36}$$

u_p being the positive input and u_m the negative input.

A_0 is the amplifier gain that is assumed as very high value. The simplest way to model this function is to apply both input signals to the same mesh, in opposite senses. At the same time we want to control the input impedances. A solution is to use two meshes as input stages and to couple them with the differential mesh through unidirectional cords. If the impedance of this last mesh is 1 Ω, the current in the mesh is

$$J = \frac{u_p - u_m}{1} \text{ [A]} \tag{10.37}$$

A last mesh of output impedance equal to the output impedance of the OA is coupled with the previous one through the cord A_0. The whole graph associated with the OA is shown in Figure 10.6.

The impedance operator associated with this graph is

$$\zeta = \begin{bmatrix} z_g + \frac{R_i}{R_iC_ip+1} & 0 & 0 & 0 \\ 0 & z_d + \frac{R_i}{R_iC_ip+1} & 0 & 0 \\ 1 & -1 & G_i & 0 \\ 0 & 0 & A_0 & R_o + R_s \end{bmatrix} [\Omega] \tag{10.38}$$

where z_g and z_d are the common mode output impedances of the differential mode signal. R_s represents the load.

A classical mounting consists of connecting the (+) input to zero and to link the output with the (−) input through a resistance RC. Figure 10.7 shows a typical amplifier mounting.

As the first positive input is short-circuited, the impedance operator of this circuit can be reduced to the last four meshes, including the reaction one:

$$\zeta' = \begin{bmatrix} \beta & 0 & 0 & -z_d \\ -1 & G_i & 0 & 0 \\ 0 & A_0 & R_{os} & -R_s \\ -z_d & 0 & -R_s & z_{dsc} \end{bmatrix} [\Omega] \tag{10.39}$$

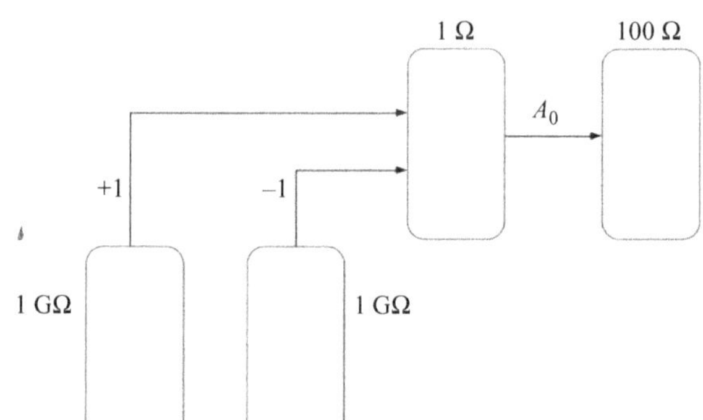

Figure 10.6 Operational amplifier graph

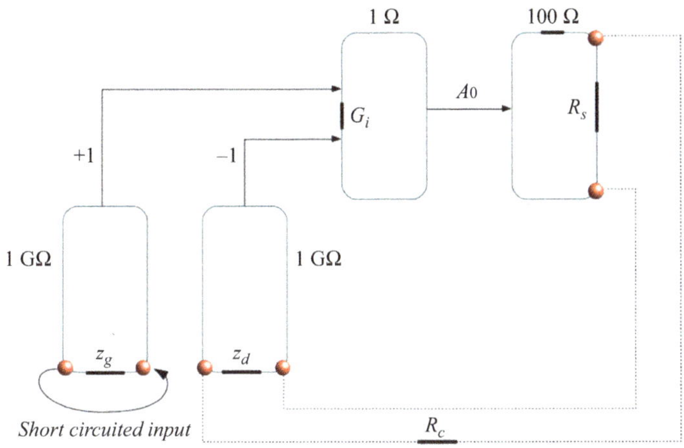

Figure 10.7 Amplifier mounting

with $\beta = R_i/(R_iC_ip + 1)$, $R_{os} = R_0 + R_s$ and $z_{dsc} = z_d + R_c + R_s$. We can inverse formally this matrix using Maxima software, for example, to obtain the matrix shown in Figure 10.8.

We know that there is just the signal input generator in the circuit. So $T_1 = u_i$ and $T_\alpha = 0, \ \forall\alpha \neq 1$. We can write

$$J^3 = \left(\zeta'_{13}\right)^{-1} T_1 = y^{31} T_1 \ \ [\mathrm{A}] \tag{10.40}$$

and coming from Maxima,

$$y^{31} = \frac{G_iR_sz_d - A_0z_{dsc}}{G_i\left(R_{os}z_{dsc} - R_s^2\right)b - z_d\left(G_iR_{os}z_d - A_0R_s\right)} \ \ [\mho] \tag{10.41}$$

The output voltage is R_sJ^3. With $R_c \gg R_s, z_d$ we find

$$v_s = R_sJ^3 = -\frac{R_c}{z_d} \ \ [\mathrm{V}] \tag{10.42}$$

which is the well-known approximated relation for this setup.

Now as we have understood how the model works, we can use it for any application involving an OA. In particular, we can use it for modelling a comparator. In this case, the source covector has two components different from zero: the two OA inputs. Looking for the fourth mesh current, we find (always using Maxima for inverting the matrix given in (10.38)):

$$R_sJ^4 = y^{41}T_1 + y^{42}T_2 = \begin{cases} \frac{A_0}{G_iR_{os}b}\left(u_p - u_m\right) \ \text{if} \ < V_{cc} \\ V_{cc} \ \text{if} \ > V_{cc} \end{cases} \tag{10.43}$$

where V_{cc} is the power supply of the OA.

$$
\begin{bmatrix}
\frac{Gi\,(Ros\,Zdsc - Rs^2)}{Gi\,(Ros\,Zdsc - Rs^2)\,b - zd\,(Gi\,Ros\,zd - A_0\,Rs)} & -\frac{A_0\,Rs\,zd}{Gi\,(Ros\,Zdsc - Rs^2)\,b - zd\,(Gi\,Ros\,zd - A_0\,Rs)} & \frac{Gi\,Rs\,zd}{Gi\,(Ros\,Zdsc - Rs^2)\,b - zd\,(Gi\,Ros\,zd - A_0\,Rs)} & \frac{Gi\,Ros\,zd}{Gi\,(Ros\,Zdsc - Rs^2)\,b - zd\,(Gi\,Ros\,zd - A_0\,Rs)} \\
\frac{Ros\,Zdsc - Rs^2}{Gi\,(Ros\,Zdsc - Rs^2)\,b - zd\,(Gi\,Ros\,zd - A_0\,Rs)} & \frac{(Ros\,Zdsc - Rs^2)\,b - Ros\,zd^2}{Gi\,(Ros\,Zdsc - Rs^2)\,b - zd\,(Gi\,Ros\,zd - A_0\,Rs)} & \frac{Rs\,zd}{Gi\,(Ros\,Zdsc - Rs^2)\,b - zd\,(Gi\,Ros\,zd - A_0\,Rs)} & \frac{Ros\,zd}{Gi\,(Ros\,Zdsc - Rs^2)\,b - zd\,(Gi\,Ros\,zd - A_0\,Rs)} \\
\frac{Gi\,Rs\,zd - A_0\,Zdsc}{Gi\,(Ros\,Zdsc - Rs^2)\,b - zd\,(Gi\,Ros\,zd - A_0\,Rs)} & \frac{A_0\,zd^2 - A_0\,Zdsc\,b}{Gi\,(Ros\,Zdsc - Rs^2)\,b - zd\,(Gi\,Ros\,zd - A_0\,Rs)} & \frac{Gi\,Zdsc\,b - Gi\,Zd^2}{Gi\,(Ros\,Zdsc - Rs^2)\,b - zd\,(Gi\,Ros\,zd - A_0\,Rs)} & \frac{Gi\,Rs\,b - A_0\,zd}{Gi\,(Ros\,Zdsc - Rs^2)\,b - zd\,(Gi\,Ros\,zd - A_0\,Rs)} \\
\frac{Gi\,Ros\,zd - A_0\,Rs}{Gi\,(Ros\,Zdsc - Rs^2)\,b - zd\,(Gi\,Ros\,zd - A_0\,Rs)} & -\frac{A_0\,Rs\,b}{Gi\,(Ros\,Zdsc - Rs^2)\,b - zd\,(Gi\,Ros\,zd - A_0\,Rs)} & \frac{Gi\,Rs\,b}{Gi\,(Ros\,Zdsc - Rs^2)\,b - zd\,(Gi\,Ros\,zd - A_0\,Rs)} & \frac{Gi\,Ros\,b}{Gi\,(Ros\,Zdsc - Rs^2)\,b - zd\,(Gi\,Ros\,zd - A_0\,Rs)}
\end{bmatrix}
$$

Figure 10.8 ζ^{-1}

If both input signals come to the OA through cables then microstrips, we can estimate what is the level of noise transmitted on each OA input. It means that rather than applying the functional signals on the input, we carry forward the noise coupled on the cables outside the equipment.

It is clear that if the two-wire line is perfectly symmetric, both emfs induced on the wires are perfectly equal and $u_p - u_m = 0$. Unfortunately, there are always differences between the lines, and the induced emfs are not perfectly compensated. Let us consider, for example, the induction coming from the bulk current injection EMC test (BCI test) on equipment. We connect our OA to two lines going on the other side to two loads associated with some sensors for example. The BCI test is not so simple to simulate. Rather than considering the whole OA, we can simply consider the two input impedances. After that, we will be able to postpone on the OA circuit the voltage obtained across the loads.

Figure 10.9 shows a typical BCI test setup. R_l are the input impedances of the sensors. R_a are the input impedances of the OA. e_0 and R_0 are the characteristics of the generator source of the current injected on the lines, while R_m is the output impedance of a second current sensor used to measure the injected current and associated with the ferrite M_2. The ferrite M_1 is employed to transmit the power coming from the generator to the lines. It is used as a transformer.

To simplify our illustration, we consider only the magnetic coupling between the two lines (see Chapter 2 for more details on the complete coupling). The system is made (under this assumption and for a limited frequency band) of four meshes: one mesh for each transformer and two meshes for the two lines. We can easily write the impedance operator:

$$\zeta = \begin{bmatrix} R_0 + L_1 p & -M_1 p & -M_1 p & 0 \\ -M_1 p & b_g + b_d + L_c p + R_G + (L_1 + L_2)p & -M_{II} p & -M_2 p \\ -M_1 p & -M_{II} p & b_g + b_d + L_c p + R_G + (L_1 + L_2)p & -M_2 p \\ 0 & -M_2 p & -M_2 p & R_m + L_2 p \end{bmatrix} [\Omega] \tag{10.44}$$

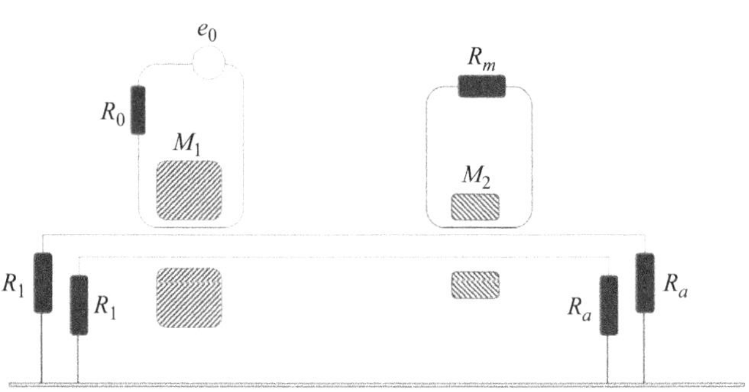

Figure 10.9 BCI EMC test

with

$$\begin{cases} b_g = \frac{R_l}{R_l(C_g/2)p+1} \\ b_d = \frac{R_a}{R_a(C_g/2)p+1} \end{cases} \tag{10.45}$$

where R_G is the ground resistance, L_c the line inductance and C_g the line capacitance. M_{ll} is the mutual inductance between the two lines.

In the test, we increase the generator level until the current measured on the second transformer reaches a defined value. The measurement is given by $R_m J^4$. The voltage obtained is proportional to the line currents. This ratio is linked with the transformer windings ratio on the primary and the secondary. We suppose here that $J^4 = J^2 + J^3$. At each frequency between 10 kHz and 100 MHz, we increase the generator level for reaching an induced current of 100 mA (typical current limit required by the EMC standards). If the current cannot be reached because the harness impedance is too high, another criterion limits the source power to twice or four times the power needed to reach the required current on 50-Ω load.

Listing given in Annexe 10.A in Python makes this kind of computation with a simple control.

It leads to the curves given in Figure 10.10. The current coming from the difference of the two output currents on the lines is null. Numerical errors give a noisy current near to zero.

The result is null because the system is perfectly symmetric. A small dissymmetry can be sufficient to change radically this result. If we consider, for example, an added capacitor associated with the ground plane of a PCB located above the mechanical reference, the order of value of this kind of capacitors is 10 pF. If we add this impedance

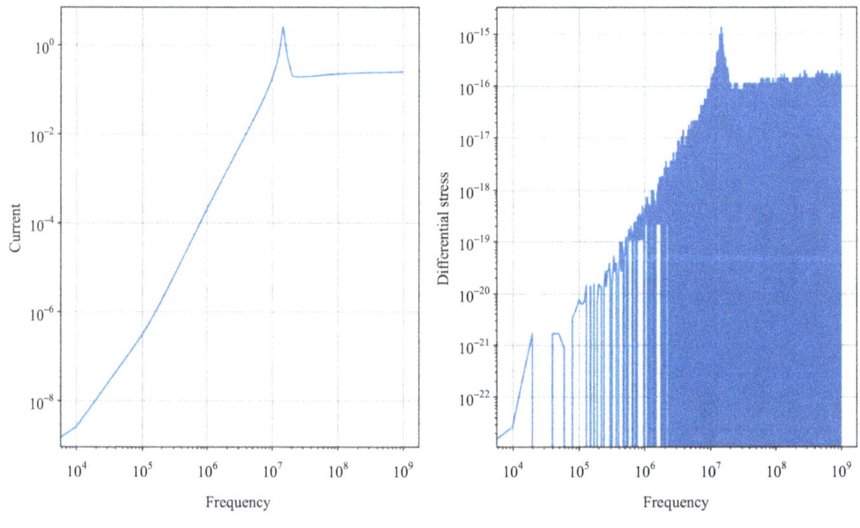

Figure 10.10 BCI EMC test on symmetric lines

to the input impedance on one line only, we obtain the new curves presented in Figure 10.11.

In that case the stress can reach more than 100 mA. This example shows us that the robustness of an analogue function against conducted disturbances goes principally by ensuring the symmetry in the impedances. Many errors are often made for that through the PCB layout management. An electrical plan layer can be, for example, a bad choice, creating this capacitance with the metallic reference which is at the base of dissymmetry. It can be preferable to use electrical 0-V tracks over a reference plan layer, the zero volt being a signal among others.

There are two criteria to define the failure threshold of components: a criterion in power and the other in voltage. The power that can be accepted by a component can be applied through various waveforms. Depending on this waveform, the component answer is not the same. When the constraint is long, the maximum power remains constant with time. If the constraint is very short, then the maximum power varies with the inverse of the time. If t is the duration of power application and if P_x is the maximum power that can be accepted by the component, we have

$$P_x = P_0\left(e^{-\alpha t} + \beta\right) \quad [\mathrm{W}] \tag{10.46}$$

This model says that for a very short duration, the applying power can be very high. But in fact this model can lead to voltages higher than the component breakdown voltage. So this equation must be accompanied by a second one pointing out the maximum voltage acceptable V_x:

$$\begin{cases} P_x = P_0\left(e^{-\alpha t} + \beta\right) \\ V < V_x \end{cases} \tag{10.47}$$

Under pulses of 1 μs of duration, the energies for degradations of usual components are (in μJ)

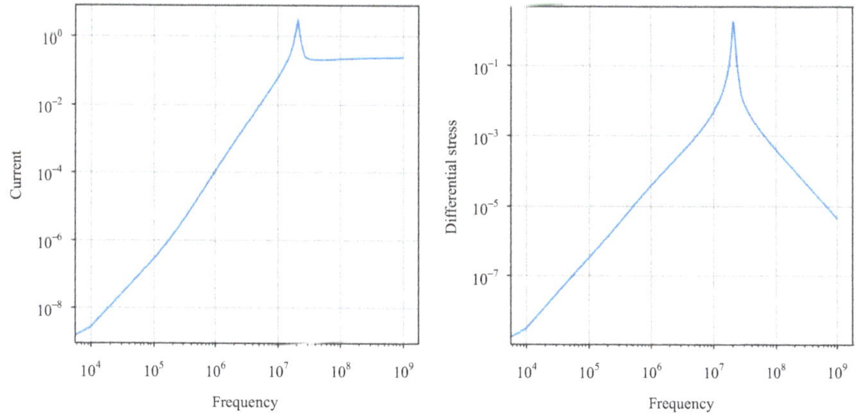

Figure 10.11 BCI EMC test on dissymmetric lines

- integrated circuit: 1
- low-power transistor: 10
- resistors (1/4 W): 10^4

10.6 Transmission to the component through the PCB

Between the outside world and the component, the PCB plays the role of an interface [2]. Once the component model is available, a synthetic model includes the external coupling mechanism, the microstrips of the PCB and the component as load. When an electrical, electronic architecture is studied, it reaches the system requirements. The requirements are enclosed in the EMC standard for the system. In this standard exists in general the BCI test and the radiated immunity test. These two tests are the sources for computations of conducted immunity. As they are well defined and quite simple, it is often very interesting to study the EMC risk in conducted immunity through these test setups rather than trying to take into account the whole system. The standard is, for the equipment supplier, the EMC reference for conducted immunity (and for other tests also).

It is important to consider the PCB interface because it creates resonances in high frequencies and change the impedance seen by the component in the whole frequency band. Topologically, we can model this structure by three submatrices. One submatrix for the outside world M_o. One of the rest of two submatrices is considered for the PCB M_p and one submatrix for the component M_c. The whole problem is defined by $M_o \oplus M_p \oplus M_c$ plus the interactions at the boundaries between these elements.

The border limit between the outside harness and the internal microstrip is often the connector. If this connector is long enough and mismatched, we must add a fourth element: this connector and its impedance operator M_n. In the case of very short connectors or direct connection, the border limit is an LC circuit. The inductance represents the length of the connection and the capacitor represents the vis-a-vis between the two wires of connection. The connector, for the condition of low ratio length over wavelength, uses a similar equivalent circuit. Finally, we can consider next matrices for our four operators:

$$\begin{cases} M_o = \begin{bmatrix} R_0 + z_c & (\frac{1}{C/2p} - z_c)e^{-\tau p} \\ (R_0 - z_c)e^{-\tau p} & \frac{1}{C/2p} + z_c \end{bmatrix} \\ M_n = \begin{bmatrix} L_p + \frac{2}{C/2p} \end{bmatrix} \\ M_p = \begin{bmatrix} \frac{1}{C/2p} + \eta_c & (z_i - z_c)e^{-\tau p} \\ (\frac{1}{Cp} - z_c)e^{-\tau p} & z_i + z_c \end{bmatrix} \\ M_c = \begin{bmatrix} z_i \end{bmatrix} \end{cases} \tag{10.48}$$

When we make the direct summation $M_o \oplus M_n \oplus M_p \oplus M_c$, the addition operation must take into account the interactions between the elements. The frontier between the outside line and the connector or between the connector and the microstrip is the capacitor $C/2$, while between the microstrip and the component it is represented by z_i. The operator ζ is then constructed like

$$\zeta = \begin{bmatrix} [M_o] & -\frac{1}{C/2p} = \zeta_{23} & [0] & [0] \\ -\frac{1}{C/2p} = \zeta_{32} & [M_n] & -\frac{1}{C/2p} = \zeta_{34} & [0] \\ [0] & -\frac{1}{C/2p} = \zeta_{43} & [M_p] & -z_i = \zeta_{56} \\ [0] & [0] & -z_i = \zeta_{65} & [M_c] \end{bmatrix} [\Omega] \quad (10.49)$$

The source comes from the transformer in the BCI test or from the field to line coupling in the radiated immunity test. It affects both meshes associated with the bran in which models the outside harness. The source covector is

$$T = \begin{bmatrix} e_1 & e_2 & 0 & 0 & 0 & 0 \end{bmatrix} \ [\mathrm{V}] \quad (10.50)$$

where e_1 and e_2 are computed from the physical description given in Figure 10.9.

It remains just to solve the system $T_\alpha = \zeta_{\alpha\beta} J^\beta$. The power across the component is defined by

$$P = \frac{1}{4}\left(z_i^* J^{4*} J^4 + z_i J^4 J^{4*}\right) \ [\mathrm{W}] \quad (10.51)$$

10.7 The failure risk

The source level *e* is always distributed between two extremes that we can estimate. If W_e is the average power value, we have to express, we can write the probability $P(e)$:

$$P(e) = 1 - \left[1 + e^{-\alpha(W - W_E)}\right]^{-1} \quad (10.52)$$

The same exercise can be conducted for the component susceptibility probability $P(c)$, around its average failure value in power P_c:

$$P(c) = \left[1 + e^{-\beta(W - P_c)}\right]^{-1} \quad (10.53)$$

The coefficients α and β give the distributions slopes stiffness, allowing one to set the minimum and maximum values pointed out by the functions.

The risk $\mathscr{R}$ is then given by the product $P(e)P(c)$:

$$\mathscr{R}(W) = \left(1 - \left[1 + e^{-\alpha(W - W_E)}\right]^{-1}\right)\left(\left[1 + e^{-\beta(W - P_c)}\right]^{-1}\right) \quad (10.54)$$

10.8 Out-band component disturbance risk

Nonlinear behaviour transforms high-frequency energies in the low-frequency one. Once a part of the transported energy coming from a parasitic source partly projected in the working frequency domain of a circuit, this circuit is disturbed following the same mechanism studied in the in-band disturbances.

The out-band disturbance is more difficult to predict because, by definition, the component datasheets do not explore this abnormal functioning. A typical example is a diode. If we power supply the diode with a sinusoïdal waveform $A \sin(\omega t)$, the current across the diode can be approximated by $A/2\,[1 + \sin(\omega t)]$. The DC component of this transformed waveform can excite a low-frequency circuit while the sinusoïdal component cannot.

In the case of signals coming from communications, the sinusoïdal waveform is modulated by a low-frequency signal. The nonlinear behaviour acts as a detector and receives a part of this modulation signal. That is through this mechanism that a smartphone can disturb the audio function of a radio or of an hi-fi chain.

As we do not know the transformation operator from high to low frequencies, we must imagine a simple technique to evaluate the transmitted level. One technique supposed here consists of taking a detector located in front of the low-frequency function. It is often the case because electronic components are equipped with electrostatic discharge protections which can be the case of diodes. A way to compute the level detected and transmitted to the low-frequency circuit means to use a diode and a capacitor. Then the demodulated signal is used to understand the out-band disturbance. Let us consider a simple digital circuit. It can be modelled by an RC circuit. If we excite this circuit by a high-frequency source (e, R_e) we have

$$u_i = \frac{1}{R_e}\left(p + \frac{1}{\beta C}\right)^{-1} \quad \beta = \frac{R_e R}{R + R_e} \quad [\mathrm{V}] \tag{10.55}$$

Bode's diagram of this transfer function is a low-pass first-order filter with a cut-off frequency at $\omega_c = (\beta C)^{-1}$. The threshold of the circuit is $u_t = (5R_e)^{-1}$. If we attack this circuit with a modulated sinusoïdal at main frequency $\omega_0 = 10^3\omega_c$ and the same amplitude ($e = u_t$), the input residual level transmitted to the circuit is $u_t/100$ which does not disturb the circuit. Now we place a diode mounted in inverse to act as a limiter and protect the circuit from high-voltage pulses like electrostatic discharges. In positive it limits the voltage to 10 V and in negative to −0.1 V. By neglecting this last level to fix it at 0 V, an $A = 5$ V sinusoïdal waveform is transformed to a signal similar to

$$v(t) = \frac{A}{2}\,[1 + \sin(\omega_0 t)] \quad [V] \tag{10.56}$$

The DC component of this signal is $A/2$ which is higher than the circuit threshold. And this DC component is not attenuated by the filter neither the inductance that limits the high frequency decreasing from the capacitance filtering. For a positive signal, if R_H is the diode impedance in blocked mode, the impedance operator becomes

$$\zeta = \begin{bmatrix} R_H & -R_H \\ -R_H & R_H + \gamma + Lp \end{bmatrix} \quad [\Omega] \tag{10.57}$$

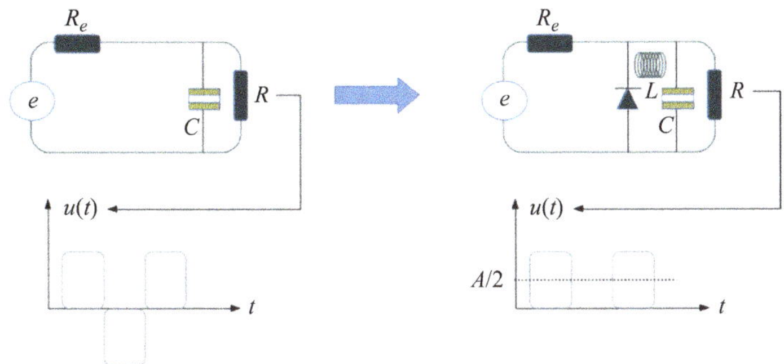

Figure 10.12 Out-band detection mechanism

with $\gamma = R/(RCp + 1)$. The current transmitted to the load is

$$J^2 = [\gamma + Lp]^{-1} e \ \text{[A]} \tag{10.58}$$

As $J^e = e/R_e$, the detection efficiency (d.e.) is defined by

$$d.e. = \frac{R_e}{\gamma + Lp} \tag{10.59}$$

So, the circuit response to the DC component must be considered notably with a functional and desired signal. For example, Figure 10.12 illustrates this mechanism of the signal DC component effect.

In the absence of any information on the studied circuit behaviour in front of out-band stress, we propose to use this subterfuge in order to estimate the out-band disturbance risk. The ratio between the sinusoïde amplitude and the in-band signal amplitude is called the d.e. The higher the d.e., the higher the disturbance risk.

Starting from the power distribution of the source $P(e)$, we multiply it by the d.e. in order to compute the failure probability for an out-band context. Then we make the product with the threshold probability $P(c)$ as before.

Remember that the source amplitude is defined by the standard. The front circuit coming from the microstrip (and connector, etc.) as an interface between the outside world and the component influence the out-band energy transfer to the component. Using the same method as previously we can easily construct the whole problem operator, just adding a diode and the high-frequency impedances required by the out-band spectrum. The high-frequency components can be estimated from the circuit dimensions and PCB technology.

10.9 Radioreceptor circuits

Disturbances of radio receptors can come from two ways as seen before. If the stress arrives in the working frequency (in-band disturbance), we speak of jamming. If the stress arrives out-band, we speak of electromagnetic interference and immunity.

10.9.1 In-band radioreceptor disturbances

If a signal $s(t)$ is modulated in amplitude, its form is

$$s(t) = A_0(t)\cos(\omega t) \quad [\mathrm{V}] \tag{10.60}$$

The receptor uses a nonlinear component to extract the modulation $A_0(t)$ from the carrier frequency:

$$o(t) = \alpha s^2(t) = \frac{A_0(t)^2}{2}[1 + \cos(2\omega t)] \leftarrow \frac{A_0(p)^2}{2} \times \left[1 + \frac{p}{p^2 + 4\omega^2}\right] \quad [\mathrm{V}] \tag{10.61}$$

Using a low-pass filter of cutting frequency f_c we can suppress all the frequencies superior to $10f_c$. If $f_c = \omega/(20\pi)$, we keep only the modulation after the filter. This is written as

$$m(p) = \left(\frac{A_0(p)^2}{2} \times \left[1 + \frac{p}{p^2 + 4\omega^2}\right]\right)\frac{\beta}{p + \omega_c} \quad [\mathrm{V}] \tag{10.62}$$

Now if a narrowband signal b comes to disturb the carrier frequency through couplings on the PCB, we can translate this by writing

$$m(p) = \left(\frac{A_0(p)^2 + b^2}{2} \times \left[1 + \frac{p}{p^2 + 4\omega^2}\right]\right)\frac{\beta}{p + \omega_c} \quad [\mathrm{V}] \tag{10.63}$$

This noise adds a constant signal to the modulation. If the ratio b/A_0 is too much low, the information included in $A_0(t)$ is lost.

Now we can have another sort of noise. If we inject a signal defined by

$$b(t) = b\sin(\omega_b t)\cos(\omega t) \quad [\mathrm{V}] \tag{10.64}$$

The filter output gives both desired and undesired components:

$$m(p) = \frac{A_0(p)^2}{2} + \frac{b}{2}\frac{\omega_b}{p^2 + \omega_b^2} \quad [\mathrm{V}] \tag{10.65}$$

In that case the signal is mixed with the noise and this makes its reception less intelligible. The mix of the two kinds of disturbances leads to the following expression:

$$m(p) = \left(\left[\frac{A_0(p)^2}{2} + b + \frac{b'}{2}\frac{\omega_b}{p^2 + \omega_b^2}\right] \times \left[1 + \frac{p}{p^2 + \omega^2}\right]\right)\frac{\beta}{p + \omega_c} \quad [\mathrm{V}] \tag{10.66}$$

This kind of argument can be used for any modulation type.

10.9.2 Out-band radioreceptor disturbances

Out-band means that the perturbator has frequencies superior to $10\omega_c$.

The signal will not be demodulated, so basically no fake information will be transmitted to the users. But the power dissipated by the filter can exceed to the limit, as an unwanted energy is added to the nominal one. So in that case the out-band signal can provoke some failure in the receptor. Often during the EMC test of receptor, immunity is verified for the out-band frequencies.

Having seen the reasons of disturbance of the radio receptor, we can address some mechanisms through which these couplings occur on the PCB.

10.9.3 Sources of disturbances of radio receptor on the PCB

We study conducted immunity. Our problem concerns the transmission of noise from an outside coupling to an internal radio-frequency function. This transmission can occur through couplings on the PCB. We consider here high-frequency couplings, as the radio-frequency functions work at high frequencies. But as we are interested in conducted immunity, we know that we cannot go higher than 100 MHz typically and 1 GHz exceptionally. Anyway, we consider 1 GHz maximum working frequency.

The couplings can be as follows:

- between neighbour microstrips;
- through the equipment cavity;
- between different layers.

Couplings between layers can be solved using similar equivalent *RLC* circuits, including vias, etc. This models family is simple and accurate. However, any tentative of modelling is doomed to fail if there are no efforts on the PCB design and the layers management.

10.9.3.1 Couplings through layers

Figure 10.13 shows the configuration we considered where the coupling can occur. For a height h and a radius a, the inductance associated with the via is given by [3]

$$L_{via} = 0.2\left[h - \log\left(\frac{h + \sqrt{h^2 + a^2}}{a}\right) + \frac{3}{2}\left(a - \sqrt{a^2 + h^2}\right)\right] \text{ [H]} \qquad (10.67)$$

Two vias neighbourings, one wearing the signal coming from the outside harness and the other wearing the radio-frequency function, can couple each other. It remains to estimate the mutual inductance between the two vias.

The mutual inductance M is defined by

$$M = \frac{\mu_0}{4\pi}\int_{h_1}\int_{h_2}\frac{d\mathbf{h}_1 \cdot d\mathbf{h}_2}{R_{12}} \text{ [H]} \qquad (10.68)$$

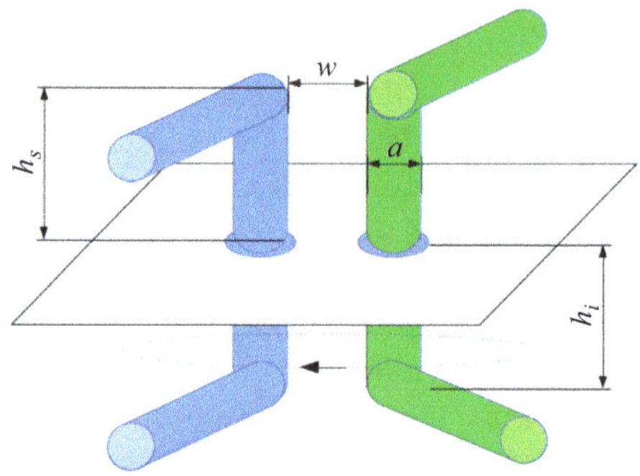

Figure 10.13 Coupling between vias

In our case, this gives

$$M = \frac{\mu_0}{4\pi}\int_0^{h_1}\int_0^{h_2}\frac{dh_1 \cdot dh_2}{\text{ffl}\sqrt{w^2 + (h_2 - h_1)^2}} \ [\mathrm{H}] \tag{10.69}$$

If two microstrips arrive to the vias, we can model them using Branin's models. The impedance operator with the coupling through the vias is

$$\zeta = \begin{bmatrix} z_c + R_0 & L_{via}pe^{-\tau p} & 0 & 0 \\ (R_0 - z_c)e^{-\tau p} & z_c + L_{via}p + z_c & 0 & -Mp \\ 0 & 0 & z_c + z_0 & L_{via}pe^{-\tau p} \\ 0 & -Mp & (z_0 - z_c)e^{-\tau p} & z_c + L_{via}p + z_c \end{bmatrix} \ [\Omega] \tag{10.70}$$

Each microstrip sees the via as one extremity, in series with the suite of the line. That is why one mesh has for impedance $L_{via}p + 2z_c$. The impedance of the radio-frequency function is z_0. Our question is: what is the noise amplitude transmitted to the radio-frequency function if a noise generator power supplies the resistance R_0 (e.g. noise coming from the conducted immunity test setup)? It means that the source vector is given by

$$T = \begin{bmatrix} e_0 & 0 & 0 & 0 \end{bmatrix} \ [\mathrm{V}] \tag{10.71}$$

Noting

- $a = z_c + R_0$
- $b = z_c + z_0$
- $RL = z_c + L_{via}p + z_c$
- $\gamma_v = L_{via}pe^{-\tau p}$
- $\gamma_R = (R_0 - z_c)e^{-\tau p}$
- $\gamma_z = (z_0 - z_c)e^{-\tau p}$

the noise measured on the radio-frequency function load z_0 is obtained inverting ζ with the Maxima giving

$$y = \frac{1}{\Delta}\begin{bmatrix} RL(RLb-\gamma_v\gamma_z)-Mp^2b & \gamma_v(RLb-\gamma_v\gamma_z) & Mp\gamma_v\gamma_z & Mpb\gamma_v \\ \gamma_R(RLb-\gamma_v\gamma_z) & a(RLb-\gamma_v\gamma_z) & Mpa\gamma_z & Mpab \\ Mp\gamma_R\gamma_v & Mpa\gamma_v & (RL^2-Mp^2)a-RL\gamma_R\gamma_v & \gamma_R\gamma_v^2-RLa\gamma_v \\ Mpb\gamma_R & Mpab & \gamma_R\gamma_v\gamma_z-RLa\gamma_z & RLab-b\gamma_R\gamma_v \end{bmatrix} [\mho] \tag{10.72}$$

with $\Delta = a(RL(RLb-\gamma_v\gamma_z)-Mp^2b)-\gamma_R\gamma_v(RLb-\gamma_v\gamma_z)$.

The noise η^3 is given by $\eta^3 = y^{3\sigma}T_\sigma = \Delta^{-1}Mp\gamma_R\gamma_v e_0$. So

$$\eta^3 = \frac{Mp\gamma_R\gamma_v e_0}{[RLb-\gamma_v\gamma_z]\,[aRL-\gamma_R\gamma_v]-aMbp^2}\ [\mathrm{A}] \tag{10.73}$$

The via structure is involved through the impedances RL and γ_v. We can study what is the influence of the via on the noise diffusion to the radio-frequency function. To do that, we calculate the partial derivative of η^3 versus RL and γ_v. Let us take the case of RL:

$$\frac{\partial\eta^3}{\partial RL} = -2\frac{abRLM\gamma_R\gamma_v p}{[RLb-\gamma_v\gamma_z]\,[aRL-\gamma_R\gamma_v]-aMbp^2}e_0\ [\mathrm{A}]\,[\mho] \tag{10.74}$$

In high frequency, this expression tends to α/L_{via}.

We have seen through this example that the tensorial analysis of networks gives the way to determine theoretically the parameters influence. After what, numerical applications can be conducted to cipher these influences.

10.9.3.2 Couplings through the equipment cavity

Coupling through cavities remains a difficult problem. If we want to cover a wide frequency band, it becomes necessary to use Branin's models adapted to the cavities modelling. As we have already seen, the first job consists of defining modes in a known empty volume, enclosed in conductive walls. These modes describe the wave surface propagating in a given direction of the cavity. The cavity is seen as a short-circuited waveguide. Then this leads to the dispersion of the waveguide which gives the propagation speed, then the wave impedance. The wave impedance finally gives the characteristic impedance of the waveguide and at this step, equivalent Branin's model of the cavity is available. It remains to determine the coupling functions with the sources embedded in the cavity.

Modes

For a PCB enclosed in a metallic box, the wave surface is described by the right section of the box, giving a rectangular profile with the PCB slice in its centre. There are two configurations similar giving the same right section seen from both sides of the box. If y is the direction of the height, z the direction of propagation and θ the curvilinear direction turning around the PCB, we can define the electric field spatial distribution by

$$\mathbf{E} = E_0 \cos\left(n\pi \frac{\theta}{\Theta}\right) e^{-\frac{z}{c}p} \mathbf{u}_y \tag{10.75}$$

The electric field must respect d'Alembert's equation (also known as Helmholtz's equation):

$$\frac{\partial^2 \mathbf{E}}{\partial y^2} + \frac{\partial^2 \mathbf{E}}{\partial \theta^2} + \frac{\partial^2 \mathbf{E}}{\partial z^2} + k^2 \mathbf{E} = 0 \tag{10.76}$$

Replacing the electric field by its expression we find

$$k = \sqrt{\left(\frac{\omega}{c}\right)^2 + \left(\frac{n\pi}{\Theta}\right)^2} \tag{10.77}$$

which gives us

$$v = \frac{\omega}{k} \tag{10.78}$$

The wave impedance is $\mu_0 v$ but it gives E/H while we want V/i. Knowing

$$V = \int_y d\mathbf{y} \cdot \mathbf{E} = YE \tag{10.79}$$

and

$$i = \oint_{[0\cdots\Theta]} d\theta \cdot \mathscr{H} \approx 2\Theta\mathscr{H} \tag{10.80}$$

we obtain

$$\eta_c = \frac{V}{i} = \mu_0 v \frac{Y}{2\Theta} \tag{10.81}$$

At this step, we have all the information to construct Branin's model representing the cavity, i.e. η_c and v. It remains to couple the source and target to the cavity fields. In this last exercise, it may be easier to compute the coupling of the cavity fields into the source or target circuits. As the coupling function is symmetric, this does not change the result.

These coupling interfaces are always described in near fields. If we consider a magnetic field B_θ and a microstrip conducting the noise coming from the outside, the emf induced in the microstrip of height h and length z_m is

$$e_m(z) = -\int_{z=a}^{b} dz h p B_\theta(z) \tag{10.82}$$

and we estimate as an example for the first mode (TEM one)

$$B_\theta(z) = \frac{\mu_0}{2\Theta} i^n(z) \tag{10.83}$$

which leads to the coupling impedance ζ_{mn}:

$$\zeta_{mn} = \frac{e_m(z)}{i^n(z)} = -h(b-a)p\frac{\mu_0}{2\Theta} \tag{10.84}$$

A similar coupling impedance can be computed between the cavity and the target microstrip. Both coupling impedances are included in the coupling tensor ζ_{mn}.

The cavity is represented by the structure:

$$\zeta_c = \begin{bmatrix} R_{cc} + \eta_c & (R_{cc} - \eta_c)\, e^{-\frac{Z}{v}p} \\ (R_{cc} - \eta_c)\, e^{-\frac{Z}{v}p} & \eta_c + R_{cc} \end{bmatrix} \tag{10.85}$$

The source has the same structure, for example,

$$\zeta_s = \begin{bmatrix} R_s + z_c & \left(R_f - z_c\right) e^{-\frac{Z_S}{v_S}p} \\ (R_s - z_c)\, e^{-\frac{Z_S}{v_S}p} & z_c + R_f \end{bmatrix} \tag{10.86}$$

and the target also (they are all successions of microstrips and lines eventually):

$$\zeta_t = \begin{bmatrix} R_t + z_c' & \left(R_u - z_c'\right) e^{-\frac{Z_t}{v_t}p} \\ \left(R_t - z_c'\right) e^{-\frac{Z_t}{v_t}p} & z_c' + R_u \end{bmatrix} \tag{10.87}$$

The problem is solved making $\zeta_s \oplus \zeta_c \oplus \zeta_t + \zeta_{mn}$, something like

$$\zeta = \begin{bmatrix} R_s + z_c & \left(R_f - z_c\right) e^{-\frac{Z_S}{v_S}p} & 0 & 0 & 0 & 0 \\ (R_s - z_c)\, e^{-\frac{Z_S}{v_S}p} & z_c + R_f & -h(b-a)p\frac{\mu_0}{2\Theta} & 0 & 0 & 0 \\ 0 & -h(b-a)p\frac{\mu_0}{2\Theta} & R_{cc} + \eta_c & (R_{cc} - \eta_c)\, e^{-\frac{Z}{v}p} & 0 & 0 \\ 0 & 0 & (R_{cc} - \eta_c)\, e^{-\frac{Z}{v}p} & \eta_c + R_{cc} & -h(k-l)p\frac{\mu_0}{2\Theta} & 0 \\ 0 & 0 & 0 & -h(k-l)p\frac{\mu_0}{2\Theta} & R_t + z_c' & \left(R_u - z_c'\right) e^{-\frac{Z_t}{v_t}p} \\ 0 & 0 & 0 & 0 & \left(R_t - z_c'\right) e^{-\frac{Z_t}{v_t}p} & z_c' + R_u \end{bmatrix} \tag{10.88}$$

Note that here we have supposed a coupling between the microstrips and the cavity at each cavity extremities. In fact in the emf and current relations with the field, it may be necessary (for the first TE mode for example) to integrate the field on z in order to obtain the expressions of e and i. This complicates the computation process but the approach remains always the same.

Annexe 10.A

```
import numpy as np
import pylab as plt
#
# values
Ro = 50. # generator self-impedance
```

```
L1 = 10E-6 # first transformer input inductance
# line parameters
D = 1E-3 # wire diameter
zc = 60.*np.log(4.*5E-2/D) # characteristic impedance
Lc = zc/3E8*1. # line inductance for a one meter line long
C = 1./(3E8*zc) # idem for the capacitance
RG = 1E-3 # typical ground resistance
M1 = 0.7*np.sqrt(L1*Lc/2.) # first transformer mutual
Rl = 50. # signal impedance
Ra = 1E5 # input impedance of the comparator
L2 = 1E-6 # current sensor inductance
M2 = 0.7*np.sqrt(L2*Lc/2.)
M11 = 0.1*Lc # mutual inductance between the two lines
Rm = 50. # measurement resistance
# computations parameters
fo = 10E3 # sampling frequency
N =100000
# tabs
ax = np.zeros(N,dtype = float)
test = np.zeros(N,dtype = float)
stress = np.zeros(N,dtype = float)
J = np.zeros((N,4),dtype = complex)
# loop
eo = 100.
for f in range(1,N):
    p = 1J*2.*np.pi*f*fo
    bg = Rl/(Rl*C/2.*p+1.)
    bd = Ra/(Ra*C/2.*p+1.)
    zeta = [[Ro+L1*p,-M1*p,-M1*p,0],\
            [-M1*p,bg+bd+(L1+L2+Lc)*p+RG,-M11*p,-M2*p],\
            [-M1*p,-M11*p,bg+bd+(L1+L2+Lc)*p+RG,-M2*p],\
            [0.,-M2*p,-M2*p,Rm+L2*p]]
    T = [[eo],[0.],[0.],[0.]]
    y = np.linalg.inv(zeta)
    J[f,:] = np.transpose(np.dot(y,T))
    test[f] = abs(J[f,3])
    ax[f] = f*fo
    stress[f] = abs(J[f,1]-J[f,2])
    if f*fo > 1E5:
        if J[f,3] < 0.1:
            eo = eo + 1.
        if J[f,3] > 0.1:
            eo = eo - 1.
# curve
plt.subplot(1,2,1)
plt.plot(ax,test)
plt.grid(True)
plt.ylabel(u'Current')
plt.xlabel(u'Frequency')
```

```
plt.yscale('log')
plt.xscale('log')
plt.subplot(1,2,2)
plt.plot(ax,stress)
plt.grid(True)
plt.ylabel(u'Differential stress')
plt.xlabel(u'Frequency')
plt.xscale('log')
plt.yscale('log')
plt.show()
```

References

[1] P. Degauque and J. Hamelin, "Compatibilité électromagnétique". Dunod, Paris 1990, ISBN 2-04-018807-X.

[2] Z. Xu, "Tensorial analysis of multilayer printed circuit boards: computations and basics for multiphysics analysis", 2019, http://www.theses.fr/2019NORMR003.

[3] M.E. Goldfarb and R.A. Pucel, "Modeling via hole grounds in microstrip", IEEE Microwave Guided Wave Letter, 1991, Vol. 1, pp. 135–137.

Chapter 11
PCB-radiated susceptibility (RS) EMC TAN modelling

Zhifei Xu[1], *Blaise Ravelo*[1], *Yang Liu*[1], *and Olivier Maurice*[2]

Abstract

The radiated electromagnetic compatibility (EMC) analyses constitute one of the major challenging issues of the electronic printed circuit board (PCB) designers and EMC engineers. A few methods are currently available for the comprehension of the EMC phenomena between the EM-field interactions with the PCBs. Once again, this chapter will propose some key solutions against this EMC mechanism misunderstanding by means of the tensorial analysis of networks approach. In this chapter, a supplementary mathematical tensorial concept of moment space will be exploited to establish the metric of the EM wave interactions and the PCB hardware components, including the electrical interconnections. Some fundamental cases of scenarios with plane wave radiation interactions and planar PCBs will be developed. The interaction between PCB and other electronic structures as interconnect wires and another PCBs will also be treated. Then, the assessment of the signal-to-noise ratio in function of the radiated emissions will be formulated and validated with numerical and practical results.

Keywords: TAN model, radiated EMC investigation, PCB modelling, radiated emission, radiated susceptibility

11.1 Far-field coupling

When two subnetworks are far apart enough, they are in far-field interaction. Different domains have their own definitions of the far-field boundaries. An automotive expert will state that the far field begins at a distance of 3λ from the source, with λ being the

[1]IRSEEM/ESIGELEC, Rouen, France
[2]ArianeGroup, Paris, France

radiation wavelength. However, an electromagnetic compatibility (EMC) engineer will claim that the far field begins at

$$d_a = \frac{5\lambda}{2\pi} \tag{11.1}$$

And an engineer from precision antennas will say that the far field begins at

$$d_b = \frac{\lambda}{2\pi} \tag{11.2}$$

In this case, we followed the automotive engineers' definitions of far field. Nevertheless, we have satisfied all these three definitions in our study.

As we have seen previously, the number of moments is equal to the number of meshes. With this interesting characteristic, we consider also an additional transformation of the types of space change. The goal is to reduce far-field interactions to a main vector [1]. For this, we will work in the network space. In the example shown in Figure 11.1, subnetworks are chosen arbitrarily, a network consisting of M meshes can act as a single point source for the other networks. This example can be assumed to be an electronic system embedded in a vehicle interacting with each other by far-field couplings.

The first subnetwork N_1 has three meshes, the second N_2 has two, and the third N_3 has only one mesh. Far-field coupling is represented by the terms α_{ij} which represent the propagation of energy between subnetworks.

Each topological subnetwork will be represented independently by its electromotive forces (EMFs) and impedance matrices in mesh space. The moment space is linking the mesh currents to their respective surface's matrix S.

We define another moment vector a that will take the network space as reference. Since the number of moments is equal to the number of meshes of the system, the term S^p_ω corresponds to meshes ω in moment space and meshes a in network space. In the

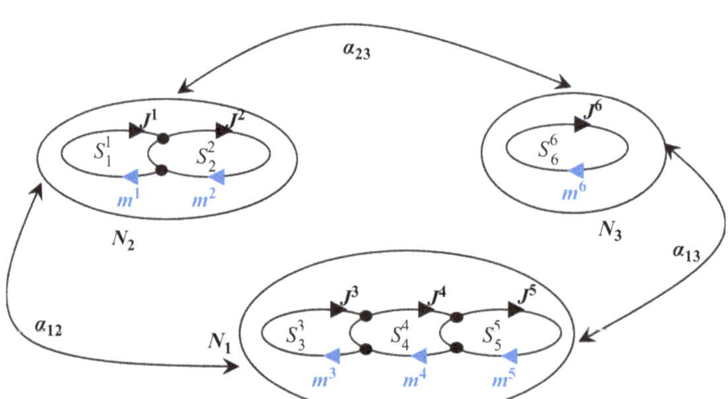

Figure 11.1 Equivalent topology of far-field coupling [2]

example, $\omega = 1, 2, \ldots, 6$ and $a = 1, 2, 3$. The matrices of the surfaces of each network are square matrices. In the example of Figure 11.1, the matrix S has sub-arrays written as follows:

$$S^1 = \begin{bmatrix} S_1^1 & 0 & 0 \\ 0 & S_2^1 & 0 \\ 0 & 0 & S_3^1 \end{bmatrix}, \quad S^2 = \begin{bmatrix} S_4^2 & 0 \\ 0 & S_5^2 \end{bmatrix} \quad \text{and} \quad S^3 = [S_6^3] \tag{11.3}$$

$$[S] = \begin{bmatrix} [S^1] & [0] & [0] \\ [0] & [S^2] & [0] \\ [0] & [0] & [S^3] \end{bmatrix} \tag{11.4}$$

The moments presented in Figure 11.1 are obviously not arranged in the same plane in the (x, y, z) coordinate. We must, therefore, take into account the geometry of the system during the transformation of the space of moments towards the space of the networks. A matrix with angles formed between a normal vector and a reference mesh, and the normal vectors with respect to their meshes.

The influence of the position of the meshes is taken into account by the sine angles formed by the vector u_ω from the mesh and the magnetic field vector B showed in the example of Figure 11.2.

The connection matrix contracting the space of moments towards the space of networks (η) [1],

$$\eta_a^\omega = \sin\left(\theta_a^\omega\right) \tag{11.5}$$

The moments vector will then be defined by expression (11.6) in the space of the networks:

$$m = u_\omega \eta_a^\omega m^a \tag{11.6}$$

where u_ω is the normal moment vector generated by the mesh current.

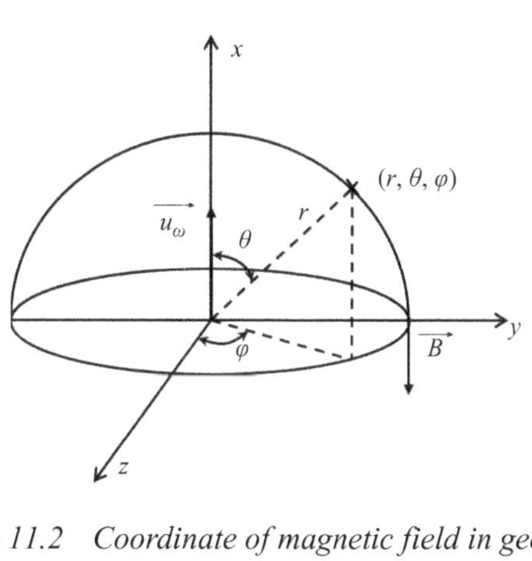

Figure 11.2 Coordinate of magnetic field in geometry

Applied to the example in Figure 11.1, the non-square connection matrix with dimension 3×6 that composes the sinus angles θ_a^ω ($\omega = 1, \ldots, 6$ moments are in the space of moments and $a = 1, 2, 3$ moments are in the network space) is formulated as

$$\eta_a^\omega = \begin{bmatrix} \sin(\theta_1^1) & 0 & 0 \\ \sin(\theta_2^1) & 0 & 0 \\ \sin(\theta_3^1) & 0 & 0 \\ 0 & \sin(\theta_4^2) & 0 \\ 0 & \sin(\theta_5^2) & 0 \\ 0 & 0 & \sin(\theta_6^3) \end{bmatrix} \tag{11.7}$$

So far, a subnetwork will interact another subnetwork as a point source or as an association of flows gathered in a single moment in the network space. The coupling in far field will then be performed by a transformation of the moment (Figure 11.3).

Considering that the meshes of the subnetworks behave as loop antennas in the example. The magnetic field is given at a distance R from the loop radiating:

$$\left|\vec{B}\right|_\theta = \mu \frac{\pi JS \sin(\theta)}{\lambda^2} \frac{e^{-jkR}}{R} = JS \sin(\theta)\alpha \tag{11.8}$$

where μ and λ are the permeability and wavelength, respectively. Or written in tensorial form with ω by taking moment space as reference or p by taking networks as reference and a or b subnetworks

$$B_b = J^p S_p^\omega \sin(\theta_\omega^a)\alpha_{ba} = \alpha_{ba}\eta_\omega^a m^\omega \tag{11.9}$$

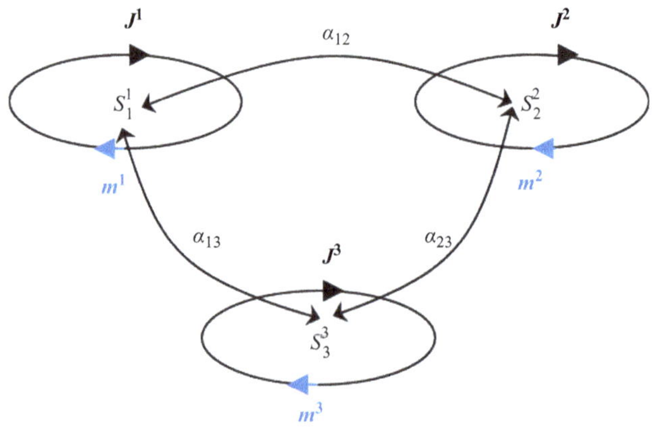

Figure 11.3 Interactions in the moment space [2]

The magnetic field in a specific geometric generated by a mesh of another network is defined through a relation:

$$e_S = S_S^q \eta_q^b B_b \tag{11.10}$$

with $s, q = 1, \ldots, M$ and $b = 1, \ldots, R$. By replacing B_b with its expression, we obtain

$$e_S = S_S^q \eta_q^b (\alpha_{ba}(\eta_\omega^a) m^\omega) = S_S^q \eta_q^b \alpha_{ba} \eta_\omega^a S_p^\omega J^p \tag{11.11}$$

We want to inject this expression directly into MKME formalism by the mutual interaction in the space of meshes [1]. In (11.11), the term α_{ba} defines the magnetic fields $\vec{B}$ generated by J^p mesh currents, S_p^ω indicates an angle θ_ω^a with respect to the reference mesh. In this case, the magnetic fields created by each of the subnetworks will generate a three-dimensional matrix α_{ba} reflecting the interactions between the three subnetworks ($a = b = 1, 2, 3$):

$$\alpha_{ba} = \begin{bmatrix} 0 & \alpha_{12} & \alpha_{13} \\ \alpha_{21} & 0 & \alpha_{23} \\ \alpha_{31} & \alpha_{32} & 0 \end{bmatrix} \tag{11.12}$$

To solve the equations of the system, we need to transfer the coupling from the space of moments to the space of meshes with the mutual coupling M_{sp}.

Note that the matrix S_p^ω has dimension 6×6, η_ω^a has dimension 6×3 and the matrix α_{ab} has dimension 3×3. Thanks to this expression, we reduced the system of six-dimensional mesh currents to a space of three-dimensional networks to define the far-field interactions between these three subnetworks. This matrix of far-field coupling is automatically defined in the space of the mesh and will be added to an impedance Z-matrix containing the description of the three independent subnetworks in the space of the meshes:

$$Z_{coupling} = Z_{sp} + M_{sp} \tag{11.13}$$

The far-field coupling defined by these equations can be easily introduced into the MKME formalism, which reduces system equations [1].

11.2 MKME for 3D multilayer PCB illuminated by I-microstrip line

The present section investigates the radiated coupling between two planar printed circuit boards (PCBs). The EM-field emitter PCB or the culprit element is constituted by a single-layer circuit. The other PCB consisted of a multilayer structure is assumed as EM-field receiver.

11.2.1 Description of system

In this case, the investigation scenario is constituted by two parallel planar PCBs with radiated EM coupling. The configuration of EMC emitter–receiver system introduced in Figure 11.4 is constituted by the following:

1. The emitter PCB Network 1 which is placed on the top is constituted by one-layer microstrip I-line connected between ports P_1 and P_2 loaded by $R_1 = R_2 = 100\ \Omega$.
2. And the receiver PCB which is placed on the bottom and assumed as Network 2. It is constituted by a four-layer PCB investigated in the previous section. It is connected to ports P_3 and P_4, which are loaded by $R_3 = R_4 = 50\ \Omega$.

The distance between the two PCBs is equal to $r = 1$ m. The emission network is modelled with the Telegrapher's structure with identical LC cells. The receiver network representing the illuminated multilayer PCB is modelled under the same approach as introduced previously. In this section, the field radiated by the emitter PCB is used instead of plane wave excitation. The voltage couplings are calculated with MKME model of Taylor's sources.

11.2.2 MKME topological analysis

The MKME graph, equivalent to the structure introduced in Figure 11.4. The coupling sources are generated by magnetic dipoles of Network 1. The coupled magnetic field for each cell of Network 2 is the sum of the magnetic field of all the H-dipoles constituting Network 1. The magnetic moment between the two networks is calculated via the MKME mesh to moment space conversion shown in the equations in

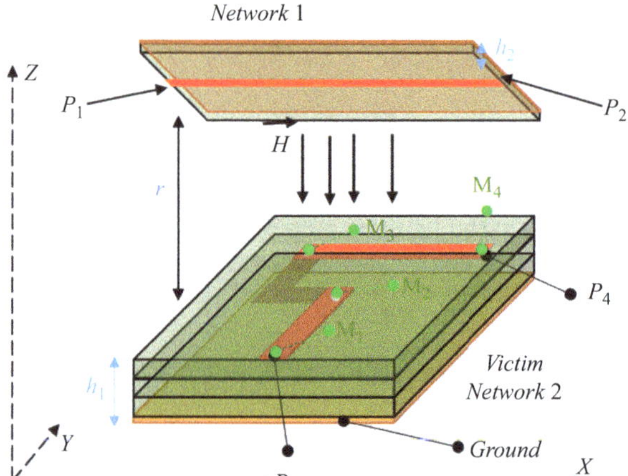

Figure 11.4 Network constitutes microstrip line (EM-field aggressor) and multilayer PCB circuit(victim) [3]

Section 11.1. The first step of the modelling is the calculation of mesh space tensors without coupling interactions based on the topology shown in Figure 11.5. After the branch to mesh space transformation, we have the mesh impedance:

$$Z_{mn} = \begin{bmatrix} [Z_{N^\circ 1}] & 0 \\ 0 & [Z_{N^\circ 2}] \end{bmatrix} \quad (11.14)$$

The mesh space impedance matrix of the entire system is denoted by Z_{mn}. It presents the dimension $M \times M$ with M is the sum of mesh number in these two systems. At the frequency f, the propagation factor is defined by

$$\alpha = \frac{\mu_0 f}{c_0 r} e^{-jk_0 r} \quad (11.15)$$

where $k_0 = 2\pi f/c$ is the free-space wave number. The interaction matrix α_{ab} with the same size as Z_{mn} between these two networks is defined by

$$\alpha_{ab} = \begin{bmatrix} 0 & 0 & 0 & \cdots & 0 & 0 & 0 \\ \vdots & \vdots & \vdots & \vdots & \vdots & \vdots & \vdots \\ 0 & 0 & \cdots & 0 & 0 & \cdots & 0 \\ 0 & 0 & \cdots & 0 & 0 & \cdots & 0 \\ 0 & \alpha & \cdots & \alpha & 0 & \cdots & 0 \\ \vdots & \vdots & \cdots & \vdots & \vdots & \cdots & \vdots \\ 0 & \alpha & \cdots & \alpha & 0 & \cdots & 0 \end{bmatrix} \quad (11.16)$$

The positions of the coefficients α_a in rows correspond to the EM coupling from the meshes in Network 2. The positions of α_b in columns represent the EM coupling to

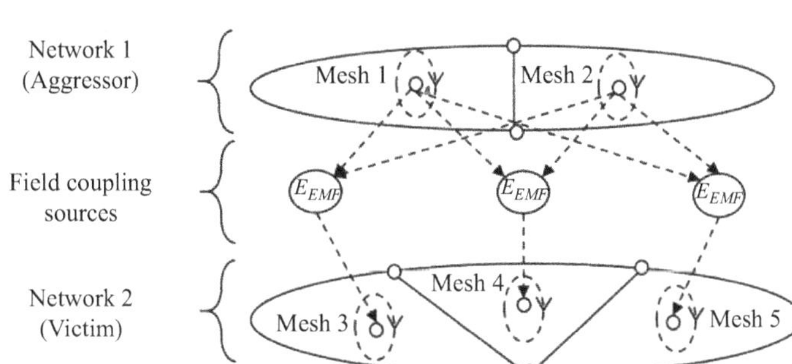

Figure 11.5 Simplified equivalent topology of the radiated EMC MKME model [3]

the meshes of Network 1. The interaction between the two networks results is given by the EMF tensor:

$$E_m = j\omega M_{mn} J^n \tag{11.17}$$

where M_{mn} is a magnetic impedance tensor expressed as

$$M_{mn} = S_m^a \alpha_{ab} S_n^b \tag{11.18}$$

The S matrix with all elements located in diagonal has the same size as Z_{mn} and expressed in (11.20) by taking

$$\left[\xi_{k=\{1,2,3\}}\right] = \begin{bmatrix} lh_k/N_{c_k} & 0 & 0 \\ 0 & \ddots & 0 \\ 0 & 0 & lh_k/N_{c_k} \end{bmatrix} \tag{11.19}$$

Quantity (11.19) is a square matrix with dimension equal to the number of Telegrapher's cells of the trace.

$$[S] = \begin{bmatrix} [\xi_1] & 0 & \cdots & 0 & 0 & 0 & 0 & 0 \\ 0 & [\xi_2] & 0 & \cdots & 0 & 0 & 0 & 0 \\ 0 & 0 & 0 & \cdots & 0 & 0 & 0 & 0 \\ \vdots & \vdots & \vdots & 0 & \cdots & 0 & 0 & 0 \\ 0 & 0 & 0 & \vdots & [\xi_3] & 0 & 0 & 0 \\ 0 & 0 & 0 & 0 & 0 & 0 & 0 & 0 \\ 0 & 0 & 0 & 0 & 0 & 0 & 0 & 0 \\ 0 & 0 & 0 & 0 & 0 & 0 & 0 & [\xi_4] \end{bmatrix} \tag{11.20}$$

The cell matrices defined by the following parameters:

1. $\xi_k = [1, 2, 3, 4]$ represents the different coupling surface of the Telegrapher's cells of different interconnects in Network 2, k represents the different interconnects in Network 2, N_k represents the number of Telegrapher's cells of different interconnects.
2. l_k are the lengths of the multilayer PCB interconnects.
3. h_k are the height of the interconnects with respect to the ground plane.
4. The diagonal zero elements represent the via surfaces in Network 2.

Knowing this mutual tensor, the general metric of the two coupled PCBs is written as

$$E_m = (Z_{mn} + j\omega M_{mn}) J^n \tag{11.21}$$

11.2.3 Validation results

11.2.3.1 POC description

Similar to the previous case, the POC circuits are implemented on FR4 epoxy dielectric substrate. Network 1 is the top PCB aggressor with the microstrip line and ground plane implemented in bottom and top, respectively. It acts as the aggressor circuit is

Table 11.1 Network 2 PCB parameters

PCB layout parameters					
Board width	**Board length**	**Board height**	**Substrate permittivity**	**Tangent**	**Metal thickness**
20 mm	80 mm	1.6 mm	4.5	0.012	0.035 mm
Via and interconnect parameters					
Interconnect width			**Length/Each**		
0.5 mm			80 mm		

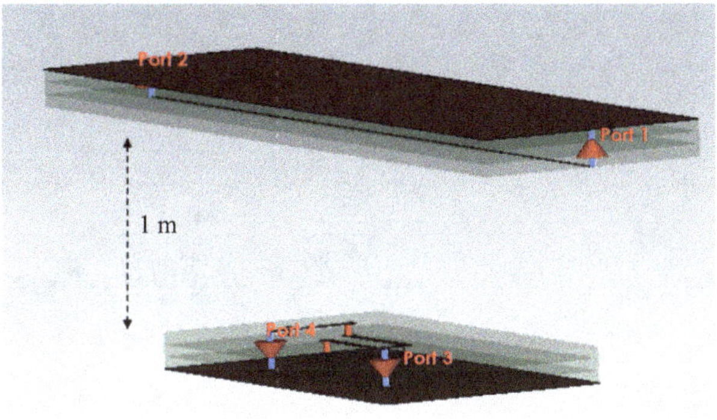

Figure 11.6 CST MWS® design of the simulated POC [3]

excited by the voltage sine wave source with amplitude 1 kV. The bottom radiated field receiver multilayer PCB constitutes Network 2. These networks' physical parameters are presented in Table 11.1 (Figure 11.6).

11.2.3.2 Discussion on the computed coupling voltages

In this case, the calculation and simulation are performed from 1 MHz to 1 GHz. The far-field source from the aggressor circuit constituted by Network 1 was established. The field coupling onto the PCB of Network 2 is assessed with voltages $V_1 = V(P_3)$ and $V_2 = V(P_4)$. They are calculated with (11.21) after the numerical implementation of radiated and interconnect structure metric expressed in (11.16) of all the two-network interactions. The comparisons between the CST MWS simulated and MKME model voltages V_1 and V_2 are displayed in Figure 11.7.

Comparisons of calculated and simulated coupled voltages on V_1 and V_2 by considering an incident wave radiated by the microstrip I-line. For the illustrative view, the results are separated into two frequency bands and plotted in Figure 11.8.

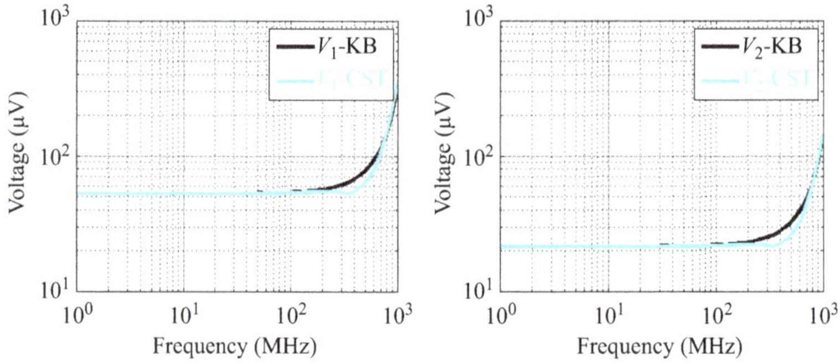

Figure 11.7 Comparisons of calculated and simulated coupled voltages on V_1 and V_2 by considering an incident wave radiated by the microstrip I-line [3]

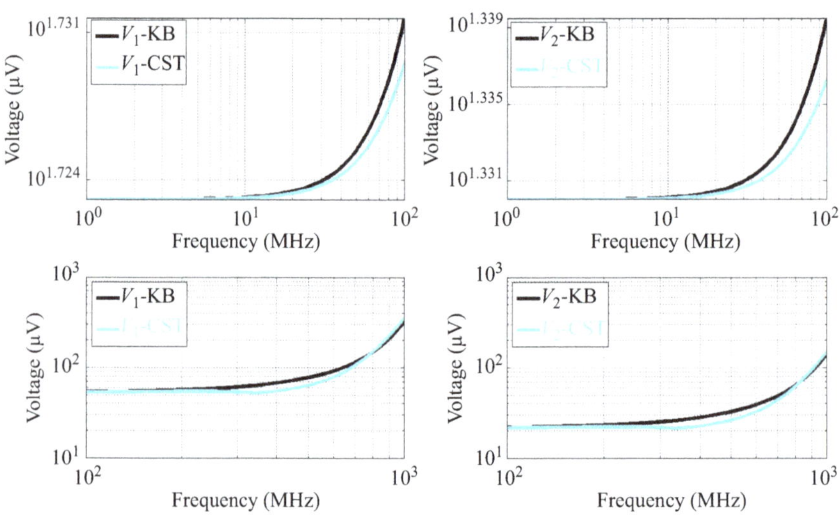

Figure 11.8 Comparisons of calculated and simulated coupled voltages on V_1 and V_2 by considering an incident wave radiated by the microstrip I-line in different frequency bands [3]

Both computed results are in good correlation. It can be noticed that the EM coupling increases with the frequency. The main differences between the MKME and CST® computations are essentially due to the assumptions of Taylor's cells and the meshing effects. The coupling voltage behaviours depend on both the intensity of the radiated field in the considered frequency band and the configuration of the interconnect line of Network 2. First, the increase of the radiated field is due to the significant magnetic field emitted at half-wavelength frequency of the culprit I-line. Then, the coupling voltage amplitude is more significant for the magnetic field interacting with the top signal line of Network 2 which must be more significant at higher frequencies. Our work is different to the work done by Rogard from Telecom and Thales [4], they only studied the radiation of a microstrip based on MKME; however, we have studied the radiation of a microstrip received by another multilayer PCB structure by MKME. Our case is more complicated.

11.3 Sensitivity analysis with MKME

In this section, we would like to discuss another advantage of MKME, the theoretical analysis can be applied for sensitivity analysis based on MKME formalism. This formalism is able to identify which parameter or subsystem plays the most important role of the system by performing the theoretical analysis, and how the parameter affects the system when it varies and how the subsystem affects. This ability can be potentially exploited for the reliability analysis. The diagnostic results can be analysed easily to repair or predict the system damages.

11.3.1 Sensitivity analysis with theoretical expression for Branin's model

11.3.1.1 Theoretical analysis

Let us start with the KB model, this analysis predicts the effects of $R_{1,2}$:

$$Z_{\mu\upsilon} = \begin{bmatrix} R_1 + Z_c & (-Z_c + R_2)e^{-\theta} \\ -(Z_c - R_1)e^{-\theta} & R_2 + Z_c \end{bmatrix} \tag{11.22}$$

By recalling equations as

$$E_\mu = Z_{\mu\upsilon} J^\upsilon \tag{11.23}$$

where E_μ is

$$E_\mu = \begin{bmatrix} V \\ Ve^{-\theta} \end{bmatrix} \tag{11.24}$$

Then, we perform the derivation of the expression in (11.24) to obtain the effects of $R_{1,2}$, where a and b represent the index of rows and columns of the branch space tensors. μ and υ represent the index of rows and columns of mesh space tensors. The derivation can be expressed as

$$V_2 = \psi(R_1, R_2) \Rightarrow dV_2 = \frac{\partial V_2}{\partial R_1} dR_1 + \frac{\partial V_2}{\partial R_2} dR_2 \tag{11.25}$$

with $Z_{\mu\upsilon}^{-1} = Y^{\upsilon\mu}$, V_2 can be derived as

$$V_2 = R_2 J^2 = R_2 (Y^{2\mu} C_\mu^a e_a) \tag{11.26}$$

Substituting (11.26) into (11.25),

$$dV_2 = R_2\left[\frac{\partial Y^{2\mu}}{\partial R_1}dR_1 C_\mu^a e_a + Y^{2\mu}C_\mu^a\frac{\partial e_a}{\partial R_1}dR_1 + \frac{\partial Y^{2\mu}}{\partial R_2}dR_2 C_\mu^a e_a + Y^{2\mu}C_\mu^a\frac{\partial e_a}{\partial R_2}dR_2\right] \tag{11.27}$$

Then, this can be simplified as

$$dV_2 = R_2\left[\left(\frac{\partial Y^{2\mu}}{\partial R_1}C_\mu^a e_a + Y^{2\mu}C_\mu^a\frac{\partial e_a}{\partial R_1}\right)dR_1 + \left(\frac{\partial Y^{2\mu}}{\partial R_2}C_\mu^a e_a + Y^{2\mu}C_\mu^a\frac{\partial e_a}{\partial R_2}\right)dR_2\right] \tag{11.28}$$

therefore,

$$\frac{dV_2}{V_2} = \frac{1}{Y^{2\mu}C_\mu^a e_a}\left[\left(\frac{\partial Y^{2\mu}}{\partial R_1}C_\mu^a e_a\right)dR_1 + \left(\frac{\partial Y^{2\mu}}{\partial R_2}C_\mu^a e_a\right)dR_2\right] \tag{11.29}$$

the relative error is

$$\frac{dV_2}{V_2} = \frac{\Delta R_1}{R_1}\psi_{R_1} + \frac{\Delta R_2}{R_2}\psi_{R_2} \Rightarrow \varepsilon_{V_2} = \psi_{R_1}\varepsilon_{R_1} + \psi_{R_2}\varepsilon_{R_2} \tag{11.30}$$

With

$$\begin{bmatrix} \psi_{R_1} = \frac{1}{Y^{2\mu}C_\mu^a e_a}\left[\left(\frac{\partial Y^{2\mu}}{\partial R_1}C_\mu^a e_a\right)\right] \\ \psi_{R_2} = \frac{1}{Y^{2\mu}C_\mu^a e_a}\left[\left(\frac{\partial Y^{2\mu}}{\partial R_1}C_\mu^a e_a\right)\right] \end{bmatrix} \tag{11.31}$$

The final equation for Branin's model sensitivity analysis is derived from (11.25):

$$\begin{aligned}\frac{\partial V_2}{\partial R_1} = {} & (R_1 e^{-2\theta}(R_2 - Z_1) \\ & (-V_{in}e^{-\theta}(R_1 - Z_1)/ \\ & (-e^{-2\theta}(R_1 - Z_1)(R_2 - Z_1) \\ & + (R_1 + Z_1)(R_2 + Z_1)) \\ & + V_{in}e^{-\theta}(R_1 + Z_1)/(-e^{-2\theta}(R_1 - Z_1) \\ & (R_2 - Z_1) + (R_1 + Z_1)(R_2 + Z_1)))/ \\ & (-e^{-2\theta}(R_1 - Z_1)(R_2 - Z_1) \\ & + (R_1 + Z_1)(R_2 + Z_1)) - R_1 e^{-\theta}(R_2 + Z_1) \\ & (-V_{in}e^{-2\theta}(R_2 - Z_1)/(-e^{-2\theta}(R_1 - Z_1) \\ & (R_2 - Z_1) + (R_1 + Z_1)(R_2 + Z_1)) \\ & + V_{in}(R_2 + Z_1)/(-e^{-2\theta}(R_1 - Z_1) \\ & (R_2 - Z_1) + (R_1 + Z_1)(R_2 + Z_1)))/ \\ & (-e^{-2\theta}(R_1 - Z_1)(R_2 - Z_1) + (R_1 + Z_1)(R_2 + Z_1)))\end{aligned} \tag{11.32}$$

$$\begin{aligned}\frac{\partial V_2}{\partial R_2} &= (R_2 e^{-\theta}(R_1 - Z_1)(-v_1 e^{-2\theta}(R_2 - Z_1)/(-e^{-2\theta}(R_1 - Z_1)(R_2 - Z_1)\\ &\quad + (R_1 + Z_1)(R_2 + Z_1))\\ &\quad + V_{in}(R_2 + Z_1)/(-e^{-2\theta}(R_1 - Z_1)(R_2 - Z_1)\\ &\quad + (R_1 + Z_1)(R_2 + Z_1)))/(-e^{-2\theta}(R_1 - Z_1)(R_2 - Z_1)\\ &\quad + (R_1 + Z_1)(R_2 + Z_1)) - R_2(R_1 + Z_1)\\ &\qquad (-v_1 x1(R_1 - Z_1)/(-e^{-2\theta}(R_1 - Z_1)(R_2 - Z_1)\\ &\quad + (R_1 + Z_1)(R_2 + Z_1))\\ &\quad + V_{in} e^{-\theta}(R_1 + Z_1)/(-e^{-2\theta}(R_1 - Z_1)(R_2 - Z_1)\\ &\quad + (R_1 + Z_1)(R_2 + Z_1)))/(-e^{-2\theta}(R_1 - Z_1)(R_2 - Z_1)\\ &\quad + (R_1 + Z_1)(R_2 + Z_1)))\end{aligned} \tag{11.33}$$

Finally, (11.25) can be solved, and the results are displayed in the next section. From (11.30) and (11.31), the outputs are the unknowns which are used as results of the system for analyses. We can get the effects of system results when the loads vary. For multiple parameters,

$$\begin{aligned} R_{o1} &= \tfrac{\partial V_o}{\partial R_1}\\ R_{o2} &= \tfrac{\partial V_o}{\partial R_2}\\ &\vdots \end{aligned} \tag{11.34}$$

The total effects can be represented as

$$Total - effects = R_{o1} + R_{o2} + \cdots \tag{11.35}$$

where $R_{o1}, R_{o2}, \ldots$ are the effects of different parameters, respectively.

11.3.1.2 Numerical analysis

For this illustrative application, let us consider the interconnect structure constituted by the microstrip I-line placed above the ground with height $h = 3$ mm, the length and width of the microstrip are equal to $l = 20$ mm, $w = 0.5$ mm. The dielectric substrate between the microstrip and ground is FR4 with $\varepsilon_r = 4.4$. We are going to study the effects of the loads on the output with an AC source at the input, the frequency bandwidth is 1 Hz to10 GHz. With the equation developed in (11.29)–(11.35), the loads R_1 and R_2 changing from 50 to 500 Ω have been studied. The results are displayed in Figure 11.9 with $\Delta R_{1,2}$ being the resistance variation.

11.3.2 Conclusion

A POC constituted by a multilayer PCB coupled with the EM-field radiated by an I-line microstrip PCB is studied. Once more, the plots of the calculated and simulated

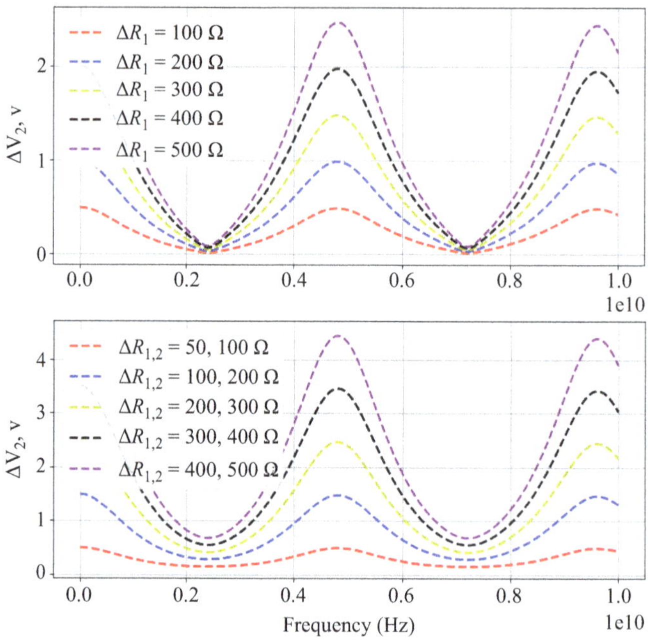

Figure 11.9 Sensitivity analysis for the loads of KB TL model

coupled voltages across the multilayer PCB terminations are in good agreement. These results confirm the MKME effectiveness to predict the R-EMC coupling onto the multilayer PCB in a wideband frequency range. At last, the sensitivity analysis has been performed both for Branin's TL model and also the R-EMC multilayer PCB network.

References

[1] Leman S. Contribution à la résolution de problèmes de compatibilité electromagnétique par le formalisme des circuits electriques de KRON. Univ. Lille; 2009.

[2] XU Z. Tensorial analysis of multilayer printed circuit boards, PhD thesis; 2019.

[3] Xu Z, Ravelo B, Maurice O, *et al.* Radiated EMC Kron's Model of 3-D Multilayer PCB Aggressed by Broadband Disturbance. IEEE Transactions on Electromagnetic Compatibility. 2019.

[4] Rogard E, Azanowsky B, and Ney MM. Comparison of Radiation Modeling Techniques Up to 10 GHz—Application on a Microstrip PCB Trace. IEEE Transactions on Electromagnetic Compatibility. 2010;52(2):479–486.

Chapter 12

TAN model of loop probe coupling onto shielded coaxial short cable

Christel Cholachue[1,2], Amélie Simoens, Olivier Maurice[3], and Blaise Ravelo

Abstract

This chapter introduces coaxial cable *S*-matrix modelling with tensorial analysis of networks (TAN). The main objective of the modelling is to determine the shielding effectiveness (SE) of shielded cable coupled with a loop probe. The TAN methodology from the equivalent graph elaboration to the Kron's branch and mesh space analyses and ended by *Z*-matrix is elaborated. The SE is formulated innovatively from the *S*-matrix. A proof-of-concept constituted by a centimetre-length braid shielded cable illuminated by a proximate millimetre-radius circular was designed and simulated with commercial full-wave simulator. The wide band computed and simulated results with parametric analysis in function of the loop probe position are in good agreement.

Keywords: Coaxial cable, circuit theory, near-field coupling, tensorial analysis of networks (TAN), modelling method

12.1 Introduction

The cables and connectors constitute indispensable key components for modern automotive and aeronautics systems [1,2]. Nowadays, the increases of the design density of electronic and electrical systems cause unneglected electromagnetic (EM) compatibility (EMC) problems which attract the EMC research engineer attention. Behind the integration density, the cable EM shielding must respect critical level of severity with respect to the EMC standard compliances [3]. Moreover, the shielded cable EMC characterizations are performed with time cost and expensive experimental techniques.

[1]Tenneco/Federal-Mogul Systems Protection SAS, Crépy en Valois, France
[2]IRSEEM/ESIGELEC, Rouen, France
[3]ArianeGroup, Paris, France

To analyse the shielded cables, an EM theory was established based on the transmission line (TL) approach [4,5]. In addition, measurement techniques of coaxial cable EM shielding effectiveness (SE) and transfer impedance (Z_t) were deployed [6,7]. The most of existing SE and Z_t measurement techniques are very expensive, present important time costs, limited to the EMC-conducted approaches and relatively limited in frequency bandwidth. For this reason, we are investigating on techniques allowing one to reduce the EMC test costs. Therefore, more appropriated modelling is needed for the new EM shielding products [8] and to optimize the complex structures as cable braid shielding [9]. An improved calculation method of shielded coaxial cable transfer impedance was proposed [10].

Nevertheless, the analytical method to model the radiated braided coaxial cable opens challenges with respect to the operating frequency band and also the structure geometrical parameters [11,12]. With the current golden age of computer-aided design, different numerical computational methods [14–21] and commercial simulation tools [22–25] have been developed for the 1D/2D/3D structure analyses. Since the 1980s, various solvers based on the familiar numerical methods as finite element method [13,19], Simulation Program with Integrated Circuit Emphasis (SPICE) [14–16], TL matrix [17], method of moments [18], finite difference time domain [20] and partial element equivalent circuit [21] have emerged in the EMC engineering area. However, most of these familiar methods necessitate complex pre- and post-processing and are time-consuming. For example, the computations of the interaction between the printed circuit board (PCB) and EM radiation with full-wave meshing are still time-consuming. It may take 1 day for pre-processing and several hours of simulations. So far, the most popular tools for the coaxial cable analyses are using the SPICE-lumped circuit simulators [14–16].

In the present chapter, we would like to analyse shielded coaxial cable structures interacting with the near-field radiated by a loop probe. Mostly, the EMC-radiated emission analyses with the software commercial tools as HFSS® [22], CST Cable Studio® [23], ADS®/Momentum® [24] and ICEy/SPEAG® [25] modules may take several hours. Alternative models [26–28] of TL interacting with EM fields as BLT model were introduced to predict rapidly the coupling effects onto a metallic wire. These radiated EM coupling calculations are basically developed with Taylor's [29], Agrawal's [30] and Rachidi's [31] equations. The relevance of these radiated EM coupling methods stands with the knowledge of the illuminating field around the cables. In [32], a measurement technique of a PCB EM near field is proposed. Despite the existing computer-aided designers and analytical models, the published research works are essentially focused on the nude wire placed above the ground plane without considering the shielded structure. Further research work must be realized for predicting the EMC-radiated coupling onto the shielded coaxial cable. The modelling method explored in the present chapter is aimed to be a wide band frequency transfer function by taking into account the EM radiation from a loop probe. To do this, we would proceed with the multi-port S-parameter calculations [33]. Most of the existing S-parameter calculations are limited to two-port microwave circuits [34–37]. These familiar analytical approaches can be complicated and fastidious for the case of the shielded coaxial cable. For this reason, more efficient and innovative EMC modelling methods must be developed.

For this reason, the tensorial analysis of networks (TAN) formalism by using Kron's method is exploited in the present chapter. The TAN was initiated the first time in the 1930s [38]. So far, this formalism is not familiar to the EMC and signal integrity (SI) engineers. Few decades earlier, the TAN formalism has been extended to treat the EMC problem of complex system [39–43]. The TAN model of radiated EMC coupling proposed in [41–43] is limited to the PCB structures. Moreover, Kron's method was deployed and applied to the planar PCB interconnect modelling for the SI analysis [44]. It was found that the TAN formalism enables us to perform PCB fast SI analyses. However, so far, few research works are available for the S-parameter modelling of shielded cable with radiated field interaction. In the continuation of the research work conducted in [44], the present chapter is originally focused on the frequency-dependent analytical modelling of typically braided shielded coaxial cable under near-field coupling from a circular loop probe.

This chapter is organized in four principal sections. Section 12.2 will describe physically and geometrically the considered problem represented by the probe–coaxial structure placed above the ground (GND) plane. After the problem formulation, the structure electrical parameters will be indicated. Section 12.3 will indicate an easy open way about the proposed TAN modelling methodology, in this case, applied to the EMC cable analysis. The graph topology will be easily and explicitly elaborated. An affordable problem metric formulation will be developed with respect to the TAN formalism. Hence, a particularly innovative computation of the full-port S-parameters and SE will be introduced. In Section 12.4, the probe–cable TAN model will be validated by commercial tool simulations. The proof-of-concept (POC) constituted by a 'I'-shape shielded coaxial cable coupled with a loop probe will be designed in 3D and in an electrical scheme. The chapter will be concluded in Section 12.5.

12.2 Formulation of problem constituted by shielded cable under loop probe radiated field aggression

This section is focused on the description of the EMC problem under study. The considered structure consists of a braided shielded coaxial cable placed above a planar ground (GND) plane coupled with metallic wire loop. The equivalent circuit of the overall system is described.

12.2.1 Geometrical definition of the problem

Figure 12.1 depicts the scenario of the shielded 'I'-shape coaxial cable under study. The structure is placed in the Cartesian reference system ($Oxyz$). The metallic probe is a circular loop having radius r and centred at the point $M_0(x_0, y_0, z_0)$. The probe is connected to the matching resistance load R_m and terminated by port P_1, which is physically positioned at $M_1(x_1, y_1, z_1)$. The mutual coupling between the probe and the coaxial cable is denoted by ζ.

As explained in Figure 12.1, the d-length metallic cable is placed at the distance h above the GND plane between the points $M_2(x_2, y_2, z_2)$ and $M_3(x_3, y_3, z_3)$. The cable port terminations are denoted by P_2 and P_3. In the function of the probe coordinate x_0, the structure under study can be assumed as cascaded pieces of TLs as illustrated in Figure 12.2. The external TL constituted by the braid shield can be split into three elementary cells $Cell_{e1}$, EM-coupled $Cell_{e2}$ and $Cell_{e3}$ having physical lengths

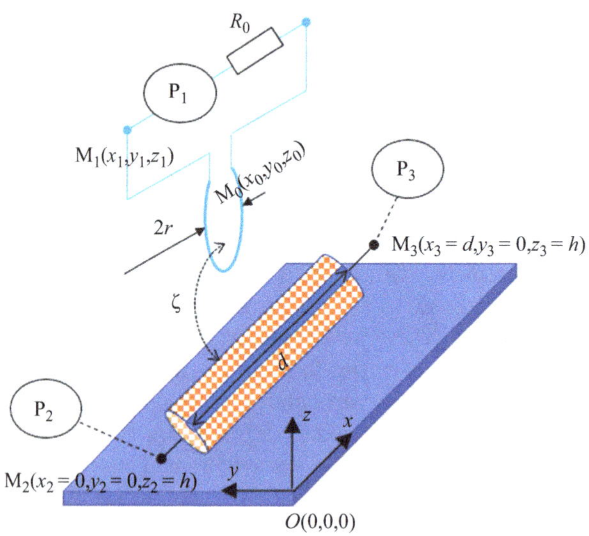

Figure 12.1 Scenario of loop probe coupled with shielded coaxial cable placed above the GND plane

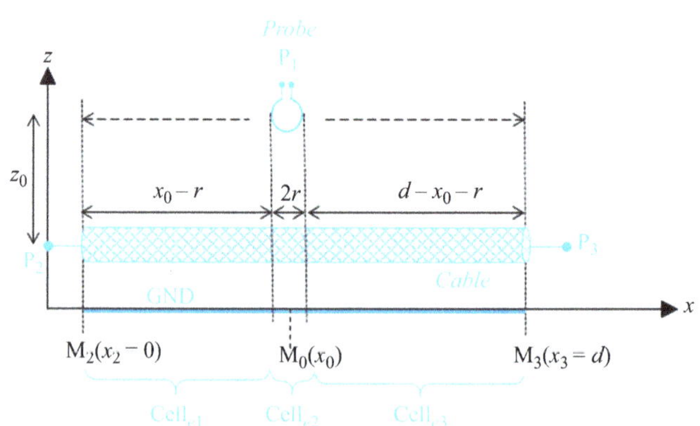

Figure 12.2 Profile view of the structure introduced in Figure 12.1 constituted by three cascaded elementary cells

$d_1 = x_0 - r$, $d_2 = 2r$ and $d_3 = d - x_0 - r$, respectively. The equivalent electrical circuit of the structure will be developed in the next section by considering this TL configuration.

12.2.2 Electrical description of the problem

As highlighted by Figure 12.3, the proposed structure can be assumed as a three-port system. We underline that this structure is particularly unconventional compared to the existing works [1–13] about the cable EMC shielding analyses because the excitation ports do not have the same references.

12.2.2.1 Circuit model

To model the circuit, each port P_k must be fed by voltage source V_k loaded by the impedance Z_k for $k = \{1, 2, 3\}$. Substantially, the circuit must be given input by the current I^k. The present study is limited to the low-frequency band where the coaxial cable presenting a physical length $d < \lambda_{\min}/10$, denoting $\lambda_{\min}$ the minimal wavelength in the considered operating frequency band.

Therefore, the equivalent circuits can be modelled by lumped R, L and C components. From the electrical point of view, the structure exposed in Figure 12.2 is constituted by lumped circuit-based three networks as detailed in Figure 12.4.

12.2.2.2 Impedance description of constituting network elements

Each constituting network can be described as follows:

- Network$_1$: By denoting the angular frequency complex variable $j\omega$, the loop probe connected to Port$_1$, represented by the RL-series impedance [32],

  $$Z_l(j\omega) \approx R_l + j\omega L_l, \tag{12.1}$$

  matched by the matching resistance R_m.
- Network$_2$: The external TL representing the braid shield is schematized by three cascaded π-cells $(Z_{pe1}–Z_{se1}–Z_{pe1})$, $(Z_{pe2}–Z_{se2}–Z_{pe2})$ and $(Z_{pe3}–Z_{se3}–Z_{pe3})$. It is coupled with the probe via the mutual inductance presenting impedance:

  $$Z_{mutual}(j\omega) = -j\omega\zeta. \tag{12.2}$$

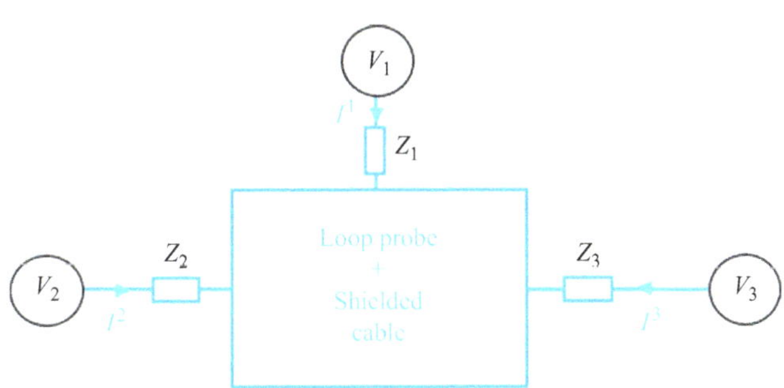

Figure 12.3 Block diagram of the structure shown in Figure 12.1

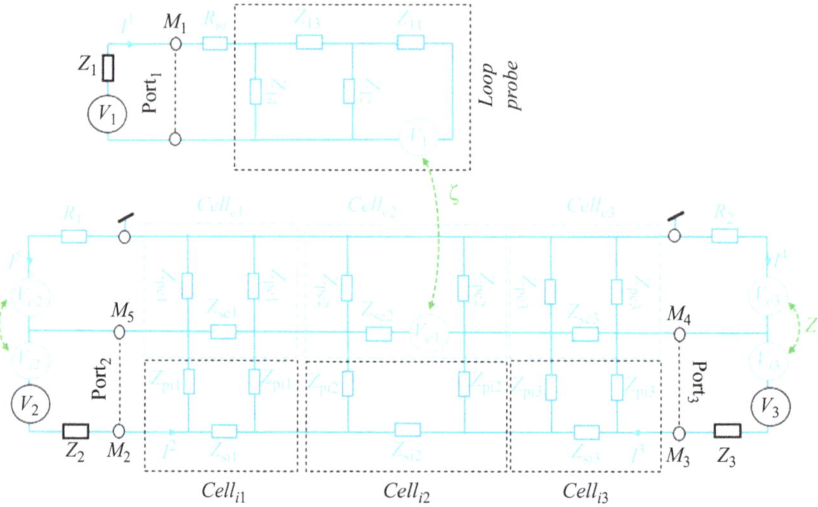

Figure 12.4 Equivalent circuit of the coaxial cable structure under study shown in Figure 12.1

The terms Z_s and Z_p designate lumped series and parallel impedances.

- Network$_3$: The internal TL is mainly built with the network of three cascaded π-cells (Z_{pi1}–Z_{si1}–Z_{pi1}), (Z_{pi2}–Z_{si2}–Z_{pi2}) and (Z_{pi3}–Z_{si3}–Z_{pi3}). By taking the external and internal coupled voltages (V_e, V_i) and currents (I^e, I^i), the internal and external TL coupling can be formulated via the simplified equivalent model of the coaxial cable shielding and inner conductor transfer impedance [5–7,9–12]:

$$Z_t(j\omega) \approx R_t + j\omega L_t, \tag{12.3}$$

which is implemented as a transfer impedance matrix:

$$\begin{bmatrix} V_e(j\omega) \\ V_i(j\omega) \end{bmatrix} = \begin{bmatrix} 0 & Z_t(j\omega) \\ Z_t(j\omega) & 0 \end{bmatrix} \begin{bmatrix} I^e(j\omega) \\ I^i(j\omega) \end{bmatrix}. \tag{12.4}$$

12.2.3 Formulation of shielding effectiveness (SE)

Acting as p-port system, the S-parameters of the structure under study represented in Figure 12.3 can be expressed as

$$[S(j\omega)] = \begin{bmatrix} S_{11}(j\omega) & \cdots & S_{1p}(j\omega) \\ \vdots & \ddots & \vdots \\ S_{p1}(j\omega) & \cdots & S_{pp}(j\omega) \end{bmatrix}. \tag{12.5}$$

The EMC coupling can be quantified with the transmission parameters $S_{k1}(j\omega)$ for $k = \{2, \ldots, p\}$. By considering the input injection from the probe referred to at Port 1, we postulate the EM SE at the cable extremities (port k) from the transmission coefficients:

$$SE_k(j\omega) = \frac{S_{k,1_{unshielded\ cable}}(j\omega)}{S_{k,1_{shielded\ cable}}(j\omega)}. \tag{12.6}$$

The TAN model enabling one to determine this SE from the electrical equivalent circuit of the probe–cable structure will be developed in the next section.

12.3 Theoretical investigation of SE modelling with TAN approach

The present section develops the implementation principle and the mathematical formulation of the TAN modelling under study. The routine process of the fundamental methodology is described. Then, the analytical way enabling one to establish the metric of the probe–cable structure is established.

12.3.1 Methodology of the S-parameter modelling of coaxial modelling under probe EM radiation

Before the exploration of the TAN formalism analytical expression, it is worth to recall the proposed modelling indicating a routine way. Indeed, the TAN modelling methodology can be merely fulfilled as follows:

- **Step 1**: The proposed TAN concept begins with the identification of the problem physical, geometrical and electrical parameters. This key step is necessary for the reformulation of the structure under study by setting the master parameters of the model.
- **Step 2**: The introduction of the classical electrical equivalent circuit must depend on the addressed problem application context. For example, in the present chapter, the study is, particularly, focused on the low-frequency bands where the coaxial cable can be assumed equivalent to lumped circuit. The electrical nature and characteristics of the problem as the mutual inductance coupling and cable *RLCG* equivalent parameters must be attentively tabulated. By dealing with the EMC problem, the electrical equivalent circuit may behave differently in function of the operation frequency band and the structure's physical sizes.
- **Step 3**: This central step enables us to carefully transcribe the TAN mathematical model of the problem under study. In the other words, the problem can be rewritten in the Kron universe. So far, it is noteworthy that the TAN graph topological presentation remains rather unfamiliar, unconventional and uncommon [38–45], for most of electrical and EMC engineers. In a nutshell, the considered problem

equivalent graph topology is built as a network constituted by elementary Kron's objects as nodes, branches, meshes, cords, etc.

- **Step 4**: The analytical abstraction of the tensorial equation aims to translate the graph topology into tensorial algebraic equations. The graph primitive elements can be translated into tensor objects by means of a typical dictionary between the physical structure–electrical circuitries–tensor objects. All the tensor algebra properties can be exploited at this stage by operating mathematically from the graph based on the branch and mesh space analyses.
- **Step 5**: The calculations of the impedance or admittance matrices can be done quickly as a function of the structure's electrical behaviour. This calculation will be explicitly and clearly described in Section 12.3.2.3 with an innovative fast linear matrix reduction method.
- **Step 6**: The final step of our modelling method is the determination of the full (over all the access port) S-parameters and the SE. This S-matrix extraction step can be quickly performed with the Z-to-S matrix transform.

12.3.2 Elaboration of equivalent graph

The TAN graph equivalent to the classical circuit presented in Figure 12.4 is depicted in Figure 12.5. Certain branches of this graph are replaced by equivalent impedance of parallel branches.

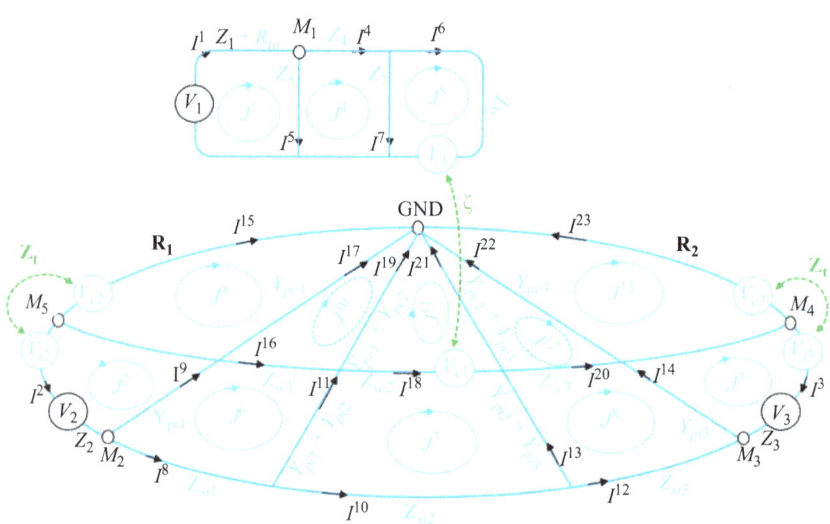

Figure 12.5 TAN equivalent graph of the coaxial cable structure shown in Figure 12.4

12.3.2.1 TAN graph index parameters

Certain branches of this graph are replaced by equivalent impedance of parallel branches. This graph is defined by index parameters:

- $B = 23$ branches,
- $M = 13$ meshes,
- $p = 3$ ports.

The analysis of the problem is transformed to the tensorial representation with mathematical laws governing the system. The solution from the TAN modelling is constituted by the determination of the branch and mesh currents I^b and J^m with $b = \{1, 2, \ldots, B\}$ and $m = \{1, 2, \ldots, M\}$, respectively.

12.3.2.2 Branch space analysis

First and foremost, the branch space analysis elaborated in the present chapter will be referred to the branch indices $(a, b) = \{1, 2, \ldots, B\}$. The graph topology is built with the given covariable voltage source E_a (corresponding to the external excitation voltages $V_{1,\ldots,3}$ for the graph under study shown in Figure 12.5), branch current I^a and the branch impedance Z_{ab}. To succeed automatically in a simple way, the algebraic mechanism of the proposed matrix reduction approach of full S-parameter extraction, it is particularly capital to set the B-size source voltage vector as follows:

$$[E_a] = \begin{bmatrix} V_1 & \cdots & V_p & 0 & \cdots & 0 \end{bmatrix}, \tag{12.7}$$

in the dual space with the contravariable B-size current:

$$[I^a] = \begin{bmatrix} I^1 \\ I^2 \\ \vdots \\ I^B \end{bmatrix}. \tag{12.8}$$

The associated twice covariable impedance is written as

$$[Z_{ab}] = [Z_{bb}] + [Z_{coupling}], \tag{12.9}$$

where Z_{bb} represents the diagonal impedance elements with branch index number b:

$$[Z_{bb}] = \begin{bmatrix} Z_{11} & 0 & \cdots & 0 \\ 0 & \ddots & \ddots & \vdots \\ \vdots & \ddots & \ddots & 0 \\ 0 & \cdots & 0 & Z_{bb} \end{bmatrix}, \tag{12.10}$$

and the coupling impedance represents the extra-diagonal elements:

$$Z_{coupling}(a,b) = \begin{cases} -j\omega\zeta \quad if \quad (a,b) = \{(6,18),(18,6)\} \\ Z_t(j\omega) \quad if \quad (a,b) = \begin{Bmatrix} (2,15),(15,2), \\ (3,23),(23,3) \end{Bmatrix} \\ 0 \quad \text{elsewhere} \end{cases} . \tag{12.11}$$

12.3.2.3 TAN connectivity dedicated to full S-parameter modelling

The mesh space analysis enables one to treat the considered problem in a reduced tensorial size compared to the branch space. This size reduction can be understood via the branch-to-mesh current connectivity defined by the Einstein mute index tensorial notation:

$$I^b = C^b_m J^m. \tag{12.12}$$

By denoting the mesh indices $m = \{1, 2, \ldots, M\}$ and $n = \{1, 2, \ldots, M\}$, the connectivity matrix is expressed in (12.13).

$$[C^a_m] = [C^b_n] = \begin{bmatrix}
1 & 0 & 0 & 0 & 0 & 0 & 0 & 0 & 0 & 0 & 0 & 0 & 0 \\
0 & 1 & 0 & 0 & 0 & 0 & 0 & 0 & 0 & 0 & 0 & 0 & 0 \\
0 & 0 & 1 & 0 & 0 & 0 & 0 & 0 & 0 & 0 & 0 & 0 & 0 \\
0 & 0 & 0 & 1 & 0 & 0 & 0 & 0 & 0 & 0 & 0 & 0 & 0 \\
1 & 0 & 0 & -1 & 0 & 0 & 0 & 0 & 0 & 0 & 0 & 0 & 0 \\
0 & 0 & 0 & 0 & 1 & 0 & 0 & 0 & 0 & 0 & 0 & 0 & 0 \\
0 & 0 & 0 & 1 & -1 & 0 & 0 & 0 & 0 & 0 & 0 & 0 & 0 \\
0 & 0 & 0 & 0 & 0 & 1 & 0 & 0 & 0 & 0 & 0 & 0 & 0 \\
0 & 1 & 0 & 0 & 0 & -1 & 0 & 0 & 0 & 0 & 0 & 0 & 0 \\
0 & 0 & 0 & 0 & 0 & 0 & 1 & 0 & 0 & 0 & 0 & 0 & 0 \\
0 & 1 & 0 & 0 & 0 & 0 & -1 & 0 & 0 & 0 & 0 & 0 & 0 \\
0 & 0 & 0 & 0 & 0 & 0 & 0 & 1 & 0 & 0 & 0 & 0 & 0 \\
0 & 0 & 0 & 0 & 0 & 0 & 1 & -1 & 0 & 0 & 0 & 0 & 0 \\
0 & 0 & 0 & 0 & 0 & 0 & 0 & 0 & 1 & 0 & 0 & 0 & 0 \\
0 & 0 & 0 & 0 & 0 & 0 & 0 & 1 & -1 & 0 & 0 & 0 & 0 \\
0 & 0 & 0 & -1 & 0 & 0 & 0 & 0 & 1 & 0 & 0 & 0 & 0 \\
0 & 0 & 0 & 0 & 0 & 0 & 0 & 0 & 0 & -1 & 0 & 0 & 0 \\
0 & 0 & 0 & 0 & 0 & 0 & -1 & 0 & 0 & 0 & 1 & 0 & 0 \\
0 & 0 & 0 & 0 & 0 & 0 & 0 & 0 & 0 & 1 & -1 & 0 & 0 \\
0 & 0 & 0 & 0 & 0 & 0 & 0 & -1 & 0 & 0 & 0 & 1 & 0 \\
0 & 0 & 0 & 0 & 0 & 0 & 0 & 0 & 0 & 0 & 1 & -1 & 0 \\
0 & 0 & 0 & 0 & 0 & 0 & 0 & 0 & 0 & 0 & 0 & 1 & -1 \\
0 & 0 & 0 & 0 & 0 & 0 & 0 & 0 & 0 & 0 & 0 & 0 & 1
\end{bmatrix} . \tag{12.13}$$

It is capital to mention that the present TAN model of S-parameter requires to choose the p-first branch-to-mesh current relationship:

$$I^{1...p} = \pm J^{1...p}. \tag{12.14}$$

12.3.2.4 Mesh space analysis

By adopting the connectivity matrix transposes,

$$\left[C_a^m\right] = {}^t\left[C_m^a\right], \tag{12.15}$$

and the branch voltage respecting Kirchhoff's second law,

$$C_m^a V_a = 0, \tag{12.16}$$

the covariable source voltage in mesh space is

$$U_m = C_m^a E_a. \tag{12.17}$$

Knowing the source voltage introduced in (12.7), this mesh M-size voltage must be given by

$$[U_m] = \begin{bmatrix} U_1 & \cdots & U_p & 0 & \cdots & 0 \end{bmatrix}, \tag{12.18}$$

with

$$U_{1...p} = \pm V_{1...p}. \tag{12.19}$$

The twice covariable branch and mesh impedances Z_{ab} and Z_{mn} which are assumed as B- and M-size square matrices are linked by the tensorial equation:

$$Z_{mn} - C_a^m Z_{ab} C_n^b. \tag{12.20}$$

Therefore, the compact equation of the problem metric can be reformulated as

$$U_m = Z_{mn} J^n, \tag{12.21}$$

or

$$J^n = Y^{nm} U_m, \tag{12.22}$$

by denoting the mesh admittance which must be twice contravariable:

$$[Y^{nm}] = [Z_{mn}]^{-1}. \tag{12.23}$$

When solving this tensorial equation, the main unknowns are the mesh currents J^n.

12.3.3 Equivalent equation of multi-port black box

The proposed cable coupled with probe system can be modelled as a Z-matrix based on the method introduced in Chapter 3. The following sections describe extraction of this impedance matrix and also the associated S-matrix.

12.3.3.1 Z-matrix extraction

According to the circuit and system theory, the structure under study can be assumed as a three-port diagram introduced in Figure 12.3. This system is usually represented by the impedance $[\Xi_{\varpi\vartheta}]$ or admittance:

$$\left[\Upsilon^{\varpi\vartheta}\right] = [\Xi_{\varpi\vartheta}]^{-1} \tag{12.24}$$

with $\varpi, \theta = \{1, 2, \ldots, P\}$ being linked by the classical definition:

$$[V_{\varpi}] = [\Xi_{\varpi\vartheta}]\left[I^{\vartheta}\right] = [\Xi_{\varpi\vartheta}]\left[C_{n_1}^{\vartheta}\right][J^{n_1}] \tag{12.25}$$

With $n_1 = \{1, 2, \ldots, P\}$, the equivalent impedance matrix relation is given by

$$[J^{n_1}] = \left[C_{\vartheta}^{n_1}\right]\left[\Xi^{\varpi\vartheta}\right][V_{\varpi}] \tag{12.26}$$

with $p_{1,2,3,4} = \{1, 2, \ldots, P\}$ and $n_2 = \{P+1, 2, \ldots, M\}$, the mesh metric given in (12.24) can be rewritten as

$$^{t}\begin{bmatrix}[U_{p_1}] & 0\end{bmatrix} = \begin{bmatrix}[Z_{p_1n_1}] & [Z_{p_2n_2}] \\ [Z_{p_3n_1}] & [Z_{p_4n_1}]\end{bmatrix}\begin{bmatrix}[J^{n_1}] \\ [J^{n_2}]\end{bmatrix} \tag{12.27}$$

where

$$\begin{cases}[Z_{p_1n_1}] = [Z_{mn}]_{(1\ldots P,1\ldots P)} \\ [Z_{p_2n_2}] = [Z_{mn}]_{(1\ldots P,P+1\ldots M)} \\ [Z_{p_3n_1}] = [Z_{mn}]_{(P+1\ldots M,P+1\ldots M)} \\ [Z_{p_4n_2}] = [Z_{mn}]_{(P+1\ldots M,1\ldots P)}\end{cases}. \tag{12.28}$$

Emphatically, we have the following subtensorial relations:

$$\begin{cases}[U_{p_1}] = [Z_{p_1n_1}][J^{n_1}] + [Z_{p_2n_2}][J^{n_2}] \\ [0_{p_3}] = [Z_{p_3n_1}][J^{n_1}] + [Z_{p_4n_2}][J^{n_2}]\end{cases}. \tag{12.29}$$

Substantially, the second equation can be transformed as

$$[J^{n_2}] = -[Y^{n_2p_3}][Z_{p_3n_1}][J^{n_1}] \tag{12.30}$$

with

$$[Y^{n_2p_3}] = [Z_{p_3n_2}]^{-1}. \tag{12.31}$$

Substituting this last expression into the first equation of (12.30), we have the innovative mesh impedance to impedance matrix formulation:

$$\left[\Xi_{p_1n_1}\right] = \left[Z_{p_1n_1}\right] - \left[Z_{p_2n_2}\right]\left[Y^{n_2p_3}\right]\left[Z_{p_3n_1}\right]. \tag{12.32}$$

12.3.3.2 Extraction of *S*-matrix impedance

As forecasted in Figure 12.3, the *S*-parameters of the structure under study are a *P*-size square matrix. Let us denote

$$\left[\Lambda^{\varpi\vartheta}\right] = R_0\left[1_{PP}\right] \tag{12.33}$$

with $R_0 = 50\ \Omega$, the general reference impedance of access ports. The three-port *S*-parameters can be deduced from the *Z*-to-*S* matrix transform. Subsequently, the targeted *S*-parameter of the structure depicted in Figure 12.1 is written as

$$S_{\varpi} = \left(\left[\Xi_{\varpi\vartheta}\right] - \left[\Lambda_{\varpi\vartheta}\right]\right)\left[\Upsilon^{\vartheta}\right] \tag{12.34}$$

where

$$\left[\Upsilon^{\vartheta}\right] = \left(\left[\Xi_{\vartheta}\right] + \left[\Lambda_{\vartheta}\right]\right)^{-1}. \tag{12.35}$$

To validate this theoretical approach, a simulation and computed results will be discussed in the next section.

12.4 Validation results

The present section investigates the feasibility of the proposed TAN *S*-parameter modelling. The aforementioned POC initiated in Figure 12.1 was designed and simulated in two commercial tools ADS from Keysight Technologies® and HFSS from ANSYS®. The obtained results are described in the next sections.

12.4.1 Description of the POC structure

The following sections indicate the POC parameters and the design methods in the different simulation environments.

12.4.1.1 Description of POC HFSS design

The 3D EM design of the circular loop probe coupled to the 'I'-shape coaxial cable simulated in a full-wave solver in the HFSS environment is presented in Figure 12.6(a). The HFSS designed structure was simulated with three active lumped ports. Ports 4 and 5 are necessary in order to connect the loads Z_4 and Z_5. The detailed design of the coaxial cable braid shield with periodical cubic holes is illustrated in Figure 12.6(b). The 3D design of unshielded cable is presented in Figure 12.7.

The *S*-parameter touchstone model taking into account the lumped loads $Z_{k=\{1,\ldots,5\}}$ and R_m was extracted from the HFSS circuit design environment as depicted in Figure 12.8. The physical and geometrical parameters of the material constituting the POC structure are addressed in Table 12.1.

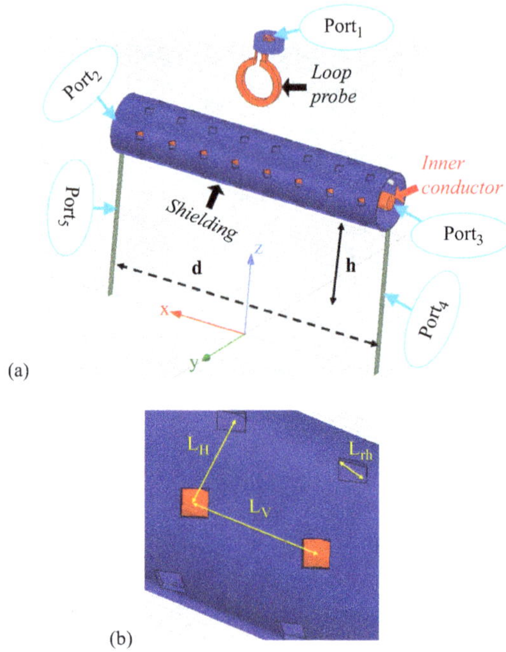

Figure 12.6 (a) 3D design of the shielded coaxial cable coupled with a circular loop probe designed in the HFSS environment and (b) zoom-in of shielded surface

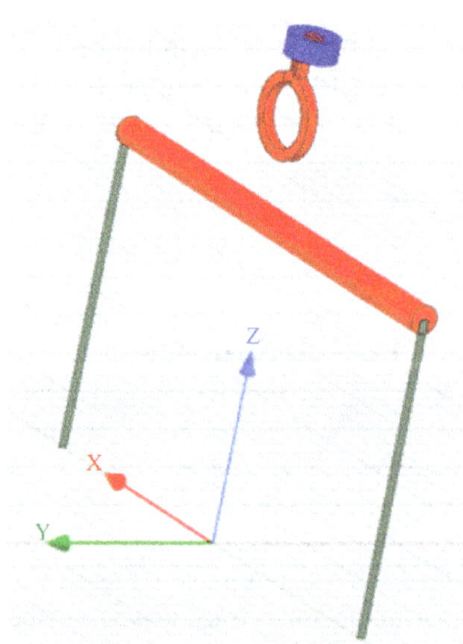

Figure 12.7 3D design of the unshielded coaxial cable coupled with a circular loop probe designed in the HFSS environment

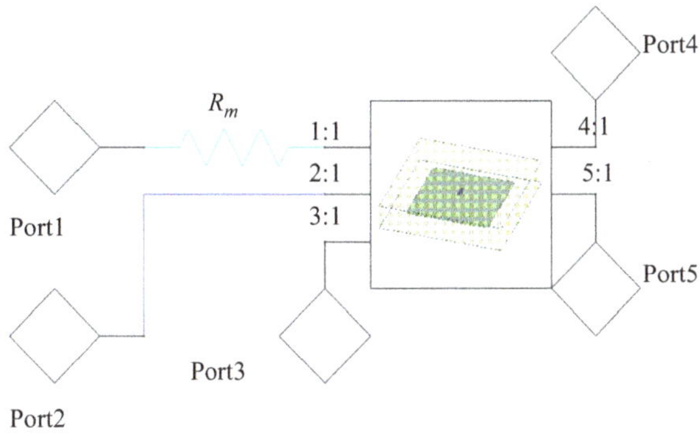

Figure 12.8 Design of co-simulation between lumped elements and 3D structure with the touchstone model of the cable structure

Table 12.1 HFSS design physical parameters

Constituting structure	**Description**	**Parameter**	**Value**
Loop probe	Material	Copper	–
	Conductor diameter	d_{Cu}	1 mm
	Radius	r	3 mm
GND plane	Material	Copper	–
	Thickness	t	35 µm
	Cable–GND distance	h	3 cm
Braid shielding	Material	Copper	–
	External radius	r_e	4.94 mm
	Internal radius	r_i	4.7 mm
	Physical length	d	50 mm
	Hole length	L_{rh}	1.3 mm
	Horizontal distance between holes	L_H	3.78 mm
	Vertical distance between holes	L_V	5.98 mm
Inner conductor	Material	Copper	–
	Radius	r_{ic}	1.34 mm
	Physical length	d	50 mm

12.4.1.2 Description of POC ADS design

The circuit schematic of the POC designed and simulated in the ADS SPICE environment is presented in Figure 12.9. This circuit was built considering the *RLC* network circuit of the loop probe, the cable external shield and the inner conductor.

The probe-shielding mutual inductance Z_{mutual} and the shielding-inner conductor transfer impedance Z_t [5–7,9–12] was calculated with the formula available in the literature. The considered circuit parameters are presented in Table 12.2.

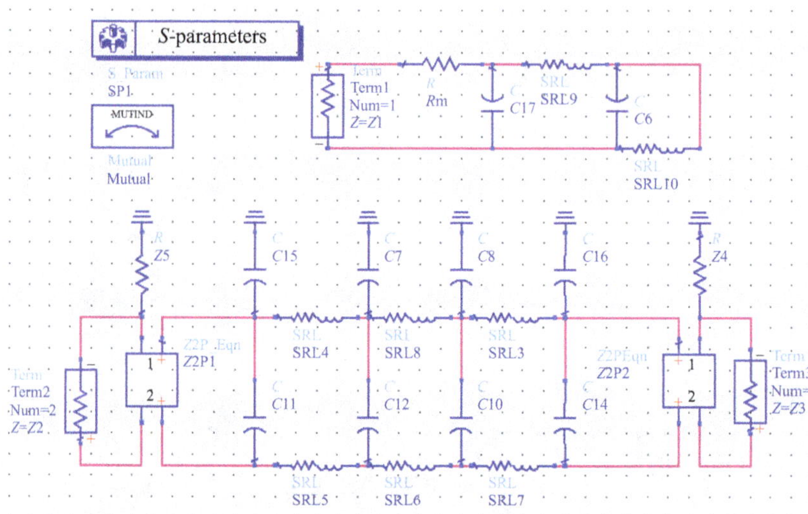

Figure 12.9 Circuit of the coaxial cable coupled with a circular loop probe designed in the ADS schematic environment

Table 12.2 ADS circuit design physical parameters

Constituting structure	Branch impedance designation	Parameters	Values
Loop probe	Matching resistance	R_m	50 Ω
	Z_4	R	1 Ω
		L	1 nH
	Z_5	C	1 pF
	Z_6	R	50 Ω
		L	1 nH
	Z_7	C	1 pF
Braid shielding	Mutual inductance	ζ	0.2 pH
	Per unit length parameters	R_{ue}	0.1 Ω/m
		L_{ue}	0.1 nH/m
		C_{ue}	0.1 nF/m
	Transfer impedance Z_t	R_t	0.5 Ω
		L_t	0.1 nH
Inner conductor	Per unit length parameters	R_{ui}	0.1 Ω/m
		L_{ui}	0.1 nH/m
		C_{ui}	0.17 pF/m

12.4.2 Comparisons of computed and simulated S-parameters

In this subsection, the POC S-parameters from 0.1 to 0.6 GHz with samples $N_{\max} = 251$ are presented. The computed from MATLAB® implementation of the TAN model routine described in Section 12.2.1 and simulated with HFSS and ADS are discussed. The influences of the probe on to the probe–cable coupling are also investigated in function of

- the parameters $x_0 = \{-20, -10, 0, 10, 20\,\text{mm}\}$ and
- the parameters $z_0 = \{5, 10, 15, 20, 25\,\text{mm}\}$.

12.4.2.1 *S*-parameter-based validation results

Figure 12.10 displays the modelled and simulated results from the shielded and unshielded cables, respectively. The *S*-parameters from TAN computation, ADS and HFSS simulations are plotted in black solid, red dashed and red dotted lines. The results reveal a good agreement between the developed fast *S*-parameter TAN modelling and simulations and highlight the trend of coupling increasing with the frequency. Acting as a passive structure, the *S*-parameters present a symmetrical aspect. The transmission coefficients which incorporate naturally the probe–cable coupling are increasing with the frequency. The TAN-simulation (TAN-ADS and TAN-HFSS)

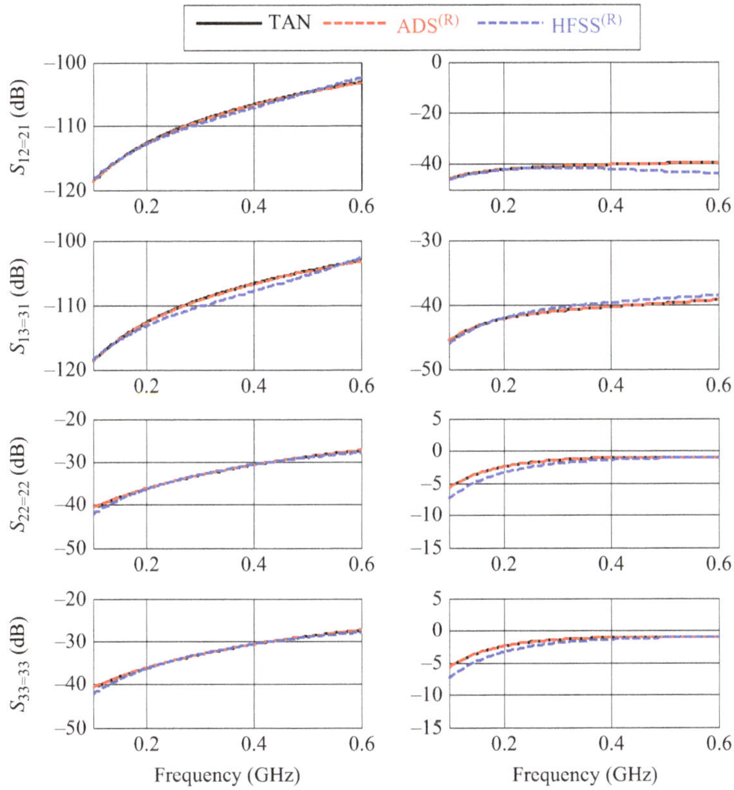

Figure 12.10 Comparisons of the reflection and transmission parameters computed with the TAN model, ADS and HFSS simulations

discrepancies between the S-parameters with indices $\xi = \{12, 13, 22, 33\}$ are assessed with the following relative error vector magnitude (REVM) formulated as

$$\varepsilon_{\xi} = \frac{\sum_{N_f=0}^{N_f=N_{\max}} \left| |S_{\xi_{TAN}}(N_f)|_{dB} - |S_{\xi_{simu}}(N_f)|_{dB} \right|}{\sum_{N_f=0}^{N_f=N_{\max}} |S_{\xi_{TAN}}(N_f)|_{dB}} \tag{12.36}$$

Table 12.3 addresses the computed REVM. In contrast to the full-wave simulation, the TAN-ADS REVM are lower than 0.1% against the TAN-HFSS, which is lower than 1.5%.

12.4.2.2 SE analyses

Figure 12.11 displays the plot of the SE derived from S-parameters via (12.6). This parameter follows literally the correlation between the transmission coefficients exposed in Figure 12.10. The SE decreases from about 84 to 58 dB with the frequency under the probe-braided cable structure analysis condition.

12.4.2.3 Study of influence of x_0

Figure 12.12(a) and (b) displays the (x_0,f)-dependent cartographies of HFSS transmission coefficients S_{12dB} and S_{13dB}, respectively, when the probe abscissa position x_0 varies from −20 to 20 mm. The coupling level corresponding to S_{12dB} increases when the probe is positioned at the side $x_0 < 0$. The inverse effect appears with S_{13dB}

Table 12.3 REVM ε_ξ (%) between TAN–ADS and TAN–HFSS results

Cable	*REVM(|S_ξ|)*	ε_{12}	ε_{13}	ε_{22}	ε_{33}
Shielded	TAN–ADS	0.08	0.08	0.07	0.07
	TAN–HFSS	0.31	0.63	1.48	1.50
Unshielded	TAN–ADS	0.01	0.01	0.04	0.04
	TAN–HFSS	3.48	1.32	20.61	20.61

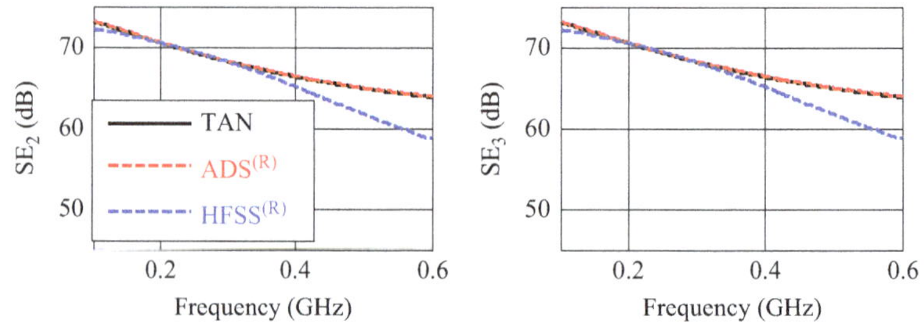

Figure 12.11 Comparisons of SE at cable Port$_2$ and Port$_3$ computed with the TAN model, ADS and HFSS simulations

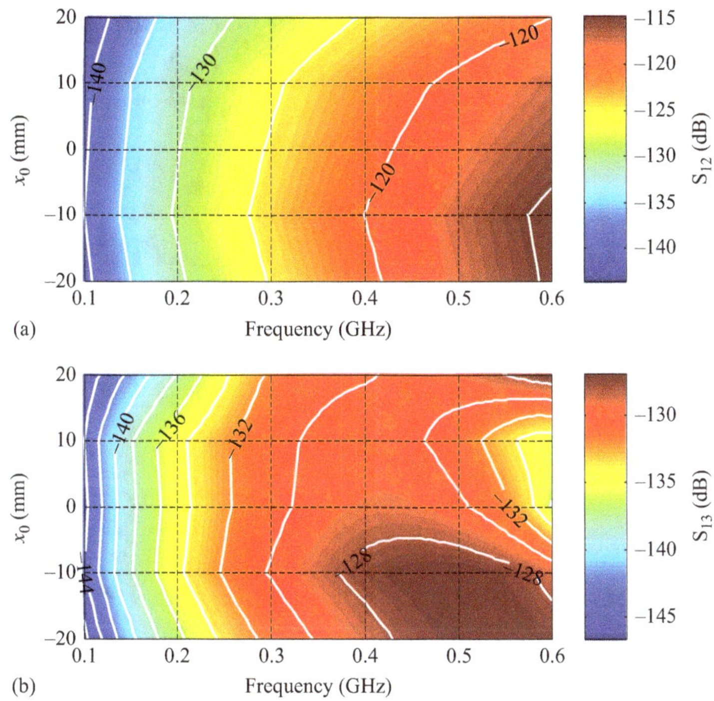

Figure 12.12 Cartographies of the transmission parameters: (a) S_{12dB} and (b) S_{13dB}, versus the geometrical position-frequency (x_0,f)

when the probe is positioned at the side $x_0 > 0$. At the frequencies $f > 0.5$ GHz, the transmission coefficient presents about 10 dB differences from −20 to 20 mm.

12.4.2.4 Study of influence of z_0

In this case of study, the probe is placed at the centre $x_0 = 0$. Figure 12.13(a) and (b) displays the (z_0,f)-dependent cartographies of HFSS transmission coefficients S_{12dB} and S_{13dB}, respectively, when z_0 increases from 5 to 25 mm. In this case, as expected naturally, the coupling levels S_{12dB} and S_{13dB} decrease when the probe position z_0 increases. The variation is less than 10 dB at $f < 0.2$ GHz and can be more than 15 dB at $f > 0.5$ GHz.

12.4.3 Discussion on the advantages and drawbacks of the TAN model

The computation time of the results depicted in Figure 12.10 was about 10 ms with the TAN model compared to more than 1 h with HFSS full-wave simulations. The computations are carried out by HFSS software using a PC equipped with

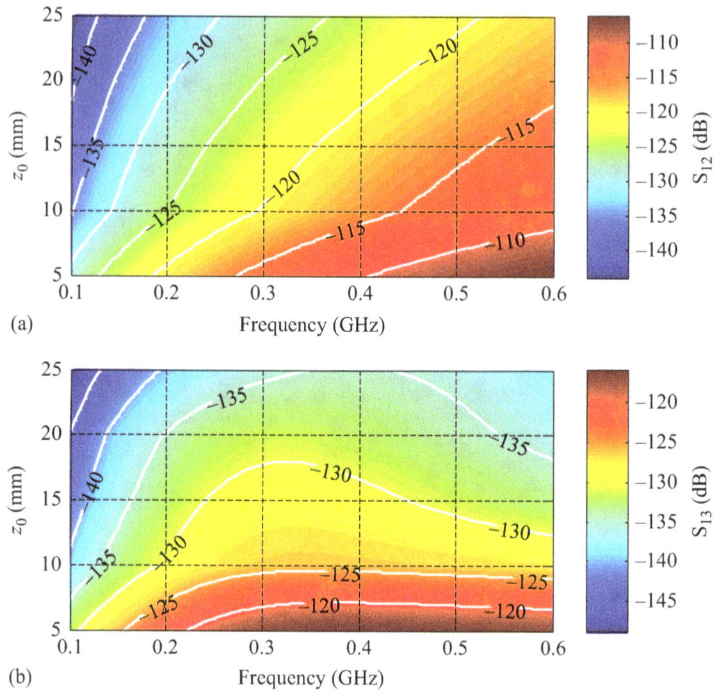

Figure 12.13 Cartographies of the transmission parameters: (a) S_{12dB} and (b) S_{13dB}, versus the geometrical position-frequency (z_0,f)

Table 12.4 Comparison between the advantages and drawbacks of the TAN model, HFSS and ADS commercial tools

Method	HFSS	ADS	TAN model
Pre-processing	2 h	10 min	1 min
Post-processing	2 min	0 s	0 s
Computation time	20 min	Some ms	Some ms
Flexibility	Good for the 3D design, not good for the lumped circuits but need to set accurate meshing accuracies	Not good for the 3D approach but well matched to the lumped circuit simulations	Possibilities to adapt to 3D structures and well matched to the lumped circuit simulations

a single-core processor Intel® Xeon® CPU E5-1620 v4 @ 3.50 GHz and a 32 GB physical RAM with 64-bits Windows 10.

The result comparisons uphold the effectiveness of the TAN model for the coaxial braid shielded cable EMC pre-analysis. As summed up in Table 12.4, the TAN concept presents considerable advantages and weaknesses.

12.5 Conclusion

An innovative TAN model of braided shielded coaxial cable coupled with a loop probe coupling is investigated. The modelling methodology is described. After the problem formulation and the associated parameter definitions, the classical equivalent circuit is introduced. Then, the equivalent graph of the probe–coaxial cable structure is elaborated. The mathematical description of the problem is established by the means of the TAN formalism. The TAN modelling is developed following Kron's method which is based on the branch and mesh space analyses. The problem solution is expressed with the full S-parameter in function of the frequency extracted from the three-port structure impedance matrix.

The relevance of the coaxial cable TAN model is approved with a POC-simulated with two different commercial tools well known in EMC engineering. As expected, a very good agreement between the S-parameters and the associated SE from some to several hundred megahertz is realized. The main advantages and weaknesses of the TAN model are listed.

References

[1] E. P. Fowler, "Cables and connectors-Their contribution to electromagnetic compatibility," in Proc. of IEEE Int. Symp. Record. Electromagnetic Compatibility, 1992, Anaheim, CA, USA, 17–21 Aug. 1992, pp. 329–333.

[2] A. Mushtaq, S. Frei, K. Siebert, and J. Barenfanger, "Analysis of shielding effectiveness of HV cable and connector systems used for electric vehicles," in Proc. of Int. Symp. Electromagnetic Compatibility (EMC EUROPE), 2013, Brugge, Belgium, 2–6 Sep. 2013, pp. 241–246.

[3] IEC 62153-4-15:2015, Metallic communication cable test methods: Part 4-15: EMC – Test method for measuring transfer impedance and screening attenuation or coupling with triaxial cell, Edn. 1, 2015.

[4] S. A. Schelkunoff, "The electromagnetic theory of coaxial transmission lines and cylindrical shields," Bell Syst. Tech. J., vol. 13, no. 4, pp. 532–579, 1934.

[5] R. B. Schulz, V. C. Plantz, and D. R. Brush, "Shielding theory and practice," IEEE Trans. Electromagn. Compat., vol. 30, pp. 187–201, 1988.

[6] B. Demoulin and L. Kone, "Shielded cable transfer impedance measurements high frequency range 100 MHz–1 GHz," IEEE-EMC Newsletter, pp. 42–50, 2011.

[7] A. Morriello, T. M. Benson, A. P. Duffy, and C. F. Cheng, "Surface transfer impedance measurement: A comparison between current probe and pull-on braid methods for coaxial cables," IEEE Trans. Electromagn. Compat., vol. 40, no. 1, pp. 69–76, 1998.

[8] Tubular Structure Shielding Products. [ONLINE] Available at: http://www.federalmogul.com/en-US/Pages/Home.aspx. [Accessed 1 February 2019]

[9] T. Kley, "Optimized single-braided cable shields," IEEE Trans. Electromagn. Compat., vol. 35, no. 1, pp. 1–9, 1993.

[10] S. Sali, "An improved model for the transfer impedance calculations of braided coaxial cables," IEEE Trans. Electromagn. Compat., vol. 33, no. 2, pp. 139–143, 1991.
[11] L. K. Warne, W. L. Langston, L. I. Basilio, and W. A. Johnson, "First principles cable braid electromagnetic penetration model," Prog. Electromag. Res. (PIER) B, vol. 66, pp. 63–89, 2016.
[12] P. Xiao, P.-A. Du, and B. Zhang, "An analytical method for radiated electromagnetic and shielding effectiveness of braided coaxial cable," IEEE Trans. Electromagn. Compat., vol. 61, no. 1, pp. 121–127, 2019.
[13] L. W. Nagel and D. O. Pederson, SPICE (Simulation Program with Integrated Circuit Emphasis), Memorandum No. ERL-M382, University of California, Berkeley, 1973.
[14] L. W. Nagel, SPICE2: A Computer Program to Simulate Semiconductor Circuits, Memorandum No. ERL-M520, University of California, Berkeley, 1975.
[15] S. Caniggia and F. Maradei, "SPICE-like models for the analysis of the conducted and radiated immunity of shielded cables," IEEE Trans. Electromagn. Compat., vol. 64, no. 4, pp. 606–616, 2004.
[16] W. J. R. Hoefer, "The transmission-line matrix method – Theory and applications," IEEE Trans. Microwave Theory Tech., vol. 33, no. 10, pp. 882–893, 1985.
[17] M. Ney, "Method of moments as applied to electromagnetic problems," IEEE Trans. Microwave Theory Tech., vol. 33, no. 10, pp. 972–980, 1985.
[18] J. Jin, The Finite Element Method in Electromagnetics, John Wiley & Sons, New York, USA, 1993.
[19] M. Laour, R. Tahmi, and C. Vollaire, "Experimental evaluation and FDTD method for predicting electromagnetic fields in the near zone radiated by power converter systems," Turk. J. Electr. Eng. Comput. Sci., vol. 25, pp. 1460–1471, 2017.
[20] A. E. Ruehli, G. Antonini, J. Esch, J. Ekman, A. Mayo, and A. Orlandi, "Nonorthogonal PEEC formulation for time- and frequency-domain EM and circuit modeling," IEEE Trans. Electromagn. Compat., vol. 45, no. 2, pp. 167–176, 2003.
[21] HFSS – 3D Electromagnetic Simulation Software. [ONLINE] Available at: https://www.ansys.com/fr-fr/products/electronics/ansys-hfss. [Accessed 1 February 2019]
[22] CST – Computer Simulation Technology. 3D Electromagnetic Simulation Software. [ONLINE] Available at: https://www.cst.com/. [Accessed 1 February 2019]
[23] ADS – RF/microwave Electronic Circuit Simulation Software. [ONLINE] Available at: https://www.keysight.com/main/software.jspx?cc=FR [Accessed 1 February 2019]
[24] ICEy – Interference & Compatibility Evaluation System. [ONLINE] Available at: https://speag.swiss/products/icey/ICEy-overview/. [Accessed 1 February 2019]

[25] R. Otin, J. Verpoorte, and H. Schippers, "Finite element model for the computation of the transfer impedance of cable shields," IEEE Trans. Electromagn. Compat., vol. 53, no. 4, pp. 950–958, 2011.
[26] F. M. Tesche, "Development and use of the BLT equation in the time domain as applied to a coaxial cable," IEEE Trans. Electromagn. Compat., vol. 49, no. 1, pp. 3–11, 2007.
[27] H. Z. Xu and W. J. Wang, "Application of the BLT equation for radiation field calculation for double transmission line," Adv. Mater. Res., vols. 756–759, pp. 4292–4295, 2013.
[28] E. R. Rajkumar, B. Ravelo, M. Bensetti, and P. Fernandez-Lopez, "Application of a hybrid model for the susceptibility of arbitrary shape metallic wires disturbed by EM near-field radiated by electronic structures," PIER B, vol. 37, pp. 143–169, 2012.
[29] C. D. Taylor, R. S. Sattewhite, and C. W. Harrison, "The response of a terminated two-wire transmission line excited by a non-uniform electromagnetic field," IEEE Trans. Antennas Propag., vol. E, pp. 987–989, 1965.
[30] A. K. Agrawal and H. J. Price, "Transient response of multiconductor transmission lines excited by a non-uniform electromagnetic field," IEEE Trans. Antennas Propag., vol. 18, pp. 432–435, 1980.
[31] F. Rachidi, "Formulation of the field to transmission line coupling equations in terms of magnetic excitation field," IEEE Trans. Electromagn. Compat., vol. 35, no. 3, 1993, pp. 404–407.
[32] Y. Liu and B. Ravelo, "Fully time-domain scanning of EM near-field radiated by RF circuits," Progress In Electromagnetics Research (PIER) B, vol. 57, pp. 21–46, 2014.
[33] K. Kurokawa, "Power waves and the scattering matrix," IEEE Trans. Microwave Theory Tech., vol. 13, no. 2, pp. 194–202, 1965.
[34] R. Mittra, A. Suntives, M. S. Hossain, and J. Ma, "A systematic approach for extracting lumped circuit parameters of microstrip discontinuities from their S-parameter characteristics," Int. J. Numer. Model., vol. 5, no. 1, Special Issue: Network Methods in Field Modelling, pp. 59–72, 2002.
[35] I. Timmins and K.-L. Wu, "An efficient systematic approach to model extraction for passive microwave circuits," IEEE Trans. Microwave Theory Tech., vol. 48, no. 9, pp. 1565–1573, 2000.
[36] T. L. Moss and Y. Chen, "Mesh analysis for extracting the S-parameters of lumped element RF and microwave circuits," Int. J. Electr. Eng. Educ., vol. 51, no. 4, pp. 330–339, 2014.
[37] F. Fesharaki, T. Djerafi, M. Chaker, and K. Wu, "S-Parameter deembedding algorithm and its application to substrate integrated waveguide lumped circuit model extraction," IEEE Trans. Microwave Theory Tech., vol. 65, no. 4, pp. 1179–1190, 2017.
[38] G. Kron, Tensor Analysis of Networks, Wiley, New York, Chapman & Hall, London, 1939.
[39] O. Maurice, Elements of Theory for Electromagnetic Compatibility and Systems, Bookelis, Aix en Provence, France, 2017.

[40] O. Maurice, A. Reineix, P. Hoffmann, B. Pecqueux, and P. Pouliguen, "A formalism to compute the electromagnetic compatibility of complex networks," Advances in Applied Science Research, vol. 2, no. 5, pp. 439–448, 2011.
[41] E. Rogard, B. Azanowsky, and M. Ney, "Comparison of radiation modeling techniques up to 10 GHz-application on a microstrip PCB trace," IEEE Trans. EMC, vol. 52, no. 2, pp. 479–486, 2010.
[42] S. O. Land, R. Perdriau, M. Ramdani, O. Maurice, and M. Drissi, "Kron simulation of field-to-line coupling using a meshed and a modified Taylor cell," in Proc. of 9th Int. Workshop on Electromagnetic Compatibility of Integrated Circuits (EMC Compo), Nara. Japan, 15–18 Dec. 2013, pp. 15–20.
[43] Z. Xu, B. Ravelo, O. Maurice, J. Gantet, and N. Marier, "Radiated EMC Kron's model of 3D multilayer PCB aggressed by broadband disturbance," IEEE Trans. EMC, Early View, 2019, pp. 1–9.
[44] B. Ravelo and O. Maurice, "Kron–Branin modeling of Y-Y-tree interconnects for the PCB signal integrity analysis," IEEE Trans. EMC, vol. 59, no. 2, pp. 411–419, 2017.
[45] R. Casagrande, O. Maurice, and A. Reineix, "High frequency bundles modeling," in Proc. of Int. Symp. Electromagnetic Compatibility (EMC EUROPE), 2013, Brugge, Belgium, 2–6 Sep. 2013, pp. 264–269.

Chapter 13

Nonlinear behaviour conduced EMC model of an ADC-based mixed PCB under radio-frequency interference (RFI)

Fayu Wan[1]

Abstract

This chapter develops a modelling of nonlinear (NL) behaviour of a mixed circuit consisted of an analogue-to-digital converter (ADC). Under normal operation, the test circuit is disturbed by radio-frequency interference (RFI). The NL model of the electromagnetic compatibility (EMC) behaviour is established with the consideration of memory effect and nonlinearity. The developed NL EMC model enables one to predict the direct shift behaviour of ADCs. As proof-of-concept (POC), an ADC demo board is realized and tested. The calculated behavioural model and measurement results are in good agreement. The model helps one to better understand the behaviour of the digital and mixed electronic circuits under RFI. The present chapter is organized in three main sections. The second section of this chapter is focused on the proposed mixed circuit as a microcontroller (μC) NL-conducted EMC with electromagnetic interference signal analysis. For better understanding, the study is focused essentially on the case of RFI higher than the ADC's sampling rate. The relationship between DC offset and sampling rate will be established analytically. The mathematical model enables one to predict the μC behaviour. The analytical approach is described with the NL model represented by the output voltage expressed in polynomial function of input. The third section of this chapter examines the validation results with μC-based mixed circuit POC. Comparison between simulations and measurements by considering the NL RFI model is discussed. Last, the fourth section of this chapter is the conclusion.

Keywords: Mixed PCB, conducted EMC, analogue-to-digital convertor, radio-frequency interference (RFI), nonlinear behaviour modelling, direct shift

[1]School of Electronic and Information Engineering, Nanjing University of Information Science & Technology (NUIST), Nanjing, China

13.1 Introduction

Because of tremendous integration density, the influence of electromagnetic interference (EMI) in modern electronic systems remains an attractive topic for the electromagnetic compatibility (EMC) research engineers [1,2]. Characterization techniques highlight the parasitic noises from different component sources of EMI emissions [1,3,4]. The EMI issues affect all electronic devices [5–11]. The EMI control was applied to the boat systems [5]. Generally, the radio-frequency interference (RFI) from cellular phones may affect undesirably implant and wireless medical devices [6,8]. Moreover, it was verified that the electronic wireless devices can interfere critically to nuclear power plants [9]. In electrotechnical engineering, it was confirmed from EMC study that the induction motor drive systems generate significantly EMIs [10]. Added to the power converter EMIs, the electric drives can generate EMC issues in the car [11]. In the same area, the DC link inverters create conducted EMI noises in the electric vehicles [12]. In the most of cohabitation, the digital electronic devices are fragilized by the electronic power systems. This critical case concerns the EMI disturbances from wireless power transfer system of an automotive charger on the analogue-to-digital converters (ADCs) [13].

Because of the EMI immunity low level, the analogue, digital and mixed low-voltage networks are the most vulnerable to EMI aggressions [14,15]. The conducted EMI can be emitted during the switching activities of digital (field-programmable gate array (FPGAs), microcontrollers (μCs), ADC–DAC, …) and power circuits (DC–DC converters, transistors …) [16–18]. To assess the EMI emission, time-domain characterizations were introduced [19,20]. But because of the experimental cost, alternative investigations based on simulations and analytical modellings have been proposed to predict the immunity and failures of chip and packaged integrated circuits [21–25].

Because of EMI issues, the mixed circuits must be analysed with electrical behavioural modelling [26,27]. The mixed electronic components are used widely in closed-loop control systems. Among the most critical mixed circuits, further studies are needed to predict the influences of RFIs onto the ADCs [27,28]. The ADC bridges digital components and analogue components in a system, and ADC errors usually lead to degradation of the system by transmitting wrong massages or breaking the balance of closed loops. The ADC also affects the stability and control performance of the system directly [14,29,30]. In the automotive field, the complex EM environment (EME) causes EMC problems of the ADC. The focus on ADC as research target is made due to the fact that ADC is the one that gets affected primarily in the EME [31,32]. A bunch of researches have been done to figure out the performance of ADC with EMI [33,34]. However, researches pertaining to the analysis of the effects of EME on the ADC are rare. Thus, the effects of the ADC in the presence of EMIs remain a challenging problem for EMC research engineers. The existing works on this topic refer to the immunity measurement and the immunity model of the ADC [35,36]. Nevertheless, the models are only compared with the measurement immunity result and the output signal nonlinear (NL) distortion of the ADC. So far, such distortions are still not easy to predict correctly. It is clear that the RFI could lead to the offset of ADCs' output signal. Several parameters influence the ADC performance in the presence of RFI, such as sampling rate, NL behaviour and memory effect of ADC. A model that contains these features is proposed to predict the ADC outputs. According to the results of these models, it is possible to figure out which parameter influences the ADC performance mostly.

This chapter focuses on the NL distortion of mixed commercial circuit output signal caused by an out-of-band interference. The originality of the model concerns the consideration of NL effects and the complexity to fit the memory effect during the mixed circuit under test (CUT) represented by the ADC conversion process.

The present chapter is organized in three main sections. Section 13.2 is focused on the proposed mixed circuit as μC NL conducted EMC with EMI signal analysis. For better understanding, the study is focused essentially on the case of RFI higher than the ADC's sampling rate. The relationship between DC offset and sampling rate will be established analytically. The mathematical model enables one to predict the μC behaviour. The analytical approach is described with the NL model represented by the output voltage expressed in polynomial function of input. Section 13.3 examines the validation results with a μC-based mixed circuit proof-of-concept (POC). Comparison between simulations and measurements by considering the NL RFI model is discussed. Last, Section 13.4 is the conclusion.

13.2 Description of the NL model of a mixed circuit under study

The present section is focused on the description of the proposed modelling methodology of the electronic mixed IC referenced PIC18F458 as the CUT under study. The NL behaviour modelling of this mixed circuit will be investigated in the remainder part of the chapter.

13.2.1 EMC problem formulation

As aforementioned the present theory is built with the S-parameter modelling. To do this, the inter-combination of the hybrid coupler and the feedback chain S-parameters will be considered. Then, the innovative topology will be characterized from the extraction of transmission and group delay.

The present study is mainly focused on the conducted RFI circuit modelling. The scenario of the problem about the proposed NL EMC effect is illustrated by the diagram shown in Figure 13.1.

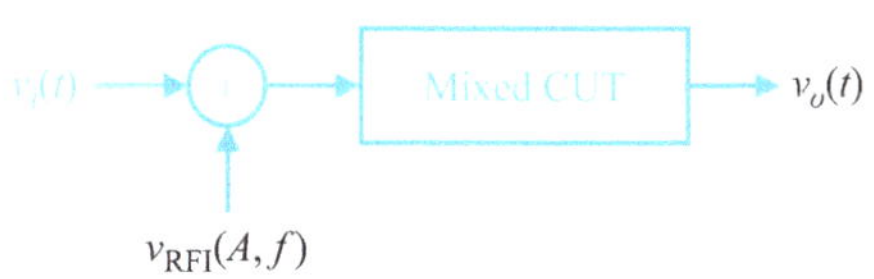

Figure 13.1 Illustrative diagram of RFI affection to the CUT

In the analytical aspect, the overall circuit must be analysed with consideration of time-dependent input test signal v_i, added by the RFI perturbation:

$$v_{\mathrm{RFI}}(t) = V\sin(\omega t) \tag{13.1}$$

specified by its angular frequency $\omega = 2\pi f$ and the amplitude V. Then, the observed output denoted by v_o is generated from the victim PCB. The ADC status of the CUT aggressed by the RFI perturbation must be detected by the μC.

13.2.2 Analytical definition of RFI

Based on the classical circuit theory, the Volterra series can be used to represent the relationship between the ADC input signal and the output signal. The circuit NL behaviour can be formulated analytically by using a Taylor series. Because of NL effect, the time-dependent output test voltages can be written as

$$v_o(t) = \sum_{k=1}^{\infty} a_k v_{\mathrm{RFI}}^k(t) \tag{13.2}$$

where $a_k (k = \{1,2,3,\ldots\})$ are the real coefficient of NL effect. Substituting (13.1) into (13.2), it yields the output expression:

$$v_o(t) = V_{\mathrm{DC}} + \sum_{k=1}^{\infty} A_k \sin(k\omega t + \theta_k) \tag{13.3}$$

with the harmonic phase shifts θ_k and amplitudes A_k of harmonics:

$$\omega_k = k\omega \tag{13.4}$$

having DC component

$$V_{\mathrm{DC}} = a_2 \times \frac{V^2}{2} + a_4 \times \frac{3V^4}{8} + \cdots \tag{13.5}$$

It is seen that the second-order coefficient and other even coefficients give rise to the signal at 0 Hz, which is called DC shift signal. The DC shift signal is proportional to even-order coefficient and the even square of the input amplitude V. Meanwhile, because of the RFI effect, the parameters of the output v_o are frequency dependent. In certain frequency ranges, the output signal can have a positive DC shift with the reference signal.

13.2.3 Output voltage analytical expression

By denoting T the sampling interval, the discrete output at the instant time $t = nT$, with n being an integer, is related to the sampled input, which is defined by

$$v_o(nT) = \sum_{k=1}^{\infty} a_k v_i^k(nT) \tag{13.6}$$

Figure 13.2 represents the equivalent circuit seen at the input pin of μC under study. In this circuit, C_{pin} is the input capacitance, and C_{hold} is the charge-holding capacitor.

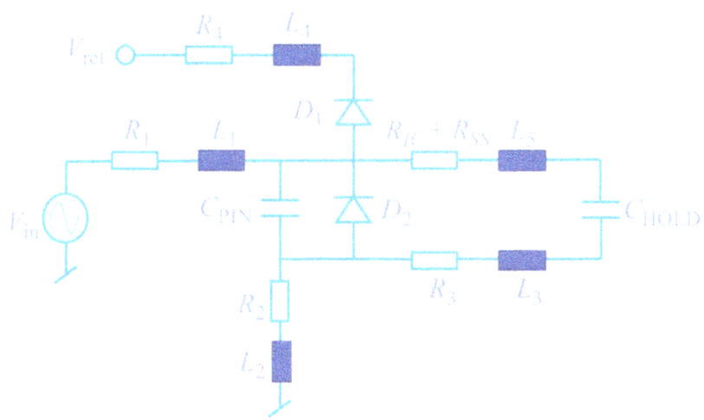

Figure 13.2 Input equivalent circuit model of the CUT

Table 13.1 Parameter of the cut equivalent input circuit

Parameters	Type	Value
V_{ref}	Voltage	5 V
R_1	Resistor	14 Ω
R_2		12.51 Ω
R_3		0.6 Ω
R_4		0.5 Ω
R_{IC}		IC7841
R_S		IC7841
L_1	Inductor	7.4 nH
L_2		5.24 nH
L_3		0.56 nH
L_4		10 nH
L_5		0.86 nH
C_{PIN}	Capacitor	5 pF
C_{HOLD}		120 pF
D_1	Diode	–
D_2	Diode	–

The two diodes are the clamp diode structure. L_1 and C_1 describe package passive effect. L_4 and R_4 represent the substrate behaviour. L_5 and R_5 stand for the pad effect and the components R_2, R_3, L_2 and L_3 give the effect of the rail.

Considering the effects of capacitor and inductor components contained in the CUT, it is obvious that the memory effect exists in ADC model. To reproduce the memory effect, this chapter hypothesizes that the output signal is related to a period of signal before the sampling point and the length of the period is supposed to be τ, and the output expression is shown as

$$v_o(n) = \frac{\int_{nT-\tau}^{nT} \left[\sum_{k=1}^{\infty} a_k v_i^k(t)\right] dt}{\tau} \tag{13.7}$$

To reproduce the additional DC shift, we proposed a black box model describing the relationship between the additional DC shift and the RFI. The black box model is considered as a Taylor series given in (13.6), and knowing the amplitude of RFI, we can determine the additional DC shift by the following relation:

$$V_{\text{add}} = \sum_{k=0}^{\infty} b_k V^k \tag{13.8}$$

We integrated the additional DC shift equations into the nonlinearity model. Therefore, we get the following expression:

$$v_o(n) = V_{\text{ref}} + \frac{\int_{(n-1)T}^{nT} \left[b_0 + \sum_{k=1}^{5} \left[a_k v_i^k(t) + b_k V^k \right] \right] dt}{\tau} \tag{13.9}$$

The memory length τ is set to be the sampling period T, and parameters vary with the RFI frequency. However, it is hard to calculate the output signal while the coefficients a and b change at the same time. Compared to the example in Figure 13.2, the DC shift caused by NL behaviour is considered to be smaller than additional DC shift. So, we fixed parameters a and b to get the model shown in the following equation, which shows higher accuracy:

$$v_o(n) = V_{\text{ref}} + \frac{\sum_{k=1}^{5} a_k \left[\int_{(n-1)T}^{nT} v_i^k(t) dt \right] + T \sum_{k=0}^{5} b_k V^k}{\tau} \tag{13.10}$$

13.3 Methodology of the EMC NL modelling of a mixed circuit

The present subsection describes the methodology of an EMC NL modelling mixed circuit.

13.3.1 Nonlinear model flow design and an input–output equivalent transfer circuit

As shown in Figure 13.3, the NL behaviour model of ADC contains the memory effect and Taylor series. The input signal is composed of a reference DC signal and an RFI, and the model gives the DC shift component. The input signal is composed of a reference DC signal and an RFI, and the model gives the DC shift component.

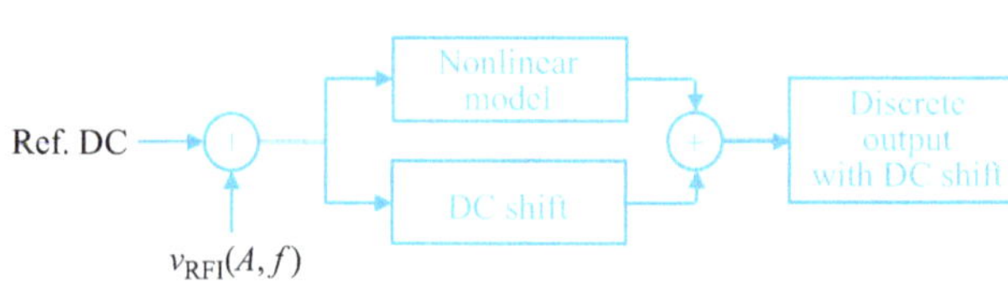

Figure 13.3 Proposed nonlinear behaviour model of ADC

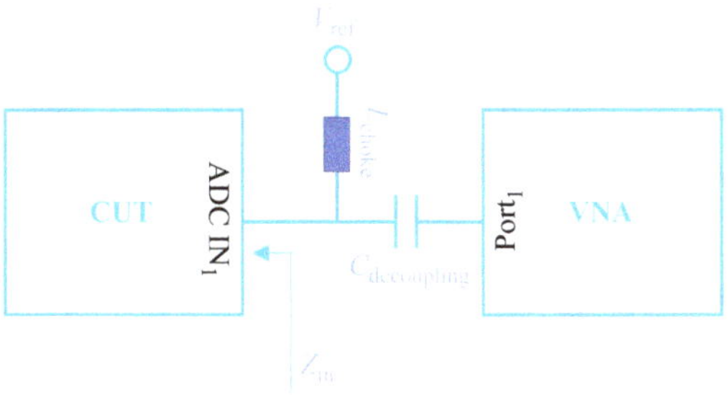

Figure 13.4 S_{11} *measurement method with VNA*

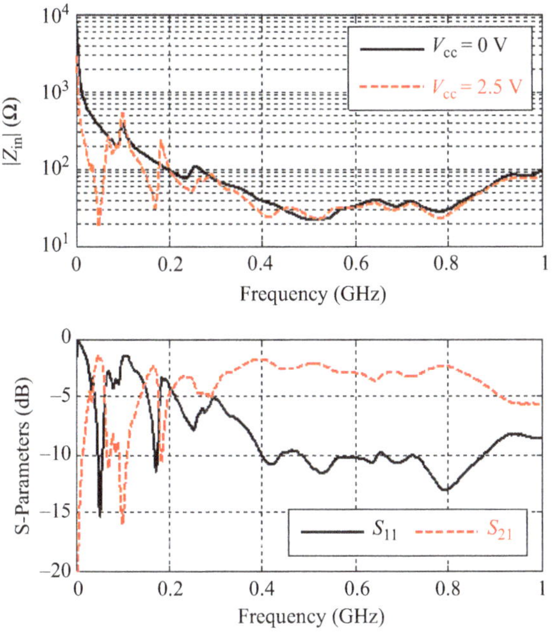

Figure 13.5 Frequency domain results

The CUT input equivalent model enables one to simulate the input impedance up to several GHzs and can be verified experimentally with the set-up shown in Figures 13.4 and 13.5 by using a vector network analyser (VNA).

13.3.2 Description of monitoring code implemented in MATLAB®

To perform this conducted EMC susceptibility, innovative interface of MATLAB program was developed and implemented in the µC. This program functioning will be described in the following sections.

13.3.2.1 Embedded software implemented into the μC

To run the tested μC software, MPLAB IDE V8.10, which is the Windows Integrated Development Environment for development systems tool, is used in this application. If the program is a C-language instead of an assembler language, Microchip C-compilers C18 is also needed. In-Circuit Debuggers help one to debug and programmer the program into μC.

The basic idea of the μC software is to convert an input analogue signal. This signal may come from a sensor or feedback signal into a digital signal, calculate and send it to PC by RS232. The main functions of the embedded software are as follows:

- Function 1: Convert the sample analogue signal into digital signal.
- Function 2: Compare the digital signal with the reference signal (DC 2.5 V), and calculate the result.
- Function 3: Send the data representing the results to a PC for display.

To increase the robustness of the software, some redundant information is added. The third function is realized by an ADC_USART module. The ADC USART module flowchart is shown in Figure 13.6. The data format is '0x aa 55 HH LL', HH is the high 2 bits, LL is the low 8 bit of the 10 bit data. The redundancy of '0x aa 55' (10101010 01010101) before the data HH LL helps one to identify the high bit from the low-bit data and also increase the reliability of the communication with a PC. The RS232 could be disturbed by the emission interference constituted by the signal generator. A practically simple format could help one to avoid the faulty data. The

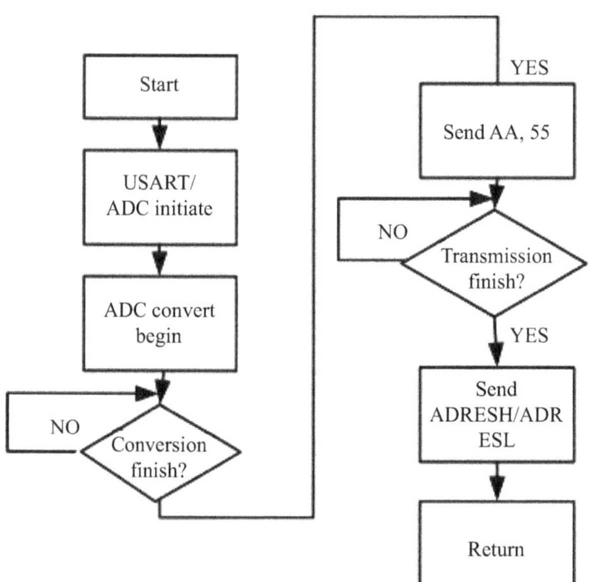

Figure 13.6 Flowchart of ADC–USART

defensive program to increase the reliability of a μC-based system will be introduced in the following section.

13.3.2.2 Implementation of automatic program

Automatic control and measurement allow one not only to save time but are to perform accurate calculations. To realize the automatic measurement, a MATLAB program is used to communicate with the signal generator and the μC test board. The MATLAB measurement algorithm is shown in Figure 13.7. The ADC output signal for each perturbation is saved. To establish such a data library, for each perturbation, 500 output data points are saved. Thus, this data library is composed of 117 files (the frequency is from 15 kHz to 1 GHz, the variable step is 10%, equaling 117 points). Each file contains 19 × 500 output data points. The test signal amplitude is varied from 0 to 18 dB m, equaling 19 points. For each perturbation signal which contains one period of the disturbance signal, we save 500 data points.

To check out the developed theory effectiveness, simulation and experimental application results will be discussed in the next section.

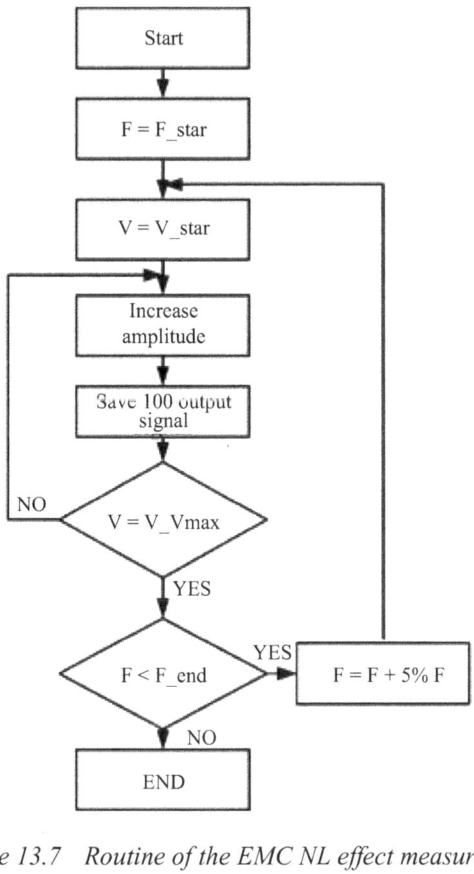

Figure 13.7 Routine of the EMC NL effect measurement

13.4 Validation results with parametric analyses

To verify the previous analytical investigation efficiency, the present section introduces the validation results with a POC. The comparisons between the calculations and experimental test results will be discussed.

13.4.1 Experimental set-up configuration

The test bench is set to confirm the developed EMC NL model. The test bench is constituted by an injection circuit, an Agilent® signal generator, a μC demon board and a PC with references indicated in Table 13.2. The test bench is shown in Figures 13.8 and 13.9. The μC ADC status under the pressure of RFI is detected by the μC. The RS232 is used to realize the communication between μC and computer, and a C-language program is used to convert the analogue signal (DC voltage and perturbation signal) into a digital signal and send it to the computer by RS232. Another MATLAB program receiving the result from the ADC is designed in the computer.

13.4.2 Empirical characteristics of RFI

The RFI is injected at the input port of the ADC. Because of the non-ideal power reference voltage (4.9 V) of the successive approximation (SAR) ADC and the reference DC signal (2.46 V), the sampling rate of the ADC is limited to about 30 kHz (by ADC conversion time); if the input interference signal is below 15 kHz, the conversion result and input signal can be considered as a linear relationship. Figure 13.10 highlights the output signal behaviours in function of the RFI frequencies.

Table 13.2 References of employed equipment

Equipment	Manufacturer	Reference
Signal generator	Agilent	N9310A
Oscilloscope	Agilent	DSO9404A
CUT	Texas Instruments®	PIC18F458

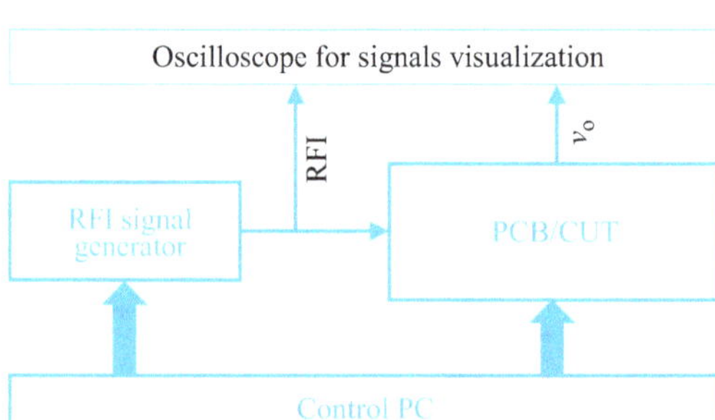

Figure 13.8 Illustrative diagram

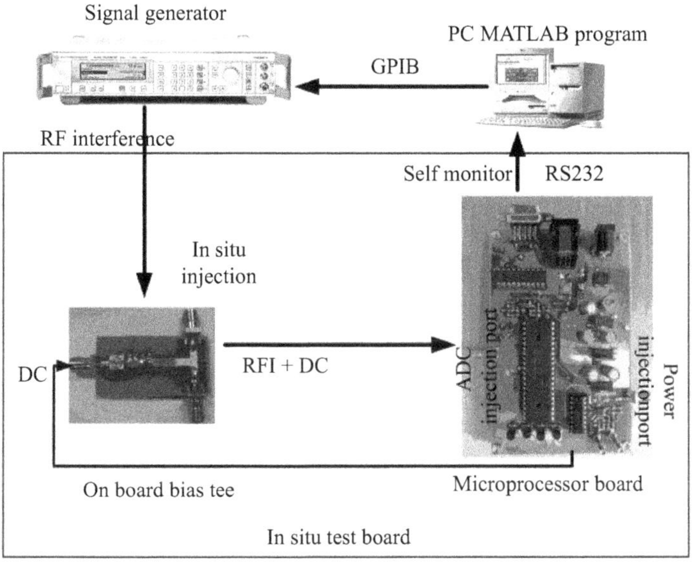

Figure 13.9 Photograph of experimental set-up of the CUT under RFI

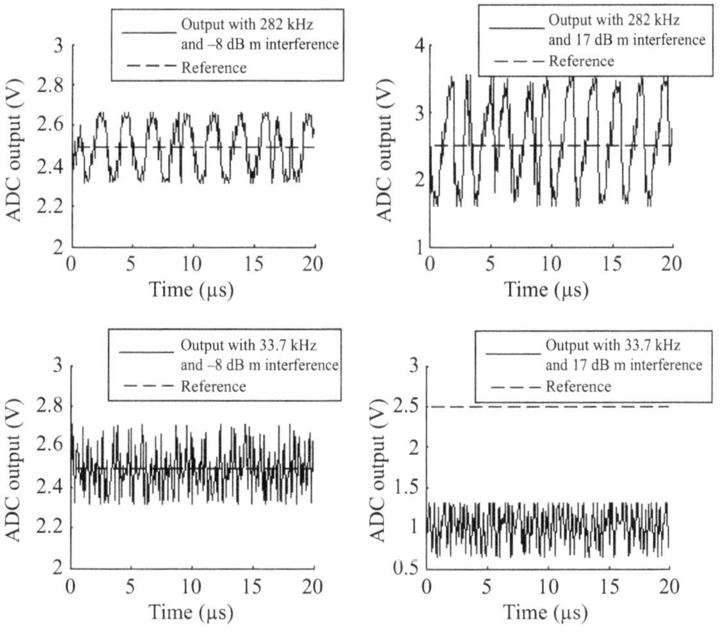

Figure 13.10 Output signals of ADC under interference in different frequencies: (a) top left, (b) top right, (c) bottom left and (d) bottom right

The input signal is a DC signal mixed with a single sinusoidal signal in this case; the DC signal is always in bandwidth of the ADC, so the single sinusoidal signal is the only consideration. The NL distortion is a harmonic distortion that is particularly harmful because the DC shift appears. The DC shift [37] is generated by the accumulation of an asymmetrically rectified signal, which is caused by the voltage clamp diode in this case. The DC shift depends on the even-order NL behaviour, and DC shift is a DC effect which is very difficult to remove or filter. As shown in Figure 13.10(c), the output signal has about 6–7 LSB DC shift compared with the reference signal. As the amplitude of the interference signal increases, the DC shift increases, as shown in Figure 13.10(d), the output signal has about 60–70 LSB DC shift compared with the reference signal.

13.4.3 Discussion on simulation and test results

After MATLAB computation of the model expressed in (13.7), the parameters were trained with input and output of the ADC and were used to get the result shown in Figures 13.11 and 13.12. Table 13.3 presents the basic coefficients used to calculate the predicted μC output.

The performed EMC NL modelling computation speed is of about tens milliseconds for 251 samples by using a PC equipped a single-core processor Intel®Core i7-4790 CPU @ 3.60 GHz and a 32 GB physical RAM with 64-bits Windows 7. The computed result allows one to claim the nonlinearity with memory. This NL effect could lead to the DC component shift. But it cannot reach such a huge DC shift in Figures 13.11 and 13.12. Then, there must be some other NL parameters need to be considered in the model.

With the model described by (13.10), we get the output results with a 33.7 MHz RFI signal, and the result is shown in Figures 13.13 and 13.14. The result matches the examples shown in Figure 13.10 well. To confirm this model, more comparisons

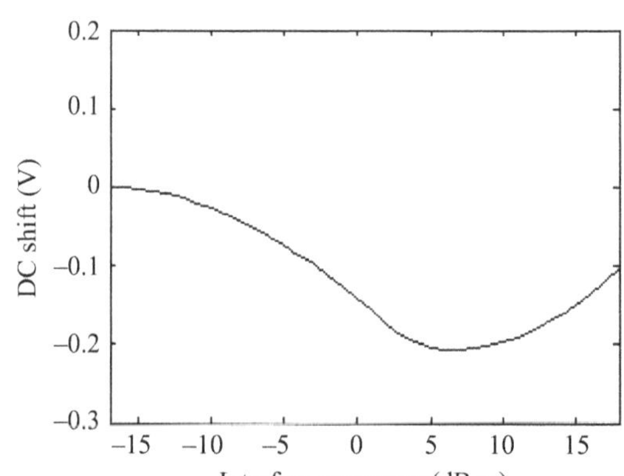

Figure 13.11 DC shift of NL ADC model

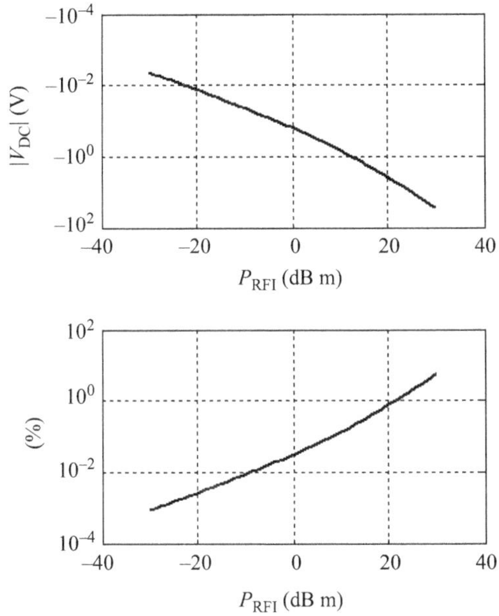

Figure 13.12 DC shift of an NL ADC model

Table 13.3 Routine of the EMC NL effect measurement

f (MHz)	a_1	a_2	a_3	a_4	a_5
12.1	1.072	−0.163	0.112	−0.027	0.002
33.7					
43.1					
f (MHz)	b_1	b_2	b_3	b_4	b_5
12.1	1	−7.2	7.2	−2.7	0.35
33.7	0.87	−8.5	9.45	−3.89	0.55
43.1	0.7	−8	9.34	−3.94	0.57

are required. The parameters of expression (13.10) are trained to compare with other measured results. Figures 13.13 and 13.14 represent the comparison between the simulated and measured DC shifts caused by 12.1 and 43.1 MHz input RFI. The simulated and measured results are very well correlated. The errors are less than 5% from kHz to several hundreds of MHzs. The results show that the proposed model predicts the behaviour of ADC very well.

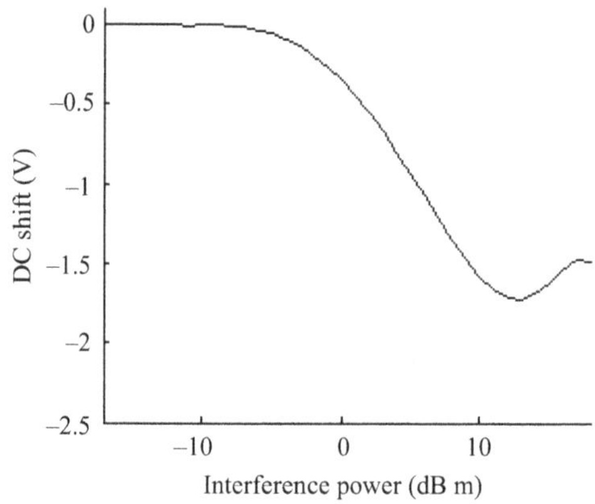

Figure 13.13 Simulated DC shift phenomenon caused by 33.7 MHz sinusoidal signal

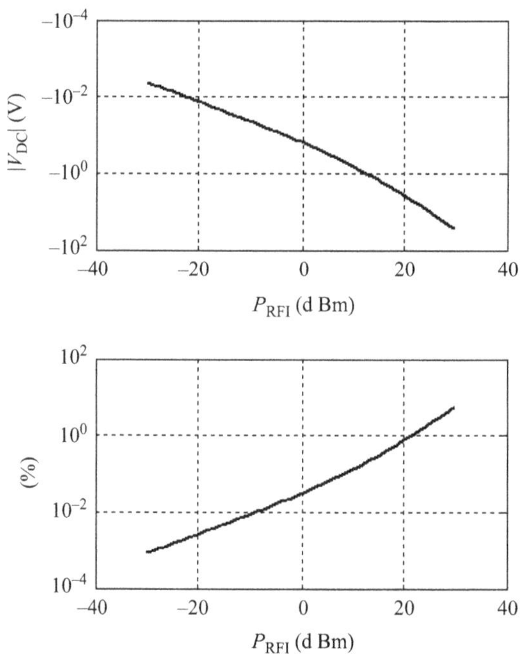

Figure 13.14 Comparison between simulated and measured results

13.5 Conclusion

An NL EMC modelling of mixed circuit is developed. The modelling method consists analytically in the combination of memory effect and Taylor series. The developed conducted susceptibility EMC model is dedicated to predict the error of discrete outputs in function of the RFI amplitude and frequency. The relevance of the model was checked by computing DC shift phenomena appeared in the typically mixed circuit. After developing a control and monitoring program specific to the demo board considered, computed and measured results in good agreement were obtained by considering of μC under the sine RFI perturbations.

In the future, this model may help us to improve the performance of mixed transceiver PCB. For example, just like the linearization of power amplifier, we can try to combine predistortion method [31] and this model to improve the PCB immunity.

Acknowledgement

This research work was supported in part by NSFC under Grant 61971230, and in part by Jiangsu Distinguished Professor program and Six Major Talents Summit of Jiangsu Province (2019-DZXX-022), and in part by the Priority Academic Program Development of Jiangsu Higher Education Institutions (PAPD) fund.

References

[1] J. P. Mills, Electromagnetic Interference Reduction in Electronic Systems, Englewood Cliffs, NJ: Prentice Hall, 1993, pp. 232–233.

[2] H. W. Denny, Grounding for Control of EMI, Gainsville, VA: Interference Control Technologies, 1993.

[3] "Military Standard MIL-STD-461F:2007. Requirements for the control of electromagnetic interference characteristics of subsystems and equipment," Department of Defense Interface Standard, MIL-STD-461F:2007, 2007.

[4] IEC62433-2 - EMC IC modelling – Part 2: Models of integrated circuits for EMI behavioural simulation – Conducted emissions modelling (ICEM-CE) – Edition 1.0, 2008.

[5] H. A. Huynh, J.-M. Jo, W. Nah and S. Y. Kim, "EMC Qualification Methodology for Semicustom Digital Integrated Circuit Design," IEEE Trans. EMC, vol. 58, no. 5, 2016, pp. 1629–1641.

[6] R. V. Carstensen, EMI Control in Boats and Ships, Gainsville, VA: Interference Control Technologies, 1979.

[7] C.-K. Tang, K.-H. Chan, L.-C. Fung and S.-W. Leung, "Electromagnetic Interference Immunity Testing of Medical Equipment to Second- and Third-Generation Mobile Phones," IEEE Trans. EMC, vol. 51, no. 3, 2009, pp. 659–664.

[8] P. M. Mariappan, D. R. Raghavan, S. H. E. Abdel Aleem and A. F. Zobaa, "Effects of Electromagnetic Interference on the Functional Usage of Medical

Equipment by 2G/3G/4G Cellular Phones: A Review," J. Adv. Res., vol. 7, no. 5, 2016, pp. 727–738.

[9] C. Lennerz, H. Pavaci, C. Grebmer, *et al.*, "Electromagnetic Interference in Cardiac Implantable Electronic Devices: Is the Use of Smartphones Safe?," J. Am. Coll. Cardiol., vol. 69, no. 1, 2017, pp. 108–110.

[10] S.-H. Ye, Y.-S. Kim, H.-S. Lyou, M.-S. Kim and J. Lyou, "Verification of Electromagnetic Effects From Wireless Devices in Operating Nuclear Power Plants," Nucl. Eng. Technol., vol. 47, no. 6, 2015, pp. 729–737.

[11] L. Ran, S. Gokani, J. Clare, K. J. Bradley and C. Christopoulos, "Conducted Electromagnetic Emissions in Induction Motor Drive Systems," IEEE Trans. Power Electron., vol. 13, no. 4, 1998, pp. 757–776.

[12] S. Guttowski, S. Weber, E. Hoene, W. John and H. Reichl, "EMC Issues in Cars with Electric Drives," in Proc. of 2003 IEEE Symp. on EMC, Symp. Record (Cat. No.03CH37446), vol. 2, Boston, MA, USA, 18–22 Aug. 2003, pp. 777–782.

[13] C. Chen, X. Xu and D. M. Divan, "Conductive Electromagnetic Interference (EMI) Noise Evaluation for an Actively Clamped Resonant DC Link Inverter (ACRDCLI) for Electric Vehicle (EV) Traction Drive Applications," in Proc. of Conf. Record of the 1997 IEEE Industry Applications Conference Thirty-Second IAS Annual Meeting (IAS'97), 5–9 Oct. 1997, New Orleans, LA, USA, pp. 1550–1557.

[14] S. Kong, B. Bae, D. Kim, H. Kim, C. Song and J. Kim, "Electromagnetic Interference on Analog-to-Digital Converters From High-Power Wireless Power Transfer System for Automotive Charger," in Proc. of 2016 IEEE Int. Symp. on EMC, Ottawa, ON, Canada, 25–29 Jul. 2016, pp. 370–373.

[15] P. Nuutinen, A. Pinomaa, J.-P. Ström, T. Kaipia and P. Silventoinen, "On Common-Mode and RF EMI in a Low-Voltage DC Distribution Network," IEEE Trans. Smart Grid, vol. 5, no. 5, 2014, pp. 2583–2592.

[16] J. P. Leca, N. Froidevaux, P. Dupré, G. Jacquemod and H. Braquet, "EMI Measurements, Modelling, and Reduction of 32-Bit High-Performance Microcontrollers," IEEE Trans. EMC, vol. 56, no. 5, 2014, pp. 1035–1044.

[17] B. Vrignon, S. Delmas Bendhia, E. Lamoureux and E. Sicard, "Characterization and Modelling of Parasitic Emission in Deep Submicron CMOS," IEEE Trans. EMC, vol. 47, no. 2, 2005, pp. 382–387.

[18] L. Ren, T. Li, S. Chandra, *et al.*, "Prediction of Power Supply Noise From Switching Activity in an FPGA," IEEE Trans. EMC, vol. 56, no. 3, 2014, pp. 699–706.

[19] H. H. Park, S.-H. Song, S.-T. Han, T.-S. Jang, J.-H. Jung and H.-B. Park, "Estimation of Power Switching Current by Chip-Package-PCB Cosimulation," IEEE Trans. EMC, vol. 52, no. 2, 2010, pp. 311–319.

[20] F. Krug and P. Russer, "The Time-Domain Electromagnetic Interference Measurement System," IEEE Trans. EMC, vol. 45, no. 2, 2003, pp. 330–338.

[21] M. S. Sarto, "Time-Domain Analysis of Electromagnetic Interference From Broadband Digital Signal Transmission on Finite-Length Power Networks," IEEE Trans. Power Deliv., vol. 13, no. 4, 1998, pp. 972–978.

[22] T. Sudo, H. Sasaki, N. Masuda and J. L. Drewniak, "Electromagnetic Interference (EMI) of System-on-Package (SOP)," IEEE Trans. Adv. Packaging, vol. 27, no. 2, 2004, pp. 304–314.
[23] T. Su, H. Zheng, Y. Yang and Z. Wang, "The Frequency Behaviour of the Intrinsic Immunity of the On-Chip Transistor Circuit," Progr. Electromagnet. Res. Lett., vol. 45, 2014, pp. 111–114.
[24] T. Steinecke, D. Hesidenz and E. Miersch, "EMI Modeling and Simulation in the IC Design Process," in Proc. of 2006 17th Int. Zurich Symp. on EMC, Singapore, Singapore, 27 Feb.–3 Mar. 2006, pp. 594–597.
[25] J. J. Laurin, S. J. Zaky and K. G. Kalmain, "EMI-Induced Failures in Crystal Oscillators," IEEE Trans. EMC, vol. 33, no. 4, 1991, pp. 334–342.
[26] F. Fiori, "Prediction of RF Interference Effects in Smart Power Integrated Circuits," in Proc. of IEEE Int. Symp. on EMC, Washington DC, USA, 21–25 Aug. 2000, pp. 345–347.
[27] W. Prodanov, M. Valle, R. Buzas and H. Pierscinski, "A Mixed-Mode Behavioral Model of a Controller-Area-Network Bus Transceiver: A Case Study," in Proc. of 2007 IEEE Int. Behavioral Modeling and Simulation Workshop, San Jose, CA, USA, 20–21 Sep. 2007, pp. 67–72.
[28] D. Dallet, D. Slepicka, Y. Berthoumieu, D. Haddadi and P. Marchegay, "[ADC Characterization in Time Domain] Frequency Estimation to Linearize Time-Domain Analysis of A/D Converters," IEEE Trans. Instrum. Meas., vol. 55, no. 5, 2006, pp. 1536–1545.
[29] F. Wan, F. Duval, X. Savatier, A. Louis and B. Mazari, "Effects of Conducted Electromagnetic Interference on Analogue-to-Digital Converter," Electron. Lett., vol. 47, no. 1, 2011, pp. 23–25.
[30] T. Liu, C. Zhou, Z. Wang, S. Zhao and Z. Liang, "Characterisation of System Clock Frequency Effects on Electromagnetic Immunity of Microcontroller Input Buffer," Electron. Lett., vol. 51, no. 23, 2015, pp. 1861–1863.
[31] F. Wan, F. Duval, X. Savatier, A. Louis and M. Belahcene, "Study of Susceptibility of an MCU Control System in the Automotive Field," in Proc. of 2010 Asia-Pacific Int. Symp. on EMC, Beijing, China, 12–16 Apr. 2010, pp. 610–613.
[32] H. H. Boo, S. W. Chung and J. L. Dawson, "Adaptive Predistortion Using a $\Delta\Sigma$ Modulator for Automatic Inversion of Power Amplifier Nonlinearity," IEEE Trans. CAS II: Express Briefs, vol. 56, no. 12, 2009, pp. 901–905.
[33] J. Wu, B. Li, W. Zhu, H. Wang and L. Zheng, "Investigations on the EFT Immunity of Microcontrollers With Different Architectures," Microelectron. Reliab., vol. 76–77, 2017, pp. 708–713.
[34] A. Ayed, T. Dubois, J. Levant and G. Duchamp, "Immunity Measurement and Modeling of an ADC Embedded in a Microcontroller Using RFIP Technique," IEEE Trans. EMC, vol. 57, no. 5, 2015, pp. 955–962.
[35] T. Su, M. Unger, T. Steinecke and R. Weigel, "Degradation of the Conducted Radio Frequency Immunity of Microcontrollers Due to Electromagnetic Resonances in Foot-Point Loops," IEEE Trans. EMC, vol. 54, no. 4, 2012, pp. 772–784.

[36] J. Gros, G. Duchamp, A. Meresse and J. Levant, "Electromagnetic Immunity Model of an ADC for Microcontroller's Reliability Improvement," Microelectron. Reliab., vol. 49, no. 9–11, 2009, pp. 963–966.

[37] X. Gao, C. Tian, L. Lao, Y. Chen and Y. Chen, "Improved Direct Power Injection Model of 16-bit Microcontroller for Electromagnetic Immunity Prediction," J. Central South Univ. Technol., vol. 18, no. 6, 2011, pp. 2031–2035.

Chapter 14

Far-field prediction combining simulations with near-field measurements for EMI assessment of PCBs

Dominik Schröder[1], *Sven Lange*[1], *Christian Hangmann*[1], *and Christian Hedayat*[1]

Abstract

Using near-field (NF) scan data to predict the far-field (FF) behaviour of radiating electronic systems represents a novel method to accompany the whole RF design process. This approach involves so-called Huygens' box as an efficient radiation model inside an electromagnetic (EM) simulation tool and then transforms the scanned NF measured data into the FF. For this, the basic idea of the Huygens' box principle and the NF-to-FF transformation are briefly presented. The NF is measured on the Huygens' box around a device under test using an NF scanner, recording the magnitude and phase of the site-related magnetic and electric components. A comparison between a full-wave simulation and the measurement results shows a good similarity in both the NF and the simulated and transformed FF. Thus, this method is applicable to predict the FF behaviour of any electronic system by measuring the NF. With this knowledge, the RF design can be improved due to allowing a significant reduction of EM compatibility failure at the end of the development flow. In addition, the very efficient FF radiation model can be used for detailed investigations in various environments and the impact of such an equivalent radiation source on other electronic systems can be assessed.

Keywords: PCB, near-field-to-far-field transformation, Huygens' box, virtual equivalent emission model for cosimulative EMI characterization, radiated EMC

Used symbols and abbreviations

Typographic markings

Vectors	letters with arrow (in this case 3D!)	e. g. $\vec{E}$
Matrices	bold capital letters	e. g. $\mathbf{A}$

[1]Department ASE, Fraunhofer ENAS, Paderborn, Germany

Indices	subscript run indices	e. g. $\vec{E}_2$
Sign indices	subscript signs, which define the calculation area	e. g. D_+
Incident field/wave	inc as indices	e. g. $\vec{E}_{inc}$
Total field	tot as indices	e. g. $\vec{E}_{tot}$
Dyad	two bars	e. g. $\bar{\bar{g}}_{\mathrm{e}}$
Derivative	∂ before that	e. g. $S = \partial D$
Reference point	apostrophe	e. g. $\vec{r}\,'$
Scalar product	$\cdot$ as arithmetic operator	e. g. $\vec{x} \cdot \vec{y}$
Vector product	$\times$ as arithmetic operator	e. g. $\vec{x} \times \vec{y}$
Convolution	$*$ as arithmetic operator	e. g. $x * y$

Most important symbols

Symbol	Meaning
A	area
$\vec{B}$	magnetic flux density
$\vec{D}$	electric flux density
D_+	inner volume of the enclosing surface S
D_-	outer volume of the enclosing surface S
δ	delta distribution
$\vec{E}$	electrical-field strength
ϵ	permittivity
ϵ_0	electrical-field constant
ϵ_{r}	relative permittivity
η	magnetic dipole moment density (auxiliary quantity)
Φ	electrical potential
Φ_{SL}	single-layer potential
Φ_{DL}	double-layer potential
$\vec{H}$	magnetic-field intensity
$G(\vec{r}, \vec{r}\,')$	Green's function
$\bar{\bar{g}}_{\mathrm{e}}$	electric Green's dyad
$\bar{\bar{g}}_{\mathrm{m}}$	magnetic Green's dyad
$\bar{\bar{I}}$	Green's unity dyad
$\vec{j}$	electrical volume flow density
$\vec{J}$	electrical surface current density
$\vec{K}$	electric displacement current density
k	wave number $k = \omega\sqrt{\epsilon\mu}$

j	imaginary unit
κ	electrical conduction
λ	wavelength
Δ	Laplace operator $\Delta = \nabla^2$
$\vec{M}$	magnetic displacement current density (auxiliary quantity)
μ	permeability
μ_0	magnetic-field constant
μ_r	relative permeability
$\vec{n}$	normal vector
∇	nabla operator $\nabla = \vec{\nabla} = \vec{e}_x(\partial/\partial x) + \vec{e}_y(\partial/\partial y) + \vec{e}_z(\partial/\partial z)$
$\vec{P}_{el}$	electrical polarization
$\vec{P}_{mag}$	magnetization
ρ_{el}	electric charge density
ρ_{mag}	magnetic space charge density (auxiliary quantity)
S	enclosing surface
σ	electrical surface charge density
ω	circular frequency $\omega = 2\pi f$
V	volume
Y	wave admittance $Y = Z^{-1}$
Y_0	wave admittance in vacuum $Y_0 = {Z_0}^{-1}$
Z	wave impedance $Z = \sqrt{(\mu/\epsilon)}$
Z_0	wave impedance in vacuum $Z = \sqrt{(\mu_0/\epsilon_0)}$

Abbreviations

Abbreviation	Meaning
DUT	device under test
EM	electromagnetic
EMC	electromagnetic compatibility
EMI	electromagnetic interferences
MoM	method of moments
NFS	near-field scanner
PCB	printed circuit board
PEC	perfect electric conductor
PMC	perfect magnetic conductor

14.1 Introduction

Due to the trend of electronic systems, modules and components getting smaller and more power efficient, current applications like (Industrial) Internet of Things, artificial intelligence in the edge and smart system in general become possible. A lot of these systems use a wireless communication interface using intentional electromagnetic waves. In addition to these intentional wireless communication, unintentional electromagnetic waves are also generated nearly everywhere [1].

Taking into account this trend, i.e., a combination of higher integration levels and smaller operating voltages, and the intentional and unintentional electromagnetic spectrum, most of the electronic modules and systems are getting more susceptible to inter- and intra-electromagnetic interferences. Inter-electromagnetic interferences includes disturbances from other modules and systems, while intra-EMI comprises influences from other parts of the same module or system. The susceptibility to electromagnetic interferences is an essential property of security/safety relevant applications, e.g., autonomous driving and human–machine interactions. Hence, all systems and especially security relevant applications have to be designed and tested on module level as well as on system level regarding electromagnetic compatibility.

When targeting an efficient development process of electromagnetic compatibility-compliant electronic modules and systems, two major methodologies can be used to accompany the whole design phase. These consist of simulations on one side and measurements on the other side. They can be utilized separately or in combination [2,3].

When utilizing simulations first, an efficient and accurate model of the system has to be generated. In the next step, this model and its results have to be validated. Once the validation is successful, several simulations with varying module parameters and properties can be performed in order to characterize the module's behaviour for different setups and architectures. The advantages of performing simulations are due to the fact that numerous diverse module setups and architectures can be tested without spending much time and resources on building test benches and on constructing different modules. In addition, simulations allow the combination of various situations. Nevertheless, during a modelling process some assumptions are made in order to get a not-too-complex model. Therefore, simulation models also exhibit a region of validity that is directly influenced by these assumptions.

To get meaningful measurement results, first the experiment itself has to be designed properly (design of experiment). At the second step, the test bench has to be implemented and calibrated before actually performing the measurements. Finally, the behaviour of the tested module can be characterized within the given architecture constraints of the module and the test bench. One essential advantage of measurements is a more realistic result also considering parameter variance and drifts. However, a characterization with measurement is more expensive and limited to the currently used model and the corresponding test bench.

In order to determine the electromagnetic interferences and electromagnetic compatibility behaviour of electronic modules and systems, different measurement methodologies can be used separately or as mentioned previously in combination with simulations. To identify emitted EMI and immunity to incident interferences on module level, measurements with TEM cells (transverse electromagnetic cells) are applicable. However, it may not be feasible to use measurements in a TEM cell when studying electromagnetic interactions between different modules and systems.

When considering interactions between modules, often conducted measurements, anechoic chambers for far-field (FF) measurements in 3- or 10-m distance and near-field (NF) scanners (NFSs) and probes for electromagnetic field measurements in close vicinity are utilized. Although FF measurement results are relevant for certification regarding EMI and EMC of all electronic components, modules and systems, they can hardly yield a detailed insight into behaviour of the device under test. That means, if, e.g., a device under test exceeds the critical value of FF measurements, it is challenging to locate the root cause of this violation by just analysing the performed FF characterizations.

NF measurements can be performed in a close vicinity of the device under test in order to obtain insight into the electromagnetic behaviour of a component, module and system allowing one to precisely detect the root of a radiation causing an immunity failure due to EMI. Nevertheless, NF scanning cannot directly be used to evaluate the EMC threshold values to pass tests for CE, FCC and/or CCC-certification.

Often it is not feasible to use only measurements or only simulation due to a lack of information or because of needed extensive analyses. Therefore, in order to efficiently accompany the design process of electromagnetic compatibility and electromagnetic interferences compliant modules and systems, this chapter combines the two mentioned domains, i.e., simulation and measurement. Moreover, an approach using the strength of both FF and NF measurement and characterization will be discussed, which is directly related to the combination of measurement and simulation.

In Section 14.2, NF measurements of magnetic and electric fields in magnitude and phase of printed circuit boards are introduced enabling a detailed analysis of its electromagnetic behaviour. Then in Section 14.3, the theoretical basics for the NF-to-FF transformation by the Huygens' principle* in three dimension (3D) are introduced and explained mathematically by using examples. Within Section 14.4, NF measurements and 3D simulations are combined in that way that these 3D simulations include the real NF measurement data in order to characterize NF behaviour within different environments. The next step is the estimation of FF behaviour based on NF measurements and NF-to-FF transformation utilizing 3D-simulations and the Huygens' principle. Finally, an example of NF-to-FF transformation based on the described combination of NF measurement and 3D simulation, including the principal of Huygens' box, is given revealing the strength of the presented method. This chapter ends up with a conclusion.

*Even if NF measurement data are not given on surfaces of an ideal cuboid, NF measurement data may be "corrected" using the method of [4].

14.2 Near-field scanning fundamentals

14.2.1 Near- and far-field definition

FF measurements are typically performed in a few meter distance from the device under test, while NF scanning is done in the close vicinity to it. So the first thing which has to be considered when talking about NF scanning is where the FF ends and the NF begins. Basically the area around an electronic system can be split into three subareas: reactive NF, radiating NF and FF [5,6].

Within the reactive NF region, the electrical structures highly interact with their surroundings. The radiating NF, also called *Fresnel* region, is the transition area between the reactive NF and the FF. Both NF regions share the characteristic that the phase angle and amplitude of the relation between the E- and H-field vectors are different for each spatial point and distance. In the FF region, also called *Fraunhofer* region, the angular field relation is independent of the distance to the device. The different regions are exemplary depicted in Figure 14.1.

The transition between those three regions is not distinct. The borders differ not only by geometry and frequency but also by the literature [5,7,8]. Within this chapter, they are defined as follows: D is the length of the radiating element, R is the distance to the device under test and λ is the wavelength of the emitted field [5]:

The reactive NF predominates for

$$R < 0.62\sqrt{\frac{D^3}{\lambda}} \tag{14.1}$$

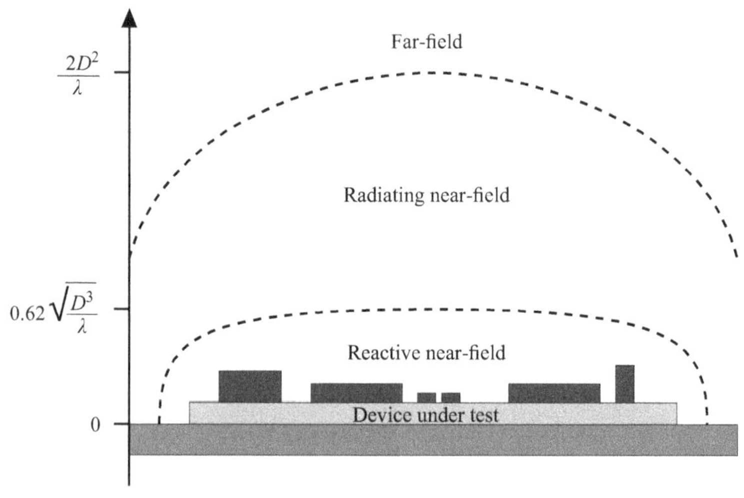

Figure 14.1 Near-field and far-field regions around a DUT

followed by the radiating NF, which is confined between

$$0.62\sqrt{\frac{D^3}{\lambda}} \leq R < \frac{2D^2}{\lambda}. \tag{14.2}$$

Hence, the FF starts at

$$\frac{2D^2}{\lambda} \leq R, \tag{14.3}$$

which is called *Fraunhofer* distance [9].

With the definition of the different regions, it can be considered in which area a field probes has to be placed for a specific purposes. When facing EMC issues which do not correspond to a wireless communication signal like Wi-Fi or Bluetooth, the radiated field strengths are usually unintended. Hence, the signals are expected to have low level in the NF of the electric device. Due to this fact and in order to have a good spatial resolution of the measured field strength, the NF probe is located very close to the device under test, so within the reactive NF region [10,11]. For antenna or radiation pattern measurements in the NF, the probe preferably has to be placed within the radiating NF region not to disturb the device under test [12]. To perform those measurements in the FF, the probe consequentially has to be placed in the FF region.

14.2.2 Radiation pattern

The result of EMC or antenna measurements whether they are done in the NF with an NFS or in the FF within an anechoic chamber often results in the aforementioned radiation pattern. In a radiation pattern or directive gain $D(\theta, \phi)^{\dagger}$, the phase of the field relation does not depend on the distance to the device. To get to a radiation pattern from an NF measurement, an NF-to-FF transformation has to be carried out, which is described in the following sections. In the case of EMC conformity measurements in an anechoic chamber in the FF, the radiation pattern is measured and (very simplified) the maximum is compared with the according regulations. To get an idea, an example of a radiation pattern for a dipole can be seen in Figures 14.2 and 14.3.

14.2.3 Near-field scanner system

The term *scanning* is related to the fact, that the electromagnetic fields are determined on various spatial points on a surface around the device under test and so the field distribution can be visualized. Therefore an NFS is equipped with a mechanical positioning system, to move the probe or the device under test itself from one measurement point to the next one. With such a system the electric field as well as the magnetic can be measured in all three room directions on equidistant or adaptive [13, pp. 35–58] surfaces around the device under test. Common surfaces are spherical [14], cylindrical [15], cubic [10] or even arbitrary [4]. While spherical or cylindrical surfaces are mainly used to characterize radiation patterns of antennas, planar or arbitrary surfaces are mainly used for EMC assessment [16]. From here on

†The radiation pattern usually is defined in spherical coordinates.

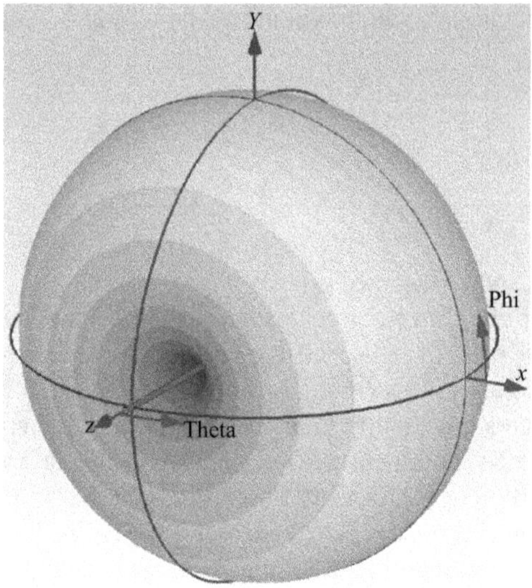

Figure 14.2 Radiation pattern of a dipole in 3D representation

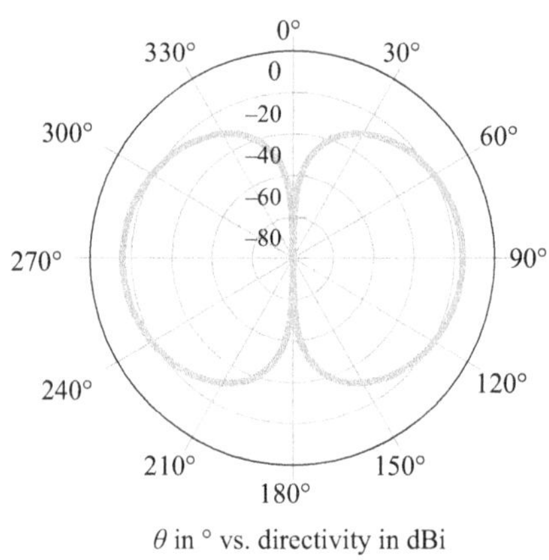

Figure 14.3 Radiation pattern of a dipole in a polar plot representation with $\phi = 90°$

it is assumed that the coordinate system is Cartesian and the measurement surfaces are planar or cubic.

Besides the positioning system, an NFS basically consists of a set of probes, the measurement receiver and a controller. The measurement receiver can have different architectures. These are compared in [7]. Since not only the amplitude but also the phase information of the NF components are needed for the purpose of an NF-to-FF transformation and any digital device could be a device under test, digital storage oscilloscopes or vector signal analysers with at least two channels are the best options. The basic setup of an NFS system is depicted in Figure 14.4. The introduced NFS

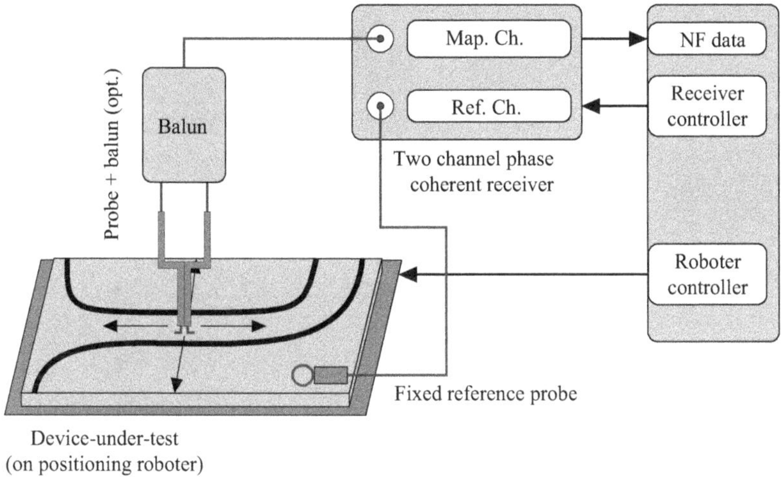

Figure 14.4 A basic near-field scanner system with a phase coherent measurement receiver

system has at least one map channel[‡] to measure the desired field on the chosen surface and one reference channel.

To obtain the reference signal, there are two basic concepts: the first one is to have a fixed reference probe spatially related to the device under test. This can be done for any kind of device under test. The second one is to feed the reference channel with a signal from the device under test itself as reference. This is only possible if the device under test signal is accessible and if its shape is linked by a known relation to the emissions measured.

14.2.4 Basic probes for near-field scanning

With the measurement receiver, the positioning system and a controller at hand, only the probes are missing for a basic NFS. In a artesian coordinate system, there is an x, y and z direction for any field component and at each spatial point. If the coordinate system and a planar surface above the DUT are defined as seen in Figure 14.5, the x and y components are the tangential (subscript t) components and the z component is the normal (subscript n). For each type of field component (tangential or normal), there exits one probe for the E-field and one probe for the H-field. The x and y components are separated by a 90° shift of the device under test or the probe.

One of the simplest ways to build the probes is to use semi-rigid cables and different probe tips, like some exemplary models depicted in Figure 14.6.

- An E_n probe has a simple monopole as tip.
- An E_t probe has a dipole as tip.

[‡]Due to the fact that it is creating a *map* of the EM fields.

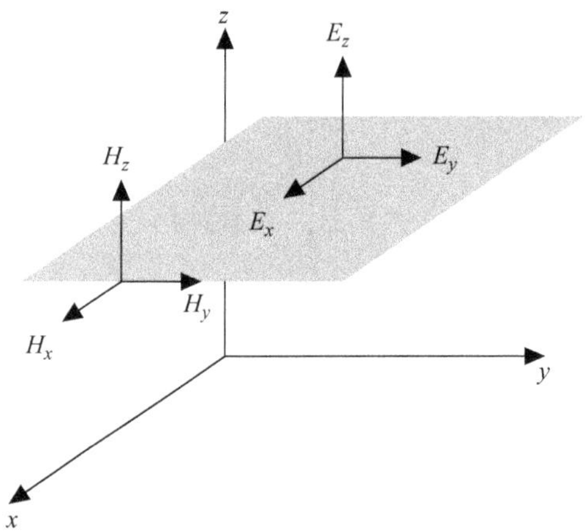

Figure 14.5 Basic Cartesian coordinate system with the according field components

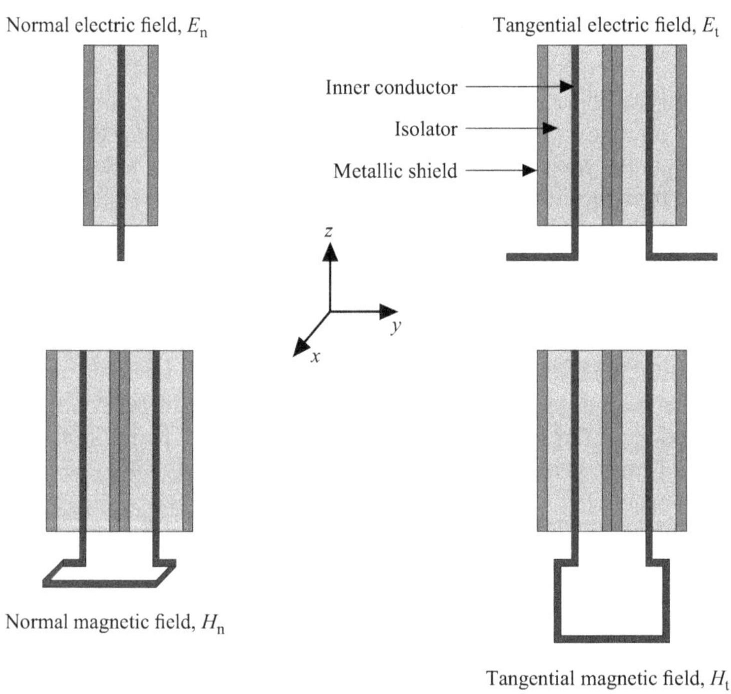

Figure 14.6 Schematic layout of a probe set with four different probes

- An H_n has a loop as tip with its normal oriented in normal direction.
- An H_t has a loop as tip with its normal oriented in tangential direction.

All probes could be connected to a balun to balance the unbalanced signals.[§] The probes cannot directly measure the desired field strength but can provide a voltage induced by the corresponding field. They need to be calibrated. In first instance a simple probe factor

$$\underline{PF} = \frac{\underline{E}|\underline{H}_{\text{meas}}}{\underline{U}_{\text{probe}}} \tag{14.4}$$

can transform the measured probe voltage $\underline{U}_{\text{probe}}$ into the according field strengths $\underline{H}_{\text{meas}}$ or $\underline{E}_{\text{meas}}$ [17]. More advanced and complex probe calibration and compensation methods can be found in [7,18,19].

14.2.5 Near-field scanner NFS3000

The measurements from the practical part of this chapter are performed using the NFS, i.e., NFS3000 [1,20]. The NFS3000 was developed at the department of Advanced System Engineering from the Fraunhofer Institute for Electronic Nano Systems ENAS. It basis builds the huge but precise positioning robot, seen in Figure 14.7.

On the one hand, device under tests up to a size of 500 mm × 800 mm × 500 mm (x, y, z) can be measured. On the other hand, a spatial step resolution of 1 μm allows the measurement of NFs of smaller device under tests like microprocessors. As a measurement receiver, a vector signal analyser is used with a dynamic range of 70 dB and a maximum input power of 20 dBm. The frequency range goes up to 6 GHz. The probes used with the NFS3000 are designed by Fraunhofer ENAS as well and match the whole system. A further feature is the optical recording of the 3-D contour of the device under test in order to avoid collisions by automatizing the scanning procedure.

14.3 Theoretical basics of near-to-far-field transformation

14.3.1 Introduction

The Huygens' principle makes it possible to calculate FF characteristics from NF measurements. Thereby the tangential electric and magnetic fields are measured, which correspond to the corresponding surface current densities and serve as source for the FF transformation. In today's computer age, simulation can be performed to calculate the radiation characteristics of antennas, for example, by inserting only the sources in the NF. These theoretical principles also form the foundation of the method of moments (MoM), which is a simulation method for solving electromagnetic problems. In the following, the theoretical basics of the Huygens' principle and MoM

[§] Hence, the balun avoids unwanted cross-polarized fields by suppressing the asymmetric part of the signal.

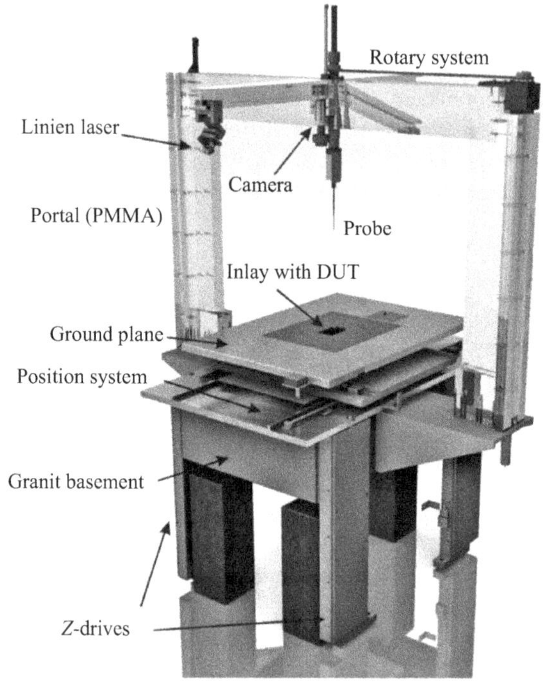

Figure 14.7 Model of the NFS3000

are described based on the primary literature [21–23], which was adapted to the FF transformation of the NFS3000 and described using some relevant examples for the NFS. Therefore, the following basics are presented in the 3D case and only for typical applications of the NFS.

14.3.2 Maxwell's equations

The Maxwell equations describe the relationship between electric and magnetic fields ($\vec{E}$ and $\vec{H}$) and their flux densities ($\vec{D}$ and $\vec{B}$), which also depends on the material parameters (permittivity ϵ and permeability μ). To derive the corresponding formulas for NF-to-FF transformation, the Maxwell equations are used as a basis and extended by imaginary auxiliary quantities ($\vec{M}$ and ρ_{mag}), which are very relevant for the further procedure. To better understand the derivation, it is assumed that the materials are homogeneous, linear, isotropic and dispersion free. The Maxwell equations in the frequency domain are [22]

$$\nabla \times \vec{E} = -\vec{M} - \mathrm{j}\omega\mu\vec{H}, \tag{14.5}$$

$$\nabla \times \vec{H} = \vec{J} + \mathrm{j}\omega\epsilon\vec{E}, \tag{14.6}$$

$$\nabla \cdot \vec{D} = \rho_{\text{el}}, \tag{14.7}$$

$$\nabla \cdot \vec{B} = \rho_{\text{mag}}. \tag{14.8}$$

The magnetic surface current density $\vec{M}$ and the magnetic space charge density ρ_{mag} have no real reference and are, therefore, often set to zero. Here, the nonphysical quantity is nevertheless used as an auxiliary quantity for calculations needs. The electric surface current density $\vec{J}$ and the electric space charge density ρ_{el} are real source variables.

14.3.3 Material equations

The material equations describe the relationship between electromagnetic flux densities ($\vec{D}$ and $\vec{B}$) and field quantities ($\vec{E}$ and $\vec{H}$), which are used often in the following [24,25]:

$$\vec{D} = \epsilon_0 \epsilon_{\mathrm{r}} \vec{E} + \vec{P}_{\mathrm{el}}, \tag{14.9}$$

$$\vec{B} = \mu_0 \mu_{\mathrm{r}} (\vec{H} + \vec{P}_{\mathrm{mag}}), \tag{14.10}$$

where $\vec{P}_{\mathrm{el}}$ and $\vec{P}_{\mathrm{mag}}$ describe the electrical and magnetic polarization, respectively. For linear, isotropic and dispersion-free materials, the two equations can be simplified to the following equations:

$$\vec{D} = \epsilon \vec{E} = \epsilon_0 \epsilon_{\mathrm{r}} \vec{E}, \tag{14.11}$$

$$\vec{B} = \mu \vec{H} = \mu_0 \mu_{\mathrm{r}} \vec{H}. \tag{14.12}$$

14.3.4 Electromagnetic boundary conditions

The transition between two materials can be explained by the boundary conditions of electromagnetism [22,24]:

$$-\vec{n} \times (\vec{E}_2 - \vec{E}_1) = \vec{M}_{12}, \tag{14.13}$$

$$\vec{n} \times (\vec{H}_2 - \vec{H}_1) = \vec{K}_{12}, \tag{14.14}$$

$$\vec{n} \cdot (\vec{D}_2 - \vec{D}_1) = \sigma_{12}, \tag{14.15}$$

$$\vec{n} \cdot (\vec{B}_2 - \vec{B}_1) = \eta_{12}, \tag{14.16}$$

where the vector $\vec{n}$ describes the normal vector between the regions 1 and 2. σ_{12} describes the electric surface charge density and η_{12}, the magnetic dipole moment density, which are interesting for further applications in electrostatics. For electrodynamics, the electric and magnetic displacement currents $\vec{K}$ and $\vec{M}$ are relevant.

In the case that region 2 is a PEC (perfect electric conductor), the equations become

$$\vec{n} \times \vec{E}_1 = 0, \tag{14.17}$$

$$-\vec{n} \times \vec{H}_1 = \vec{K}_{12}, \tag{14.18}$$

$$-\vec{n} \cdot \vec{D}_1 = \sigma_{12}, \tag{14.19}$$

$$\vec{n} \cdot \vec{B}_1 = 0. \tag{14.20}$$

This assumption is often made for metals.

For a perfect magnetic conductor (PMC) the following equations result in

$$\vec{n} \times \vec{E}_1 = \vec{M}_{12}, \tag{14.21}$$

$$\vec{n} \times \vec{H}_1 = 0, \tag{14.22}$$

$$\vec{n} \cdot \vec{D}_1 = 0, \tag{14.23}$$

$$-\vec{n} \cdot \vec{B}_1 = \eta_{12}. \tag{14.24}$$

PMC is a hypothetical material, which is relevant for some electromagnetic problems, e.g., PMC can be used to reduce numerical calculations for symmetrical field distributions.

14.3.5 Formulations for radiation

For the NF-to-FF transformation, it is necessary to establish a direct relationship between the respective current densities ($\vec{J}$ and $\vec{M}$) and the respective field sizes ($\vec{E}$ and $\vec{H}$). To do this, (14.5) and (14.6) are combined and the magnetic charges/currents are set to zero ($\vec{M} = \eta = 0$) [22,24]:

$$\nabla \times \nabla \times \vec{E} = \omega^2 \mu \epsilon \vec{E} - \mathrm{j}\omega\mu\vec{J}. \tag{14.25}$$

By using the following vector identity

$$\nabla \times \nabla \times \vec{E} = \nabla(\nabla \cdot \vec{E}) - \Delta\vec{E} \tag{14.26}$$

the substitution $k = \omega\sqrt{\mu\epsilon} = (2\pi/\lambda)$, the (14.7) and the equation of continuity

$$\nabla \cdot \vec{J} = -\mathrm{j}\omega\rho_e \tag{14.27}$$

the following expression defines

$$\Delta\vec{E} + k^2\vec{E} = \mathrm{j}\omega\mu\vec{J} - \frac{1}{\mathrm{j}\omega\epsilon}\nabla(\nabla \cdot \vec{J}) \tag{14.28}$$

and/or

$$\Delta\vec{E} + k^2\vec{E} = \mathrm{j}\omega\mu\left(\vec{J} + \frac{1}{k^2}\nabla(\nabla \cdot \vec{J})\right). \tag{14.29}$$

This expression has the form of the Helmholtz equation and can be solved by using Green's function $G(\vec{r}, \vec{r}\,')$ [22,26]:

$$\vec{E}(r) = -\mathrm{j}\omega\mu \int_V G(\vec{r}, \vec{r}\,')(\vec{J}(\vec{r'}) + \frac{1}{k^2}\nabla'(\nabla' \cdot \vec{J}(r')))dV'. \tag{14.30}$$

The previous approach is correspondingly applicable to the magnetic field

$$\nabla \times \nabla \times \vec{H} = \omega^2 \mu \epsilon \vec{H} + \nabla \times \vec{J} \tag{14.31}$$

from which the following equation for the magnetic field follows:

$$\vec{H}(r) = -\mathrm{j}\omega\epsilon \int_V G(\vec{r}, \vec{r}\,')(\vec{M}(\vec{r'}) + \frac{1}{k^2}\nabla'(\nabla' \cdot \vec{M}(r')))dV'. \tag{14.32}$$

The three-dimensional Green function for electrodynamics is [24,27]

$$G(\vec{r},\vec{r}\,') = -\frac{e^{-\mathrm{j}k|\vec{r}-\vec{r}\,'|}}{4\pi\,|\vec{r}-\vec{r}\,'|}. \tag{14.33}$$

This shows that the electric field $\vec{E}(r)$ can be represented by the electric surface current density $\vec{J}(r)$ and by the magnetic surface current density $\vec{M}(r)$.

In order to get a better understanding and overview later, variables are introduced, which combine many properties.

Equations (14.30) and (14.32) can be shortened by modifying the Green function. For this purpose, the following is defined [27–29]:

$$\bar{\bar{g}}_{\mathrm{m}}(\vec{r}) = \nabla \times (\bar{\bar{I}} \cdot G(\vec{r},\vec{r}\,')), \tag{14.34}$$

$$\bar{\bar{g}}_{\mathrm{e}}(\vec{r}) = -\mathrm{j}kZ\left(\bar{\bar{I}} + \frac{1}{k^2}\nabla'\nabla'\right)G(\vec{r},\vec{r}\,'), \tag{14.35}$$

where $\bar{\bar{g}}_{\mathrm{m}}$ is generally called the magnetic Green dyad[‖] and $\bar{\bar{g}}_{\mathrm{e}}$ the electric Green dyad. $\bar{\bar{I}}$ describes the Green unity dyad. The material properties were summarized by the wave impedance ($Z = \sqrt{(\mu/\epsilon)}$), later by the wave admittance ($Y = Z^{-1}$) and the wave number ($k = \omega\sqrt{\epsilon\mu}$).

The following properties also apply [23,27,29]

$$\bar{\bar{g}}_{\mathrm{e}}(-\vec{r}) = \bar{\bar{g}}_{\mathrm{e}}(\vec{r}) \text{ even function,} \tag{14.36}$$

$$\bar{\bar{g}}_{\mathrm{m}}(-\vec{r}) = -\bar{\bar{g}}_{\mathrm{m}}(\vec{r}) \text{ odd function,} \tag{14.37}$$

$$\bar{\bar{g}}_{\mathrm{m}}(-\vec{r}) = -\frac{1}{\mathrm{j}kZ}\nabla \times (\bar{\bar{g}}_{\mathrm{e}}(-\vec{r})), \tag{14.38}$$

$$\bar{\bar{g}}_{\mathrm{e}}(-\vec{r}) = \frac{1}{\mathrm{j}kY}(\nabla \times \bar{\bar{g}}_{\mathrm{m}}(-\vec{r}) - \bar{\bar{I}}). \tag{14.39}$$

Additionally the Green dyads can be brought into the form of (14.25) and (14.31) [23,29,30]:

$$\nabla \times \nabla \times \bar{\bar{g}}_{\mathrm{e}} = \omega^2\mu\epsilon\bar{\bar{g}}_{\mathrm{e}} - \mathrm{j}\omega\mu\bar{\bar{I}}\delta, \tag{14.40}$$

$$\nabla \times \nabla \times \bar{\bar{g}}_{\mathrm{m}} = \omega^2\mu\epsilon\bar{\bar{g}}_{\mathrm{m}} + \nabla \times (\bar{\bar{I}}\delta). \tag{14.41}$$

By these changes, (14.30) and (14.32) can be written down to the following convolution product with the electrical surface current density:

$$\vec{E}(\vec{r}) = (\bar{\bar{g}}_{\mathrm{e}} * \vec{J})(\vec{r}), \tag{14.42}$$

$$\vec{H}(\vec{r}) = (\bar{\bar{g}}_{\mathrm{m}} * \vec{J})(\vec{r}). \tag{14.43}$$

This allows the entire fields in Figure 14.8 to be defined as follows [28,31]:

$$\vec{E}_{\mathrm{tot}} = \vec{E}_1 + \vec{E}_2 + \vec{E}_{\mathrm{inc}} = (\bar{\bar{g}}_{\mathrm{e}} * \vec{J}_1)(\vec{r}) + (\bar{\bar{g}}_{\mathrm{e}} * \vec{J}_2)(\vec{r}) + \vec{E}_{\mathrm{inc}}, \tag{14.44}$$

$$\vec{H}_{\mathrm{tot}} = \vec{H}_1 + \vec{H}_2 + \vec{H}_{\mathrm{inc}} = (\bar{\bar{g}}_{\mathrm{m}} * \vec{J}_1)(\vec{r}) + (\bar{\bar{g}}_{\mathrm{m}} * \vec{J}_2)(\vec{r}) + \vec{H}_{\mathrm{inc}}. \tag{14.45}$$

[‖]Tensor second stage.

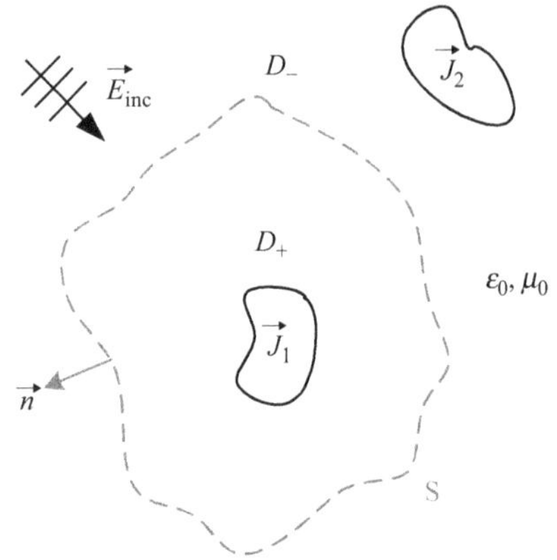

Figure 14.8 Equivalent representation of any problem with two surface current densities ($\vec{J}_1$ and $\vec{J}_2$) and an incident plane wave $\vec{E}_{inc}$ with an enveloping surface S *[23,24]*

14.3.6 Surface equivalence theorem

The surface equivalence theorem (also called Huygens' principle) states that every point of a radiating wave can be considered itself as a source. This theorem allows electromagnetic fields to be exchanged for equivalent substitute sources. Thus, on a closed surface, S it is sufficient to know the respective electric- and magnetic-field quantities and thus to calculate the further course of the field outside or inside the closed surface S. In electrodynamics, the magnetic surface current density $\vec{M}$ and the electrical surface current density $\vec{K}$ can be used as equivalent quantities. These current densities generate then the same propagating field. Thus, any surface S in the NF can be measured electromagnetically, and the field outside (also in the FF) can be calculated. These surface current densities are determined by the tangential fields through the boundary conditions of (14.13) and (14.14) [24].

Surface equivalence theorem in the electrostatic

The surface equivalence theorem can also be adapted for electrostatics. Here the equivalent quantities in the three-dimensional are the electric charge density σ and the magnetic dipole moment density η. It is important here that the magnetic component is a dipole. These quantities are determined from the normal components by the boundary conditions of the (14.15) and (14.16).

Here, the Poisson equation $\Delta\Phi = -(\rho_e/\epsilon_0)$ is used as a basis with the Green function for three-dimensional problems in statics $g_0 = G(\vec{r}, \vec{r}\,') = (1/(4\pi \mid \vec{r} - \vec{r}\,'|))$ and the boundary condition $\lim_{\vec{r}\to\infty} \Phi(\vec{r}) = 0$ used [22,23].

This results in the following equation for the electrical potential:

$$\Phi(\vec{r}) = \left(\frac{\rho_e}{\epsilon_0} * g_0\right)(\vec{r}) \tag{14.46}$$

For the example in Figure 14.9 the following calculation follows [23]:

$$\underbrace{\frac{1}{\epsilon_0}\oint_S \sigma(\vec{r}\,')g_0 \mathrm{d}A(\vec{r}\,')}_{\Phi_{SL}} + \underbrace{\frac{1}{\epsilon_0}\oint_S \mu(\vec{r}\,')\vec{n}(\vec{r}\,')\nabla' g_0 \mathrm{d}A(\vec{r}\,')}_{\Phi_{DL}} = \begin{cases} -\Phi_2(\vec{r}); & \vec{r} \in \mathrm{D}_+ \\ \Phi_1(\vec{r}); & \vec{r} \in \mathrm{D}_- \end{cases}, \tag{14.47}$$

where Φ_{SL} denotes the single layer potential with the electrical surface charge density $\sigma(\vec{r}) = -\epsilon_0 \vec{n}(\vec{r})\nabla\Phi(\vec{r})$. Φ_{DL} describes the double-layer potential with the magnetic dipole moment density $\mu(\vec{r}) = \epsilon_0 \Phi(\vec{r})$ [21,23].

By $\vec{E} = -\nabla\Phi$, the electric-field strength in electrostatics can thus be calculated. As in electrodynamics, $\vec{r} \notin \mathrm{S}$, because the fields and electromagnetic sources can be unsteady in transition [23].

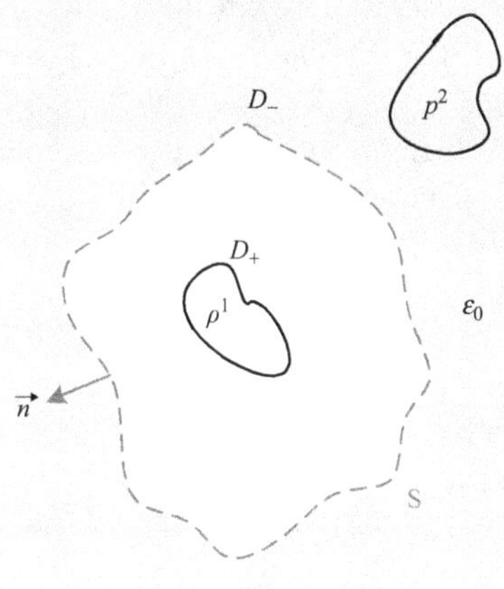

Figure 14.9 Free space with two electric space charge densities for use in electrostatics

Figure 14.10 shows an equivalent representation as an example with any closed surface S with volume D used from Figure 14.8. Here, the source was the volume $\vec{j}$, which is used for the Green theorem and its integration over the volume in the following. At first, the materials are homogeneous in the whole space, but they can be adjusted by the material parameters ϵ and μ in the Green dyads. The enveloping surface S or several enveloping surfaces S_n must enclose the complete other homogeneous medium. The second dyadic Green theorem is used as follows [23,28,32,33]:

$$\int_D (\nabla \times \nabla \times \bar{\bar{Q}}) \cdot \bar{\bar{P}} - \bar{\bar{Q}} \cdot (\nabla \times \nabla \times \bar{\bar{P}}) \mathrm{d}V$$

$$= -\oint_{\partial D = S} (\nabla \times \bar{\bar{Q}})(\vec{n} \times \bar{\bar{P}}) + \bar{\bar{Q}} \cdot (\vec{n} \times \nabla \times \bar{\bar{P}}) \mathrm{d}A. \tag{14.48}$$

Here for the dyads $\bar{\bar{P}} = \vec{E}$ and $\bar{\bar{Q}} = \bar{\bar{g}}_e$.

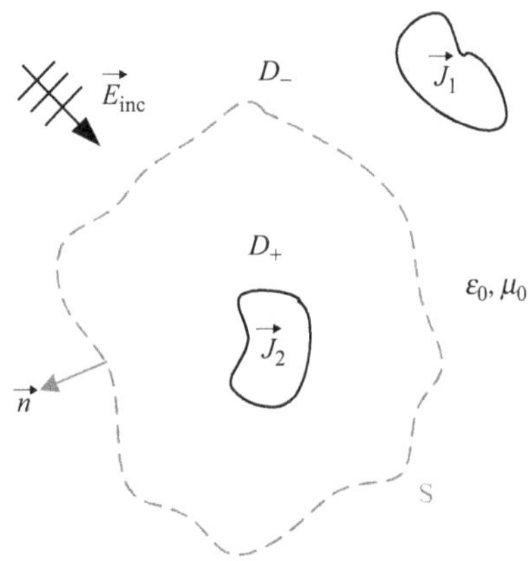

Figure 14.10 Equivalent representation of any problem with two volume flow densities ($\vec{j}_1$ and $\vec{j}_2$) and an incident plane wave $\vec{E}_{inc}$ with an enclosing surface S [23]

Now the second Green theorem (14.48) is used together with the induction law (14.7), the equation for the electric Green's dyad (14.42) and the Helmholtz equation of the electric dyad (14.40), resulting in the following formula[23]:

$$(\bar{\bar{g}}_e * \vec{j}_1)(\vec{r}) = \int_{D\mp} \vec{E}_1(\vec{r}\,')\delta(\vec{r} - \vec{r}\,')\mathrm{d}V(\vec{r}\,')$$

$$+ \oint_S (\vec{n}_\mp \times \vec{H}_1)(\vec{r}\,') \cdot \bar{\bar{g}}_e - \frac{1}{\mathrm{j}kZ}(\vec{n}_\mp \times \vec{E}_1)(\vec{r}\,') \cdot (\nabla \times \bar{\bar{g}}_e)\mathrm{d}A(\vec{r}\,') \quad \text{for } \vec{r} \notin \mathrm{S}. \tag{14.49}$$

By the filter property of the Dirac function in (14.49), it can be seen that if $r \in D_-$ is at the integral of $D_\mp$, the closed surface integral around the enveloping surface S becomes zero, since then applies

$$(\bar{\bar{g}}_e * \vec{j}_1)(\vec{r}) = \int_{D\mp} \vec{E}_1(\vec{r}\,')\delta(\vec{r} - \vec{r}\,')\mathrm{d}V(\vec{r}\,') = \vec{E}_1(\vec{r}) \quad \text{for } \vec{r} \in D_-. \tag{14.50}$$

This feature is used in the following for the three field components ($\vec{E}_{inc}$, $\vec{E}_1$ and $\vec{E}_2$), which is followed by

$$-\oint_S \bar{\bar{g}}_e \cdot (\underbrace{\vec{n}_+ \times \vec{H}_{inc}}_{\vec{K}_{inc}})(\vec{r}\,') + \bar{\bar{g}}_m \cdot (\underbrace{\vec{n}_+ \times \vec{E}_{inc}}_{\vec{M}_{inc}})(\vec{r}\,')\mathrm{d}A(\vec{r}\,') = \begin{cases} \vec{E}_{inc}; & \vec{r} \in D_+ \\ 0; & \vec{r} \in D_- \end{cases}, \tag{14.51}$$

(see Figure 14.11(a))

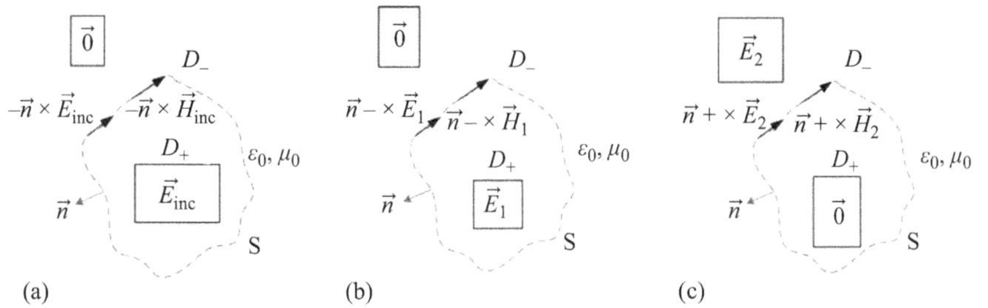

Figure 14.11 Equivalent representation of the respective electric fields and their surface current densities or tangential fields on the surface of the enveloping surface S: (a) only $\vec{E}_{inc}$, (b) only $\vec{E}_1$ and (c) only $\vec{E}_2$ [23,24]

$$\oint_S \bar{\bar{g}}_e \cdot \underbrace{(\vec{n}_- \times \vec{H}_1)}_{\vec{K}_1}(\vec{r}\,') + \bar{\bar{g}}_m \cdot \underbrace{(\vec{n}_- \times \vec{E}_1)}_{\vec{M}_1}(\vec{r}\,')\mathrm{d}A(\vec{r}\,') = \begin{cases} \vec{E}_1; & \vec{r} \in D_+ \\ 0; & \vec{r} \in D_- \end{cases}, \quad (14.52)$$

(see Figure 14.11(b))

$$\oint_S \bar{\bar{g}}_e \cdot \underbrace{(\vec{n}_+ \times \vec{H}_2)}_{\vec{K}_2}(\vec{r}\,') + \bar{\bar{g}}_m \cdot \underbrace{(\vec{n}_+ \times \vec{E}_2)}_{\vec{M}_2}(\vec{r}\,')\mathrm{d}A(\vec{r}\,') = \begin{cases} 0; & \vec{r} \in D_+ \\ \vec{E}_2; & \vec{r} \in D_- \end{cases}. \quad (14.53)$$

(see Figure 14.11(c))

The correct total field results from the superposition of the three equations ($\vec{H}_{tot} = \vec{H}_{inc} + \vec{H}_1 + \vec{H}_2$ and $\vec{E}_{tot} = \vec{E}_{inc} + \vec{E}_1 + \vec{E}_2$). Thereby $\vec{n}_\pm = \pm\vec{n}$ was set [23] as

$$\vec{E}_{tot}(\vec{r}) = \int_{D_\mp} \bar{\bar{g}}_e \cdot \vec{j}_\mp \mathrm{d}V(\vec{r}\,') \pm \oint_S \bar{\bar{g}}_e \cdot \underbrace{(\vec{n} \times \vec{H}_{tot})(\vec{r}\,')}_{\vec{K}_{tot}} + \bar{\bar{g}}_m \cdot \underbrace{(\vec{n} \times \vec{E}_{tot})}_{\vec{M}_{tot}} \mathrm{d}A(\vec{r}\,') + \begin{cases} 0; & \vec{r} \in D_+ \\ \vec{E}_{inc}; & \vec{r} \in D_- \end{cases}. \quad (14.54)$$

(see Figure 14.12)

Similarly, this can also be done with the magnetic field:

$$\vec{H}_{tot}(\vec{r}) = \int_{D_\mp} \bar{\bar{g}}_m \cdot \vec{j}_\mp \mathrm{d}V(\vec{r}\,') \pm \oint_S \bar{\bar{g}}_m \cdot \underbrace{(\vec{n} \times \vec{H}_{tot})(\vec{r}\,')}_{\vec{K}_{tot}} - Y^2 \bar{\bar{g}}_e \cdot \underbrace{(\vec{n} \times \vec{E}_{tot})}_{\vec{M}_{tot}} \mathrm{d}A(\vec{r}\,') + \begin{cases} 0; & \vec{r} \in D_+ \\ \vec{H}_{inc}; & \vec{r} \in D_- \end{cases}. \quad (14.55)$$

In Figure 14.13, only the superposition of all tangential-field components or the surface current densities is shown. This principle clearly shows the Huygens' principle. Here, the surface current densities or the tangential fields are calculated along a closed surface, and the desired outer field $\vec{E}_2$ is described in the outer area D_-:

$$\vec{E}(\vec{r}) = \oint_S \bar{\bar{g}}_e \cdot \vec{K}_{tot} + \bar{\bar{g}}_m \cdot \vec{M}_{tot} \mathrm{d}A(\vec{r}\,') + \begin{cases} -\vec{E}_1(\vec{r}) - \vec{E}_{inc}(\vec{r}); & \vec{r} \in D_+ \\ \vec{E}_2; & \vec{r} \in D_- \end{cases}. \tag{14.56}$$

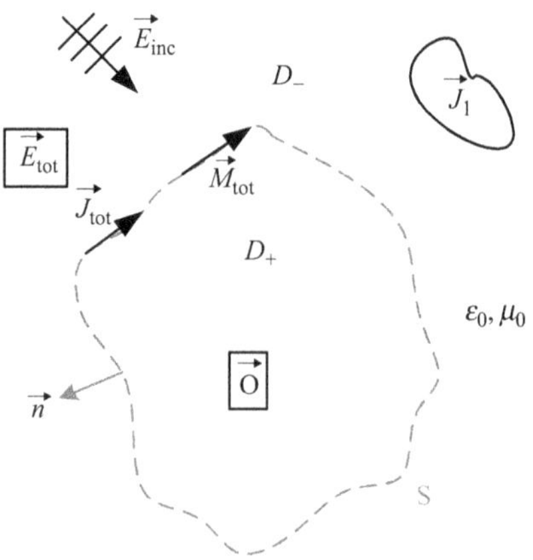

Figure 14.12 Equivalent representation [23,24]

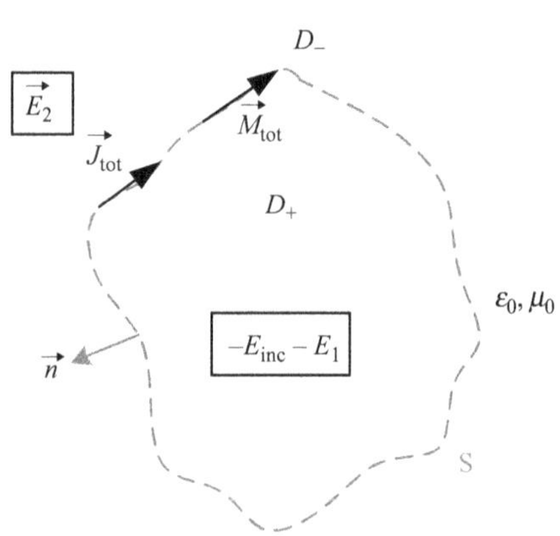

Figure 14.13 Equivalent representation for only the surface current densities [23,24]

Method of moments

The MoM (generally also often called boundary element method) is a numerical surface-based method, which is often used in radar technology and was first mentioned by Harrington in *Field Computation by Moment Methods* [21] in 1968. This allows problems that have a large volume to be solved with high priority. Therefore, MoM is often used for antenna problems in which the radiation characteristics of antennas, backscattering cross sections of objects or other possible problems of fields or waves at long distances are necessary, because the wave propagations can be operated at long ranges with less computational effort. Volume-based methods have the disadvantage there that they have to calculate the whole volume and, depending on the scope, even the latest supercomputers are overstrained.

The analytical fundamentals of MoM are the same as those of the Huygens' principle and the integral equation method described there (integro-differential equations), but these are continued there and also used for other problems. Some of the important features of the MoM are [21–23]

- surface-based method;
- good usages for antenna technology and long distances;
- also applicable for eigenvalue problems with waveguides;
- densely populated system matrices requiring a high calculation effort;
- currently only usable in the frequency domain;
- not or poorly applicable for materials that are nonlinear, anisotropic or have an inhomogeneous distribution.

14.3.7 Surface equivalence theorem for the NFS environment

In Section 14.2 the NFS3000 was presented, which is now adapted in Figure 14.14 according to the integral equations described before. $\vec{j}_1$ describes the emitted fields of the DUT. The ground plane is ideally assumed here as PEC and describes the

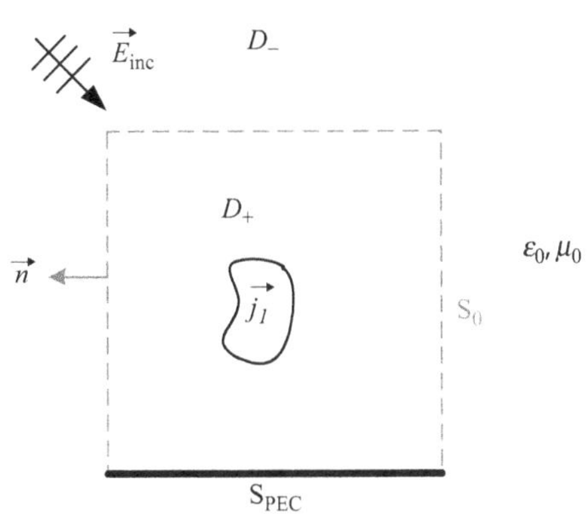

Figure 14.14 Equivalent representation of the NFS3000 measurement

area S_{PEC}. The NFS measures the area S_0 and determines the tangential electric and magnetic fields. The total necessary for the procedure is $S = S_0 + S_{PEC}$.

By using PEC and the conditions from (14.17) and (14.18), the magnetic surface current density is not present and the following equations apply to the measured surface current density or tangential fields in this arrangement:

$$\begin{aligned}\vec{E}(\vec{r}) &= \oint_S \bar{\bar{g}}_e \cdot \vec{K}_{tot} + \bar{\bar{g}}_m \cdot \vec{M}_{tot} \mathrm{d}A(\vec{r}\,') \\ &= \int_{S_0} \bar{\bar{g}}_e \cdot \vec{K}_{S_0} + \bar{\bar{g}}_m \cdot \vec{M}_{S_0} \mathrm{d}A(\vec{r}\,') + \int_{S_{PEC}} \bar{\bar{g}}_e \cdot \vec{K}_{S_{PEC}} \mathrm{d}A(\vec{r}\,') \\ &= \begin{cases} -\vec{E}_{inc}(\vec{r}); & \vec{r} \in D_+ \\ \vec{E}_1(\vec{r}); & \vec{r} \in D_- \end{cases},\end{aligned} \tag{14.57}$$

$$\begin{aligned}\vec{H}(\vec{r}) &= \oint_S \bar{\bar{g}}_m \cdot \vec{K}_{tot} + \bar{\bar{g}}_e \cdot \vec{M}_{tot} \mathrm{d}A(\vec{r}\,') \\ &= \int_{S_0} \bar{\bar{g}}_m \cdot \vec{K}_{S_0} - Y^2 \bar{\bar{g}}_e \cdot \vec{M}_{S_0} \mathrm{d}A(\vec{r}\,') + \int_{S_{PEC}} \bar{\bar{g}}_m \cdot \vec{K}_{S_{PEC}} \mathrm{d}A(\vec{r}\,') \\ &= \begin{cases} -\vec{H}_{inc}(\vec{r}); & \vec{r} \in D_+ \\ \vec{H}_1(\vec{r}); & \vec{r} \in D_- \end{cases}.\end{aligned} \tag{14.58}$$

Since the surface current density $\vec{K}_{S_{PEC}}$ is not measured on the ground plane, the following conditions must be observed during measurement:

$$\int_{S_{PEC}} \bar{\bar{g}}_e \cdot \vec{K}_{S_{PEC}} \mathrm{d}A(\vec{r}\,') \approx \int_{S_{PEC}} \bar{\bar{g}}_m \cdot \vec{K}_{S_{PEC}} \mathrm{d}A(\vec{r}\,') \approx 0. \tag{14.59}$$

This is achieved by orienting the DUT so that the primary radiation is aligned to S_0 and by grounding the ground plane. Therefore, the following formulas follow for the FF transformation procedure for the NFS3000:

$$\int_{S_0} \bar{\bar{g}}_e \cdot \vec{K}_{S_0} + \bar{\bar{g}}_m \cdot \vec{M}_{S_0} \mathrm{d}A(\vec{r}\,') = \begin{cases} -\vec{E}_{inc}(\vec{r}); & \vec{r} \in D_+ \\ \vec{E}_1(\vec{r}); & \vec{r} \in D_- \end{cases}, \tag{14.60}$$

$$\int_{S_0} \bar{\bar{g}}_m \cdot \vec{K}_{S_0} - Y^2 \bar{\bar{g}}_e \cdot \vec{M}_{S_0} \mathrm{d}A(\vec{r}\,') = \begin{cases} -\vec{H}_{inc}(\vec{r}); & \vec{r} \in D_+ \\ \vec{H}_1(\vec{r}); & \vec{r} \in D_- \end{cases}. \tag{14.61}$$

14.4 Near-field-to-far-field transformation using the Huygens' box principle

The theoretical approach from the previous chapter is now used to transform the measured NF data into an FF using a full-wave simulation tool. Within this chapter, the simulation software CST Studio Suite is used, but other similar tools could fit as well. In the following, the combination of the NF measurements and those simulation tools to perform an NF-to-FF transformation is described. The NFS system used is the NFS3000 described in Section 14.2.5.

14.4.1 Huygens' box measurement

The Huygens' box measurement within this chapter is performed on a cube around the device under test with a Cartesian coordinate system. Since the tangential-field components are of interest for the Huygens' box principle, the two right-angled tangential components are needed for each of the six surfaces of the cube, for the H- and the E-field strengths. An example is depicted in Figure 14.15. Principally, the NFs can also be measured on arbitrary, non-planar surfaces, e.g., with a fixed distance to each component of the device under test instead of a global fixed height according to device under test fixing. With the *plane wave expansion*, those non-planar measurements can be transferred to an according planar surface, needed for the Huygens' box [4,7]. The advantage of a non-planar measurement could be that if higher parts such as heat sinks, through hole transistors or capacitors, are placed on the PCB, the NF probe can still be placed close to the other parts of the device under test to achieve a better quality and sensitivity of the results.

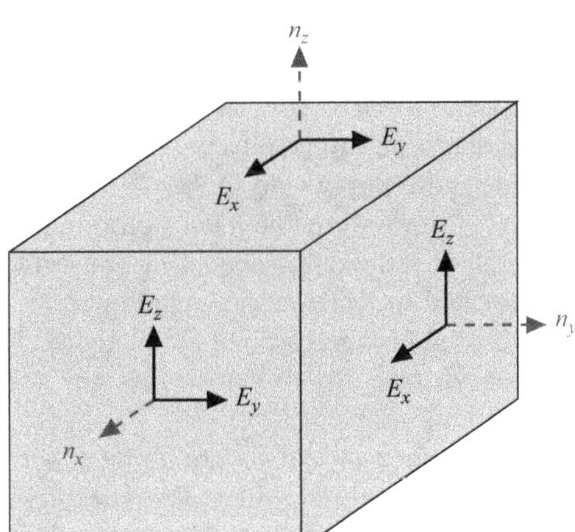

Figure 14.15 Example of a Huygens' box with the according e-field strengths

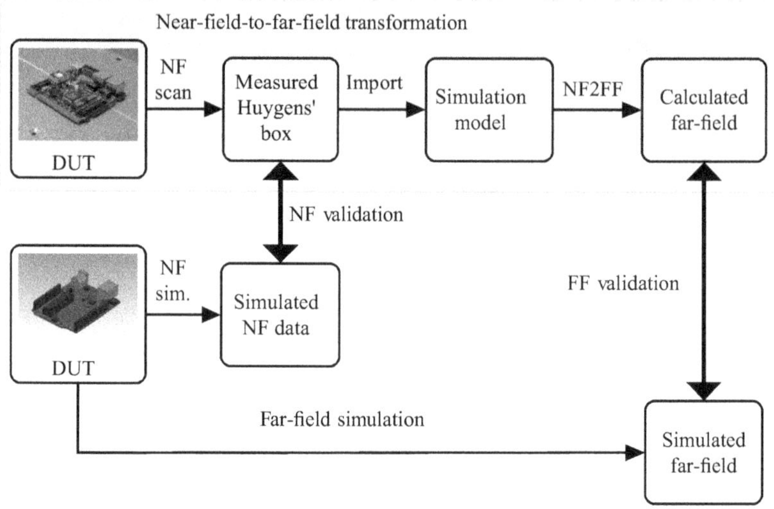

Figure 14.16 Workflow for the near-to-far-field transformation and validation with a simulation setup

The workflow for the NF-to-FF transformation using the Huygens' box is shown in Figure 14.16. At first the NF has to be measured as described. Then those measurements have to be transferred into a data format, which the simulation tool can handle. Finally, the simulation tool can calculate the FF data using the MoM Section 14.3.6.

14.4.2 Validation example and setup

Having the necessary theory and the tools, the principle now will be validated with a practical example. As validation device under test a 50 Ω microstrip line is used. The microstrip itself has a length of 111.4 mm and a width of 7.4 mm. It is fabricated on a Rogers RO4003 substrate and it has two 50 Ω SMA connectors for excitation and termination. For validation, the microstrip line is fed with a sinusoidal signal generator and is terminated with a 50 Ω terminator. The validation workflow is depicted in Figure 14.16, and the fabricated structure is shown in Figure 14.17.

The vertical and horizontal distance around the complete PCB is chosen to be 3 mm resulting in a planar box size of 240 mm × 160 mm × 15 mm (x, y, z). The stepsize of the scanning grid is 2 mm.

The microstrip is mounted on a special fixture for an easier Huygens' box measurement. Due to practical reasons, only five sides of the Huygens' box are considered. The sixth side in the bottom is the ground plane. Hence, the fields are negligible and can be left out.

The measurements and simulations are performed at three different frequencies: $f_1 = 500$ MHz, $f_2 = 1{,}000$ MHz and $f_3 = 1{,}500$ MHz.

Figure 14.17 Microstrip for validation on the NFS3000 near-field scanner

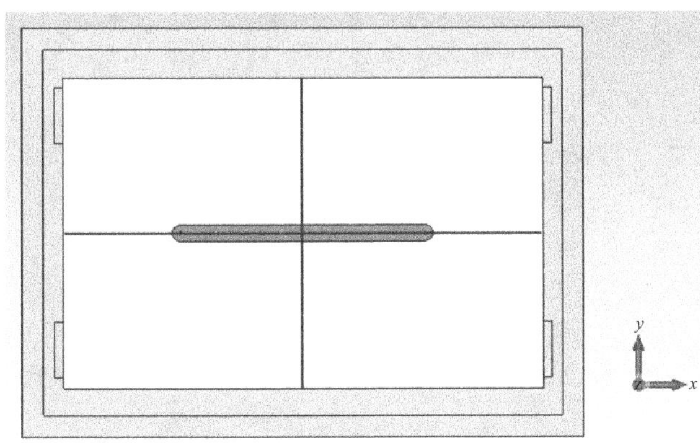

Figure 14.18 Simulation model of the microstrip and measurement line across the structure

14.4.3 Near-field results

In the following, the results from the NF measurement and the simulation of the device under test are compared at a frequency of $f_2 = 1{,}000$ MHz, to validate the measurement. To compare both results, a cross section across the microstrip in y-direction, like it can be seen in Figure 14.18, is made for the H_y and E_y components from the top side, to show the accuracy and performance of the scanning system including the probes.

From Figures 14.19 and 14.20 it can be seen that the amplitude of the measurement and the simulation show a good similarity for the E_y and H_y components. The quality as well as the quantity of the measurement points fit well. Slight differences occur due to different reasons. For instance, the probes and the device under test are connected by several adapters to the feeding and receiver which are not fully compensated. Further reasons for the differences are a limited dynamic range and measurement noise of the measurement receiver. Also the probes itself have an influence on the device under test and can disturb the fields. Finally, some portion of cross-polarized field components might influence the measurement [20]. Furthermore, the device under test in practical measurements is not perfectly planar. Hence, the distance between device under test and NF probe varies.

Besides the amplitude, the phase information is no less important to the NF-to-FF transformation. Hence, the phase is compared in Figures 14.19 and 14.20, also across the microstrip. The figures show that the phase fits quite well near the microstrip line where amplitudes of the field strengths are comparably high. At the borders, so farther away from the microstrip line, where the amplitude is lower or even at the noise floor, the phases show a noisier behaviour. This is due to the fact that the phase is extracted from a reference probe. So if the amplitude is near the noise floor, this phase extraction is no longer sufficient. But the area with those low amplitudes has a negligible contribution to the FF and, therefore, the phase error should hardly disturb the transformation.

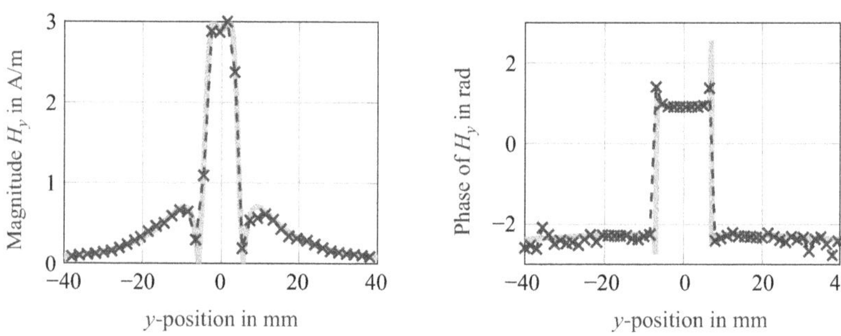

Figure 14.19 Magnitude and phase of H_y at $f = 1$ GHz for measurement −×− and simulation —

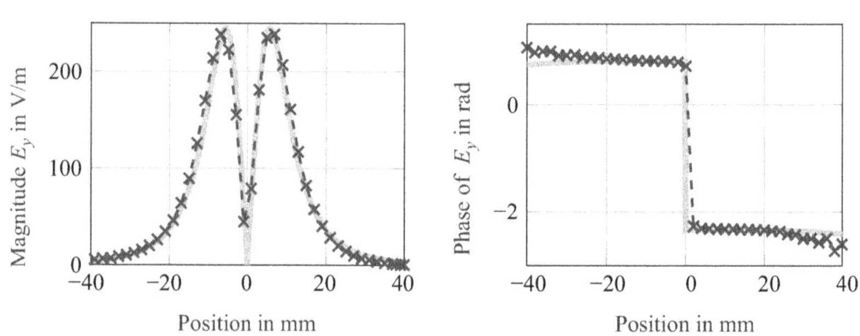

Figure 14.20 Magnitude and phase of H_y at $f = 1$ GHz for measurement −×− and simulation —

14.4.4 Near-field-to-far-field transformation

With the NF measurement results and a successful validation of those using the simulated NF, the NF-to-FF transformation can be accomplished. For this reason, the NF data are imported into the simulator as an NF source with the Huygens' box related components. For the validation of the NF-to-FF transformation, the calculated FF is compared with the full-wave simulated FF from the simulation model (see the workflow from Figure 14.16). In the following, two different examinations are done. First, the NF-to-FF transformation is done without phase relation to point out the importance of a phase coherent measurement. The second one is with the phase information from the vectorial measurement.

14.4.4.1 Without phase relation

As previously described, the following results are neglecting the phase information, so that every measurement point has the same phase. Here, only the frequency $f_2 = 1{,}000$ MHz is used. In Figure 14.21, the results are shown in a 3-D plot. The colour bar in the middle is valid for both results and shows the electric FF value in dB(V/m) at a 3 m distance.

It can be clearly seen that the shape and the amplitude of the FFs do not fit at all. The maximum amplitude for the purely simulated FF is −14 dB(V/m) while the maximum for the transformed FF is above 0 dB(V/m). The shape is also quite different. Hence the NF-to-FF is not working without the phase coherence of the measurement results.

14.4.4.2 With phase relation

As is shown that without phase information the transformed FF is completely different compared to the full-wave simulation, now the phase information from the phase coherent measurement is used for the NF-to-FF transformation. The results are

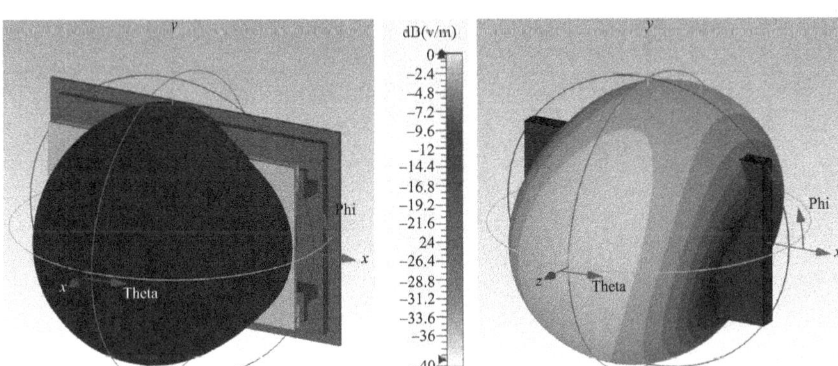

Figure 14.21 Simulated far-field and transformed far-field without phase coherence at $f_2 = 1{,}000$ MHz

depicted in Figure 14.22. It is important to note that the FF results on the left-hand side are the same as in Figure 14.21, but with a different colour bar, to be comparable to the transformed counterpart on the right-hand side.

Here, the results fit much better. Not only that the shape is quite comparable for both parts, but also the amplitudes are very close. As was stated, the amplitude in the propagation direction for the microstrip simulation model is -14 dB(V/m) and for the transformed FF using the measurements the amplitude is -15 dB(V/m). Therefore, with this example it can be shown that the NF-to-FF is working well.

Finally, the electric FF value in a 3 m distance in the $D(\phi = 0, \theta = 0)$ direction is calculated and compared over frequency in Figure 14.23. All transformed FFs for

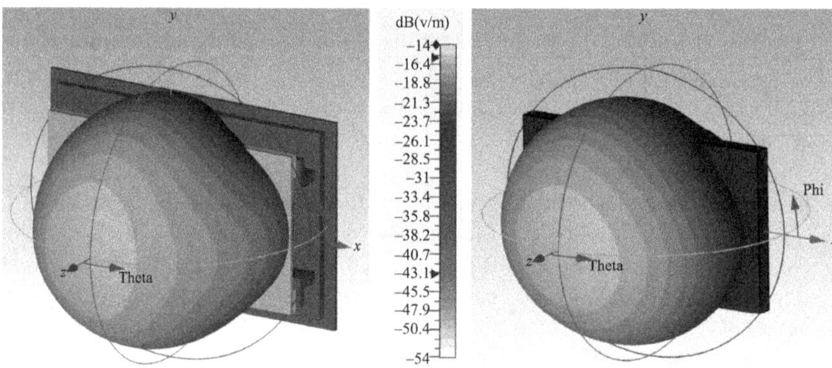

Figure 14.22 Simulated far-field and transformed far-field with phase coherence at f_2 = 1,000 MHz

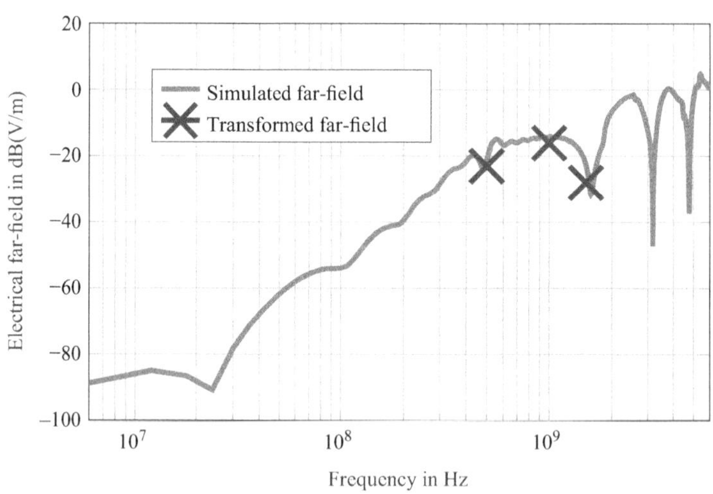

Figure 14.23 Simulated and transformed electrical far-field at $D(\phi = 0, \theta = 0)$ at 3 m distance over frequency

the three different frequencies are compared with their simulative counterpart. This value somehow goes along with the radiated EMC emission from the device under test. Here it can be clearly seen that the predicted emissions fit very well with the simulated one. Even the resonances near 500 and 1,500 MHz are hit. Summarizing the NF-to-FF transformation approach, it can be said that, by using a phase coherent NF measurement with NFSs like the NFS3000, a good prediction of the radiation pattern and EMC emission is possible.

14.5 Extended use of the near-field scan

A big advantage of NF scanning compared to an FF measurement is that the NF results can be used in different ways. While the FF measurement only gives information about the radiated emissions in some meter distance, NF data give more insights and have further scenarios to be used.

Whether to obtain the intended radiation pattern of an antenna or antenna array, or to obtain the unintended EMC emissions of a device under test, one scenario was already presented in a detailed way: the NF-to-FF transformation, so using the NF scan to predict the FF behaviour.

An even more simple use of an NF scan is to use the results as they are to have a look at the field strengths itself for debugging and noise source locating reasons [1,10,16]. For instance, if a device under test has EMC issues at a specific frequency, the engineer can look at the NF results to find out where those signals originate.

A third and very useful way to use the NF data is to utilize them as an NF source for full-wave simulations with surrounding objects. These can be objects like housings, a car or other electric devices [34]. Some examples are as follows:

- A device under test with an antenna can be measured with an NFS without any housing. In subsequent simulations, different kinds of materials or geometries for the housing can be modelled and tested. The goal would be to find a configuration with the least influence on the radiation pattern.
- With the digitization and electrification of cars, more and more electric devices find a place inside the car. By NF scanning those new devices, the NF source, for example, could be used to find the best place for this device due to EMC constraints.
- Due to the increasing amount of smart and wireless electric devices in all living areas, NF interactions between those devices increase as well. Since EMC certifications are only related to the FF behaviour, NF disturbances are not taken into account. With an NF source and a simulation tool, the NF interaction with other devices can be simulated and examined.

14.6 Conclusion

In this chapter, a technique is presented to derive the FF of electric devices by measuring the electric- and magnetic-field strength in the NF region and by combing the results with simulations using an NF-to-FF transformation approach. The basic definitions and hardware components of NF scanning are introduced to get an overview of the measurement technique. With the mathematical description of the surface equivalence theorem, the fundamentals and prerequisites for the electromagnetic simulation tools to perform the NF-to-FF transformation are shown. Combining the practical and theoretical basics, a workflow of an NF-to-FF transformation and validation is shown. A microstrip structure is chosen to compare the simulated and measured NFs in first instance. With the measurement results and the simulation model at hand, a validation of the transformation shows that the NF measurement and the FF prediction fit in a quite good way. Hence, the chosen validation setup shows that the presented approach works and that it can be utilized to predict the FFs for EMC reasons. A further advantage of NF measurements compared to a simple FF measurement is its various application scenarios. Besides the shown approach, the results can be used for radiation pattern determination, locating the source of EMC disturbances or as simulation source to investigate the interactions with the environment. Using this methodology, i.e., combining 3D-simulation, NF measurement and NF-to-FF transformation, the whole EMC-compliant design process of electronic systems can be accompanied in an efficient and productive way.

References

[1] Hangmann C, Mager T, Khan S, *et al.* Improved RF design using precise 3D near-field measurements and near-field to far-field transformations. In: Smart System Integration – International Conference and Exhibition on Integration Issues of Miniaturized Systems. Auerbach: Verlag Wissenschaftliche Scripten; 2016.

[2] Mynster AP, and Sorensen M. Validation of EMC near-field scanning amplitude and phase measurement data. In: EMC Europe 2012. IEEE; 2012.

[3] Altair. Combining Near-Field Measurement and Simulation for EMC Radiation Analysis; 2017. Available from: http://www.microwavejournal.com/articles/29215-combining-near-field-measurement-and-simulation-for-emc-radiation-analysis.

[4] Reinhold C, Hangmann C, Mager T, *et al.* Plane wave spectrum expansion from near-field measurements on non-planar lattices. In: 5th International Conference on Electromagnetic Near-Field Characterization and Imaging. Rouen, France: ESIGELEC / IRSEEM; 2011.

[5] Yaghjian A. An overview of near-field antenna measurements. IEEE Transactions on Antennas and Propagation. 1986;34(1):30–45.

[6] IEEE Standard for Definitions of Terms for Antennas. IEEE Std 145-2013 (Revision of IEEE Std 145-1993). 2014;p. 1–50.

[7] Tankielun A. Data Post-Processing and Hardware Architecture of Electromagnetic Near-Field Scanner. Hannover: Gottfried Wilhelm Leibniz Universität; 2007.

[8] Frenzel T, Rohde J, and Opfer J. Elektromagnetische Schirmung von Gebäuden. Bonn: BSI; 2007.

[9] Selvan KT and Janaswamy R. Fraunhofer and Fresnel distances: unified derivation for aperture antennas. IEEE Antennas and Propagation Magazine. 2017;59(4):12–15.

[10] Reinhold C, Mager T, and Hedayat C. Three dimensional near field scanning measurement techniques for improved RF design. In: SAME 2009, 12th Edition. Sophia Antipolis, France; 2009.

[11] Mager T, Tankielun A, Reinhold C, *et al.* Electromagnetic near-field scanning in time and frequency domain. In: Same 2008, 11th Edition. Sophia Antipolis, France; 2008.

[12] Newell AC. Current state-of-the-art in near-field antenna measurements. In: IEEE Antennas and Propagation Society International Symposium. 2001 Digest. Held in conjunction with: USNC/URSI National Radio Science Meeting (Cat. No.01CH37229). vol. 4; 2001. p. 420–423.

[13] Claeys T. Increasing the Accuracy and Speed of EMI Near-Field Scanning. Leuven: KU Leuven; 2018.

[14] Noren P, Foged LJ, and Garreau P. State of the art spherical near-field antenna test systems for full vehicle testing. In: 2012 6th European Conference on Antennas and Propagation (EUCAP). IEEE; 2012. p. 2244–2248.

[15] Qureshi MA, Schmidt CH, and Eibert TF. Adaptive sampling in spherical and cylindrical near-field antenna measurements. IEEE Antennas and Propagation Magazine. 2013;55(1):243–249.

[16] Baudry D, Bicrel F, Bouchelouk L, *et al.* Near-field techniques for detecting EMI sources. In: 2004 International Symposium on Electromagnetic Compatibility (IEEE Cat. No.04CH37559). vol. 1. IEEE; 2004. p. 11–13.

[17] Zhang J, Kam KW, Min J, *et al.* An effective method of probe calibration in phase-resolved near-field scanning for EMI application. IEEE Transactions on Instrumentation and Measurement. 2013;62(3):648–658.

[18] Spang M. Einsatz von Feldsonden mit mehreren Ausgaengen in EMV-Nahfeldmessungen von Leiterplatten. Der Technischen Fakultaet der Universitaet Erlangen-Nuernberg; 2012.

[19] Schmidt M. Methoden zur Messdatenverarbeitung und Erhoehung der Sensitivitaet für EMV-Nahfeldmessungen. Der Technischen Fakultaet der Friedrich-Alexander-Universitaet Erlangen-Nuernberg; 2018.

[20] Schroeder D, Hangmann C, Hedayat C, *et al.* Characterization of H-field probes regarding unwanted field suppression using different calibration structures. In: Smart Systems Integration; 13th International Conference and Exhibition on Integration Issues of Miniaturized Systems. VDE; 2019. p. 1–4.

[21] Harrington RF. Field Computation by Moment Methods. 0898744652. University of Michigan. New York: Krieger Publishing Co., Inc.; 1968.

[22] Gibson WC. The Method of Moments in Electromagnetics. 78-1-4200-6145-1. Boca Raton, FL: Chapman & Hall CRC; 2008.

[23] Sievers D. Anwendung finiter Gruppen zur effizienten Berechnung elektromagnetischer Felder in symmetrischen Strukturen auf Basis der Randelementmethode. Technischen Universität Darmstadt. Darmstadt; 2008.

[24] Kark KW. Antennen und Strahlungsfelder. 78-3-8348-9755-8. Wiesbaden: Vieweg+Teubner Verlag; 2010.

[25] Jackson JD. Classical Electrodynamics. New York: John Wiley & Sons, Inc.; 1999.

[26] Midori M, Kurihara H, and Aoyagi T. A fundamental study on estimation method of 10 m test-range electric field strength by near-field measurement. In: 2014 International Symposium on Electromagnetic Compatibility. Tokyo: IEICE; 2014. p. 493–496.

[27] Tai CT. Dyadic Green Functions in Electromagnetic Theory. Volume 2. New York, NY: IEEE Press series in Electromagnetic Waves; 1992.

[28] Tai CT. Generalized Vector and Dyadic Analysis. New York, NY: IEEE Computer Society Press; 1997.

[29] Havrilla M. Electric and magnetic field dyadic Greens' functions and depolarizing dyad for a magnetic current immersed in a uniaxial dielectric-filled parallel plate waveguide. In: 2011 XXXth URSI General Assembly and Scientific Symposium. Istanbul, Turkey; 2011.

[30] Sarabandi K. Dyadic Green's function. In: Theory of Wave Scattering from Rough Surfaces and Random Media. Ann Arbor, MI: University of Michigan; 2009.

[31] Yaghjian AD. Electric dyadic Greens' functions in the source region. Proceedings of the IEEE. 1980;68.

[32] Cvetkovic M, and Poljak D. Surface equivalence principle and surface integral equation (SIE) revisited for bioelectromagnetics application. International Journal of Computational Methods and Experimental Measurements. 2018;6(6):1182–1191.

[33] Kreyszig E. Advanced Engineering Mathematics. Ohio State University. Jefferson City, MO: John Wiley & Sons, Inc.; 1993.

[34] Sørensen M, Franek O, Ebert H, *et al.* How to handle a Huygens' box inside an enclosure. In: IEEE International Symposium on Electromagnetic Compatibility (EMC). IEEE; 2013.

Chapter 15
Element of information for numerical modelling on PCB: focus on boundary element method

Toufic Abboud[1] *and Benoît Chaigne*[1]

Abstract

The present chapter provides the key information about the numerical modelling of printed circuit board (PCB). The constituting key elements are defined and analytically expressed in function of the basic parameters. The purposed model is applied to the calculation of a small PCB with investigation on the meshing effect on the convergence. The main focus of the study is to analyse the inherent and critical points in function of the scale variation. The meshing effects on the PCB 3D modelling will be investigated.

Keywords: PCB, numerical modelling, convergence, 3D modelling, meshing

15.1 Boundary element method

In this section, we recall the basics of the integral formulation of Maxwell's equations in the frequency domain. *In this chapter, the harmonic time dependence* $e^{-i\omega t}$ *is used.* We consider the general equations not the eddy current approximation which assumes that the size of object is very small compared to the wavelength.

15.1.1 Integral representation formulas

This method is based on the integral representation of the solution of Maxwell's equation in a 3D homogeneous domain Ω which can be bounded or unbounded. If $\vec{E}$ and $\vec{H}$ satisfy

$$\nabla \times \vec{E} - i\omega\mu\vec{H} = \vec{0} \quad \text{and} \quad \nabla \times \vec{H} - i\omega\varepsilon\vec{E} = \vec{0} \in \Omega$$

and the Silver–Müller radiation condition in the case Ω is an exterior domain

$$\left(\sqrt{\mu}\vec{H} \times \boldsymbol{r} - r\sqrt{\varepsilon}\vec{E}\right) \to 0 \quad \text{when } r = |\boldsymbol{r}| \to +\infty$$

[1]IMACS, XTEC, École Polytechnique, Palaiseau, France

then the electromagnetic field has the following integral representation:

$$\vec{E}(\boldsymbol{r}) = i\omega\mu \int_\Gamma G(\boldsymbol{r},\boldsymbol{r}')\vec{j}(\boldsymbol{r}')d\Gamma(\boldsymbol{r}') - \frac{1}{i\omega\varepsilon}\nabla \int_\Gamma G(\boldsymbol{r},\boldsymbol{r}')div_\Gamma \vec{j}(\boldsymbol{r}')d\Gamma(\boldsymbol{r}')$$
$$-\nabla \times \int_\Gamma (\boldsymbol{r},\boldsymbol{r}')\overrightarrow{m}(\boldsymbol{r}')d\Gamma(\boldsymbol{r}')$$

and

$$\vec{H}(\boldsymbol{r}) = \nabla \times \int_\Gamma G(\boldsymbol{r},\boldsymbol{r}')\vec{j}(\boldsymbol{r}')d\Gamma(\boldsymbol{r}') + i\omega\varepsilon \int_\Gamma G(\boldsymbol{r},\boldsymbol{r}')\overrightarrow{m}(\boldsymbol{r}')d\Gamma(\boldsymbol{r}')$$
$$-\frac{1}{i\omega\mu}\nabla \int_\Gamma G(\boldsymbol{r},\boldsymbol{r}')div_\Gamma \overrightarrow{m}(\boldsymbol{r}')d\Gamma(\boldsymbol{r}')$$

where $k = \omega\sqrt{\varepsilon\mu}$ is the wavenumber in the domain Ω, $G(\boldsymbol{r},\boldsymbol{r}') = e^{ik|\boldsymbol{r}-\boldsymbol{r}'|}/4\pi|\boldsymbol{r}-\boldsymbol{r}'|$ is Helmholtz Green's kernel in the domain Ω that satisfies the Sommerfeld radiation condition at infinity $\vec{j} = \vec{H} \times \hat{n}$ is the equivalent electric current, $\overrightarrow{m} = -\vec{E} \times \hat{n}$ is the equivalent magnetic current, and $\hat{n}$ is the unit outwards normal on Γ.

In the case of a piecewise homogeneous material, typically a multilayer printed circuit board (PCB), one must use such representation in each domain.

15.1.2 Integral equation

In order to access the value of the electromagnetic field at any point $\boldsymbol{r}$ in space, we are reduced to look for the equivalent currents: the unknowns. To do this, we have to satisfy the boundary conditions. For example $\overrightarrow{E} \times \hat{n} = \vec{0}$ on the surface of a perfect electric conductor (PEC), and $\overrightarrow{E} \times \hat{n}$ and $\overrightarrow{H} \times \hat{n}$ are continuous through a dielectric interface.

This leads to a set of integral equations, whereas resolution provides the equivalent currents.

Let us develop the case of only one dielectric (exterior) domain with a PEC boundary. The boundary condition implies that the equivalent magnetic current is zero:

$$\overrightarrow{m} = \hat{n} \times \vec{E} = \vec{0}.$$

To determine the equivalent electric current, we use the tangential component of the integral representation of the electric field:

$$\overrightarrow{E}_\Gamma(\boldsymbol{r}) = i\omega\mu\Pi_r \int_\Gamma G(\boldsymbol{r},\boldsymbol{r}')\vec{j}(\boldsymbol{r}')d\Gamma(\boldsymbol{r}') - \frac{1}{i\omega\varepsilon}\nabla_\Gamma \int_\Gamma G(\boldsymbol{r},\boldsymbol{r}')div_\Gamma\vec{j}(\boldsymbol{r}')d\Gamma(\boldsymbol{r}')$$

For simplicity, let us assume a source term in the form of an ideal voltage generator located at some point $\boldsymbol{r}_g$. We obtain the integral equation:

$$\mathcal{Z}\vec{j}(\boldsymbol{r}) = \mathcal{U}_g(\boldsymbol{r}) \quad \text{for all } \boldsymbol{r} \in \Gamma \tag{15.1}$$

where

$$\mathcal{Z}\vec{j}(\boldsymbol{r}) = i\omega\mu\Pi_r \int_\Gamma G(\boldsymbol{r},\boldsymbol{r}')\vec{j}(\boldsymbol{r}')d\Gamma(\boldsymbol{r}') - \frac{1}{i\omega\varepsilon}\nabla_\Gamma \int_\Gamma G(\boldsymbol{r},\boldsymbol{r}')div_\Gamma\vec{j}(\boldsymbol{r}')d\Gamma(\boldsymbol{r}')$$
$$+ Z_g(\vec{j}(\boldsymbol{r}_g)\cdot\hat{v}_g)\hat{v}_g\delta(\boldsymbol{r}-\boldsymbol{r}_g)$$

is the sum of electric-field integral operator and the local impedance of the generator, and

$$\vec{\mathcal{U}}_g = U_g\hat{v}_g\delta(\boldsymbol{r}-\boldsymbol{r}_g)$$

is the source term or right-hand-side (RHS) of the equation. This is the so-called EFIE formulation (for electric-field integral equation). It can be generalised to handle a thin, conducting sheet with finite conductivity.

The multi-domain case is little longer to develop, but the principle is the same. On the interface Γ_{12} between two domains Ω_1 and Ω_2, we have at first one integral representation in each domain, with at this interface two couples of equivalent currents $(\vec{j}_1, \vec{m}_1)$, for the representation in the domain Ω_1 and $(\vec{j}_2, \vec{m}_2)$ in Ω_2. A first consequence of the transmission condition is that

$$\vec{j}_1 + \vec{j}_2 = \vec{0} \quad \text{and} \quad \vec{m}_1 + \vec{m}_2 = \vec{0}$$

Hence, we are reduced to look for a couple of equivalent currents, e.g. $(\vec{j}_1, \vec{m}_1)$. By equating the expressions of the integral representations of the tangential components of the electromagnetic field in each domain, we obtain a set of two integral equations. We can obtain in this way the so-called PMCHWT formulation.

In the case of a PCB, we are led to the following unknown currents:

- electric and magnetic currents on each dielectric interface,
- electric currents on electric conductors.

This is not the only choice, many alternative formulations exist. For example for the PEC problem, one can use the magnetic equation to express the boundary condition which leads to the magnetic field integral equation, or a combination of these equations to obtain the Combined Field Integral Equation. One can also use other representation formulas, for instance, via multilayer Green's tensor to obtain a new integral representation formulation with unknowns on the conducting parts.

15.1.3 Variational formulation and finite element approximation

Instead of solving (15.1) for all points $\boldsymbol{r} \in \Gamma$, we usually write an equivalent variational formulation:

$$\int_\Gamma \mathcal{Z}\vec{j}(\boldsymbol{r})\cdot\vec{j}^t(\boldsymbol{r})d\Gamma(\boldsymbol{r}) = \int_\Gamma \vec{\mathcal{U}}_g(\boldsymbol{r})\cdot\vec{j}^t(\boldsymbol{r})d\Gamma(\boldsymbol{r}) \quad \textit{for all}\ \vec{j}^t \in V(\Gamma) \tag{15.2}$$

where $V(\Gamma)$ is a functional space of admissible tangential currents. Admissible stands for a certain regularity and boundary condition (e.g. zero flux on free edges). This variational principle, known in electromagnetic literature as *Rumsey reaction principle*, states that the solution of the problem is the admissible current that equates the virtual powers for all admissible test currents.

More generally, in each domain we define its Rumsey reaction, which for any pair of test currents $(\vec{j}^t, \vec{m}^t)$ on the boundary is given by

$$\mathcal{R}_i\left((\vec{j}, \vec{m});(\vec{j}^t, \vec{m}^t)\right) = \int_{\delta\Omega_i} \left(\vec{E}\left(\vec{j}, \vec{m}\right) \cdot \vec{j}^t - \vec{H}\left(\vec{j}, \vec{m}\right) \cdot \vec{m}^t\right)$$

where $\vec{E}(\vec{j}, \vec{m})$ and $\vec{H}(\vec{j}, \vec{m})$ are the electric and magnetic fields defined by the integral representation formulas.

A finite element approximation is obtained by discretising the functional space $V(\Gamma)$ by a set of finite dimensional subspaces: $V_h(\Gamma) \subset V(\Gamma)$ where h is a parameter intended to go to zero. Such approximation can be achieved by subdividing Γ into small parts and by approximating the current on each part by a simple expression. Usually, Γ is subdivided into small triangles or approximated by the union of such triangles, Γ is then approximated by Γ_h. h represents the typical triangle diameter. The set $\mathcal{T}_h$ of all triangles is called the mesh. It can be obtained from the geometric description (3D CAD model) by using an automatic mesher. The finite element space is defined as follows:

$$V_h(\Gamma_h) = \left\{\vec{j}_h : \vec{j}_h = \vec{\alpha} + \beta \mathbf{r} \quad in\ T, \quad \text{and fluxes are continuous through edges}\right\}$$

At multiple edges, the continuity of fluxes has the form of Kirchhoff's law: the sum of outgoing fluxes is zero. This is the so-called $H(div)$ Raviart–Thomas also known as Rao–Wilton–Glisson finite element. The degrees of freedom (DoF) are the fluxes through the mesh edges: the DoF I_a associated with the edge a common to triangles T and T', oriented from T to T', represents the current (in Ampère) which flows from triangle T to triangle T':

$$I_a = \int_a \vec{j} \cdot \hat{v}_a \, da$$

The associated basis function is given by

$$\vec{j}_a^t(M) = \frac{\overrightarrow{SM}}{2|T|} \text{ if } M \in T,$$

$$\vec{j}_a^t(M) = -\frac{\overrightarrow{S'M}}{2|T|} \text{ if } M \in T', \text{ and}$$

$$\vec{j}_a^t(M) = \vec{0} \text{ elsewhere}$$

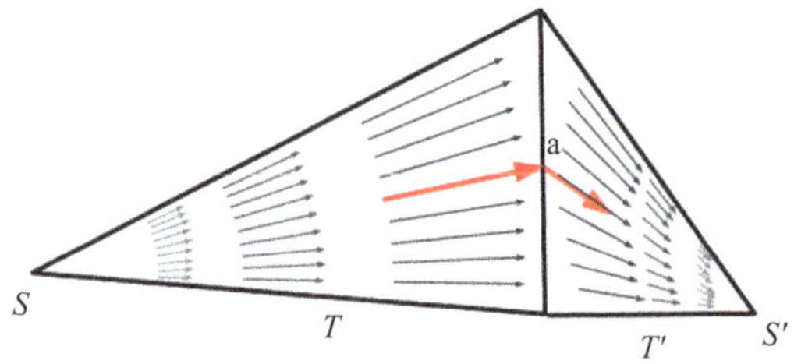

H(div) Basis function

$$div_\Gamma \vec{j}^t_a(M) = +\frac{1}{|T|} \text{ if } M \in T,$$

$$div_\Gamma \vec{j}^t_a(M) = -\frac{1}{|T|} \text{ if } M \in T', \text{ and}$$

$$div_\Gamma \vec{j}^t_a(M) = 0 \text{ elsewhere.}$$

The discretised problem is the following: we look for a current in the form of

$$\vec{j}_h(M) = \sum_{a \in \{edges\}} I_a \vec{j}^t_a(M) \tag{15.3}$$

such that

$$\int_{\Gamma_h} \mathcal{Z}\vec{j}_h(\boldsymbol{r}) \cdot \vec{j}^t_a(\boldsymbol{r}) d\Gamma_h(\boldsymbol{r}) = \int_{\Gamma_h} \vec{\mathcal{U}}_g(\boldsymbol{r}) \cdot \vec{j}^t_a(\boldsymbol{r}) d\Gamma_h(\boldsymbol{r}) \text{ for all edges } a \tag{15.4}$$

That is equivalent to solve the linear system:

$$\mathcal{Z}_h \mathcal{I}_h = \mathcal{U}_h \tag{15.5}$$

with

$$(\mathcal{Z}_h)_{i,j} = \int_{\Gamma_h} \mathcal{Z}\vec{j}^t_j(\boldsymbol{r}) \cdot \vec{j}^t_j(\boldsymbol{r}) d\Gamma_h(\boldsymbol{r}) 1 i, j \leq N\text{: the BEM matrix,}$$

$$(\mathcal{U}_h)_i = \int_{\Gamma_h} \mathcal{U}_g(\boldsymbol{r}) \cdot \vec{j}^t_i(\boldsymbol{r}) d\Gamma_h 1 \leq i \leq N\text{: the RHS, and}$$

$$(\mathcal{I}_h)_i = I_i 1 \leq i \leq N \text{ : the vector of unknown currents (} I_i \text{ through edge no. } i\text{)}$$

where N is the number of DoF.

15.1.4 Solution

The matrix $\mathcal{Z}_h$ of the linear system (15.5) is $N \times N$ complex and full. So computing this matrix costs $N \approx 10^4$, floating point operations (*flops*), and the memory requirement is $N^2 w$, where 16 bytes when using double precision and 8 bytes when single precision is sufficient. For a more robustness, double precision is recommended in practice. Classically, solving the linear system using a direct Cholesky method consists in *factorising the matrix* $\mathcal{Z}_h$:

$$\mathcal{Z}_h = \mathcal{L}_h \mathcal{L}_h^T$$

where $\mathcal{L}_h$ is a lower triangular matrix, for a given *RHS* $\mathcal{U}_h$ solving successively two triangular systems:

$$\mathcal{L}_h \mathcal{J}_h = \mathcal{U}_h \text{ forwards substitution and}$$

$$\mathcal{L}_h \mathcal{I}_h = \mathcal{J}_h \text{ backwards substitution.}$$

This solving procedure gives access to the current fluxes through all the edges of the mesh and hence, the current density. Using the integral representation formulas, one can access the electromagnetic field at any location.

Cholesky factorisation costs $\mathcal{O}(N^3)$ floating point operations, while forwards and backwards substitutions cost $\mathcal{O}(N^2)$ operations for each RHS.

Factorisation is the limiting step. In practise, using classical solvers limits the class of problems solvable on a standard desktop to $N \approx 10^4$ and to $N \approx 10^6$ on a high-performance computer (HPC).

15.2 Numerical and practical issues

15.2.1 Performance issue and fast solvers

Fortunately, there exist several algorithms allowing to handle much bigger N by dramatically reducing memory requirements, typically to $\mathcal{O}(N\log N)$, and number of flops, typically to $\mathcal{O}(N\log^2 N)$. $\mathcal{H}$-matrix method is one of the most recent and efficient of such methods. It is based on a hierarchical compressed format. For computing and assembling the boundary element method (BEM) matrix, the algorithms start by the construction of a cluster tree using nested dissections.

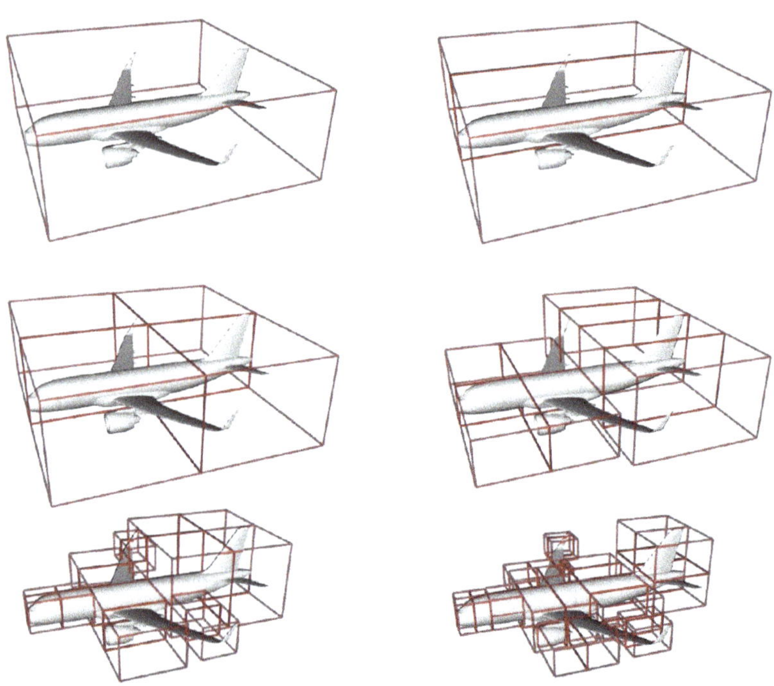

Cluster tree

Matrix block representing the interactions between two clusters τ and σ is assembled either as a dense or as a compressed block. That defines the block-cluster tree of the $\mathcal{H}$-matrix. An admissibility condition (or criterion) allows to choose between dense and compressed format.

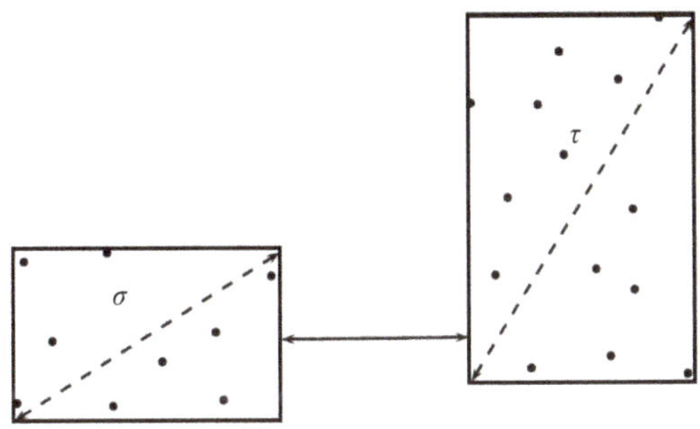

Admissibility condition based on the distance between clusters

Block compression takes advantage of the low-rank property of 'far-field' interactions to approximate an admissible $m \times n$ matrix block M as

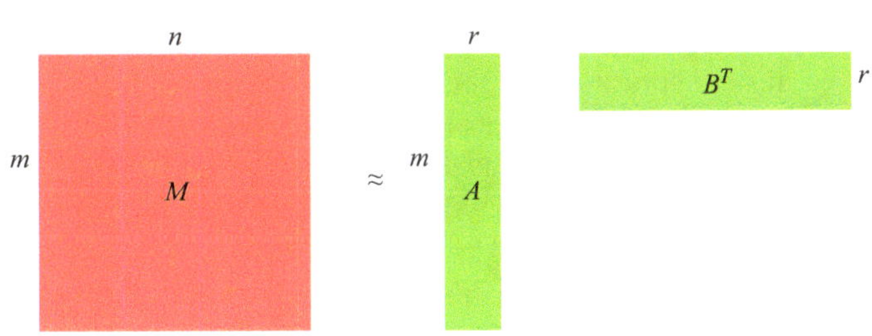

Bloc compression: $M \approx AB^T$

where A and B are, respectively, $m \times r$ and $n \times r$ full matrix blocks, r being the rank of the block, or more precisely the ϵ-rank:

$$\| M - AB^T \| \leq \epsilon \| M \|$$

Usually, r is very small compared to m and n. This compression may be achieved by algebraic techniques such as adaptive cross approximation or analytical techniques based on a low-rank approximation of the kernel.

Hierarchical matrix

A certain number of operations are defined on matrices in this format such as approximate addition and multiplication. Gauss and Cholesky methods can be adapted to factorise an $\mathcal{H}$-matrix with factors in the same format.

This allows to handle high-frequency BEM problems of the class $N \approx 10^5$ on a standard desktop and $N \approx 10^7$ on a HPC.

15.2.2 Low-frequency instability

While frequency domain BEM modelling is very popular in RCS and antenna engineering community, its application to EMC problems faces lower frequency instability problems. In fact, any full-wave numerical method has the same kind of issue because the low-frequency limit of Maxwell's equations is singular and that induces the numerical instability to the discrete system. Of course, here the concept of low frequency is not absolute but depends on the relative sizes of the wavelength, the object, and the elements of the mesh. Using classical BEM formulation, instability may appear when the local size of mesh elements h is too small compared to the wavelength, typically $h \leq 10^{-5} \times \lambda$. This is the case when the object studied is very small compared to the wavelength, or if it includes very small geometric details.

To illustrate the low-frequency instability issue, let us consider the case of the EFIE, and we refer the reader to [1] for the analysis and solution of the stability problem in this case. The EFIE matrix is the sum of these two terms:

$$(\mathcal{Z}_h)_{i,j} = ikZ((\mathcal{Z}((\tilde{\mathcal{Z}}_h^1)_{i,j} + (\tilde{\mathcal{Z}}_h^2)_{i,j}) + \text{local terms}$$

where

$$(\tilde{\mathcal{Z}}_h^1)_{i,j} = \int_{\Gamma_h}\int_{\Gamma_h} G(\boldsymbol{r},\boldsymbol{r}')\cdot \vec{j}_i(\boldsymbol{r})d\Gamma_h(\boldsymbol{r}')d\Gamma_h(\boldsymbol{r}) \text{ and}$$

$$(\tilde{\mathcal{Z}}_h^1)_{i,j} = -\frac{1}{k^2}\int_{\Gamma_h}\int_{\Gamma_h} G(\boldsymbol{r},\boldsymbol{r}')div_{\Gamma_h}\vec{j}_j(\boldsymbol{r}')div_{\Gamma_h}\vec{j}_i(\boldsymbol{r})d\Gamma_h(\boldsymbol{r}')d\Gamma_h(\boldsymbol{r})$$

Using the expressions of the degree 1 $H(div)$ finite element basis functions written in Section 15.3, we find the order of magnitude of $(\mathcal{Z}_h^1)_{i,j}$

$$(\tilde{\mathcal{Z}}_h^1)_{i,j} = O(h^2)$$

whereas

$$(\tilde{\mathcal{Z}}_h^1)_{i,j} = O(k^2)$$

So if h is too small compared to the wavelength, i.e. $kh \ll 1$, that means that

$$\begin{aligned}(\tilde{\mathcal{Z}}_h^2)_{i,j} &\gg (\tilde{\mathcal{Z}}_h^2)_{i,j} \text{ and}\\ (\mathcal{Z}_h)_{i,j} &\approx ikZ(\tilde{\mathcal{Z}}_h^2)_{i,j} + \text{ local terms}\end{aligned}$$

But matrix $\tilde{\mathcal{Z}}_h^2$ is not invertible: its kernel contains all solenoidal currents, i.e. divergence-free currents whose dimension is of the order of the number of vertices of the mesh.

In order to correctly solve the EFIE system in the low-frequency regime, loop-tree algorithm allows to decompose the finite element space as

$$V_h(\Gamma_h) = V_h^0(\Gamma_h) \oplus V_h^1(\Gamma_h)$$

where $V_h^0(\Gamma_h)$ is the subspace of divergence-free currents and $V_h^1(\Gamma_h)$ is the complementary space of classical basis functions given by the loop-tree algorithm.

The matrix of the system has the following form:

$$\mathcal{Z}_h = \begin{pmatrix} \mathcal{Z}_h^{0,0} & \mathcal{Z}_h^{0,1} \\ \mathcal{Z}_h^{1,0} & \mathcal{Z}_h^{1,1} \end{pmatrix}$$

and the unknown and RHS vectors

$$\mathcal{I}_h = \begin{pmatrix} \mathcal{I}_h^0 \\ \mathcal{I}_h^1 \end{pmatrix} \text{ and } \mathcal{U}_h = \begin{pmatrix} \mathcal{U}_h^0 \\ \mathcal{U}_h^1 \end{pmatrix}$$

The system can then be solved for any frequency using Schur complement:

$$\begin{aligned}&\text{first solve } \left(\mathcal{Z}_h^{1,1} - \mathcal{Z}_h^{1,0}(\mathcal{Z}_h^{0,0})^{-1}\mathcal{Z}_h^{0,1}\right)\mathcal{I}_h^1 = \left(\mathcal{U}_h^1 - \mathcal{Z}_h^{1,0}(\mathcal{Z}_h^{0,0})^{-1}\mathcal{U}_h^0\right)\\ &\text{then solve } \mathcal{Z}_h^{0,0}\mathcal{I}_h^0 = \left(\mathcal{U}_h^0 - \mathcal{Z}_h^{0,1}\mathcal{I}_h^1\right)\end{aligned}$$

Both systems are well conditioned. As $(kh) \to 0$

$\mathcal{Z}_h^{0,0}$ converges to the magnetodynamics limit (or eddy current approximation), and

$\mathcal{Z}_h^{1,1}$ converges to the electrostatic limit.

15.2.3 Meshing

The biggest practical difficulty encountered in modelling concerns the generation of the mesh. This is done by using standard CAD/mesher software. Generating a good mesh often involves cleaning the CAD model. The quality and size of the mesh have a direct impact on the computational time – possibly the feasibility of the numerical solving procedure – and the quality of the solution. The challenge is to generate a sufficiently fine mesh to ensure a good precision while keeping the number N of unknowns within a reasonable limit. Numerical experts have developed many a priori empirical rules depending on the application to generate a mesh 'achieving' a good compromise between precision and computational time with the available computer resources. The mesh must represent the 'relevant' geometric details with a certain precision. In each dielectric domain, the size h of the elements must resolve the wavelength, e.g. $h \leq \lambda/5$ or even smaller depending on the application and the observable. The local size of element h must also resolve singularities whether of geometrical or source origin. These rules, which come from years of experience, are difficult to transmit to junior design engineers; and usually in many industries, meshes are first validated by numerical experts. To better disseminate the designed tools, while ensuring the robustness of the results, a posteriori error estimator which indicates the zones of the mesh is to be refined. This process can be automated to obtain an adaptative mesh refinement. This procedure is often referred to as ABEM for Adaptive BEM.

15.3 Formulation and stability issues

15.3.1 LAYER formulation

While loop-tree algorithms work well to stabilise the integral equation in the case of the EFIE, it cannot be easily generalised to multi-domain case with the PMCHWT formulation. In the framework of DGA/RAPID project MOSCEM, a new formulation coined 'LAYER formulation' has been introduced that addresses this limitation. It also allows to handle thin layer of conducting material between two dielectric domains, taking into account the skin effect and sharp attenuations (see [2] for an introduction and early results).

This formulation is based on the integral representation of the electromagnetic field with independent equivalent currents in each domain. On a transparent interface between two adjacent dielectric domains Ω_1 and Ω_2, even if

$$\vec{j}_1 + \vec{j}_2 = \vec{0} \quad \text{and} \quad \vec{m}_1 + \vec{m}_2 = \vec{0}, \tag{15.6}$$

currents in each domain are approximated independently, doubling the number of unknowns on the interface. In this framework, the interface is modelled in a thin layer ranging from transparent to opaque, with model generalising (15.6). Furthermore, that allows to apply the loop-tree algorithm in each domain in order to solve the low-frequency instability issue.

On each interface, the formulation consists of writing the tangential traces of the electric and magnetic fields in each domain and expresses the boundary condition with data from the adjacent domain. For example at a transparent interface $\Gamma_{1,2}$ between the dielectric domains Ω_1 and Ω_2, we write

$$\text{in } \Omega_1: (\overrightarrow{E}_1)_t = \overrightarrow{m}_2 \times \hat{n}_2 \text{ and } (\vec{H}_1)_t = -\vec{j}_2 \times \hat{n}_2$$
$$\text{in } \Omega_2: (\overrightarrow{E}_1)_t = \overrightarrow{m}_1 \times \hat{n}_1 \text{ and } (\vec{H}_1)_t = -\vec{j}_1 \times \hat{n}_1$$

then we replace the tangential trace by its integral representation in the domain. With a thin-layer model, we express the transmission condition as an impedance condition in each domain with given data from the adjacent one. In this way, the system has the following form (in the case of two domains):

$$\mathcal{Z}_h = \begin{pmatrix} R^{1,1} & S^{1,2} \\ S^{2,1} & R^{2,2} \end{pmatrix}$$

where $R^{1,1}$ and $R^{2,2}$ are the discrete Rumsey operators, respectively, in Ω_1 and Ω_2, $S^{1,2}$ and $S^{2,1}$ are the local interactions between domains: sparse matrices.

This formulation presents many advantages:

It can handle any kind of thin-layer model ranging from transparent to opaque, it provides low-frequency (and hence all frequencies) stable numerical scheme, and it provides a highly accurate procedure for computing sharp attenuations phenomena as shielding problem.

These advantages largely compensate for the doubling of the DoF at the interfaces, because one can choose a coarser mesh while ensuring a precise solution, thanks to the choice of the formulation.

15.3.2 Validation with an analytic solution

To demonstrate these advantages and compare an analytical reference, we consider a shielding effectiveness problem. We consider a thin composite hollow sphere with a thickness of 1 mm, conductivity $\sigma = 10^5$ S/m and an average radius of 1 m.

This sphere is illuminated by an incident plane wave

$$\overrightarrow{E}^{in}(\boldsymbol{r}) = \overrightarrow{E}_0^{in} e^{ik\hat{u}\cdot\boldsymbol{r}}$$

with $\|\vec{E}_0^{in}\| = 1$ V/m. We are interested in the electric and magnetic fields inside the sphere, for example at its centre. This problem has an analytical solution based on Mie's series.

We compare Mie's series with classical BEM formulation (cs) and different versions of BEM LAYER formulations.

This simulation validates the ability to take into account a thin conductive layer taking into account the skin effect on a very large bandwidth (from 1 Hz to several hundreds of megahertz) and demonstrating a very high shielding effectiveness.

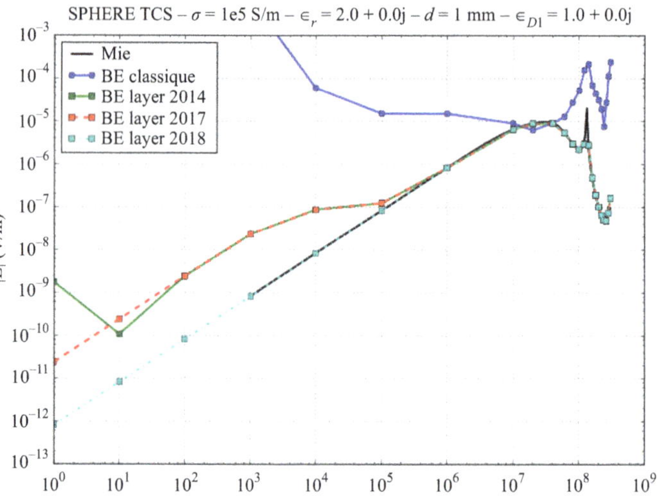

Electric field at the centre of the sphere

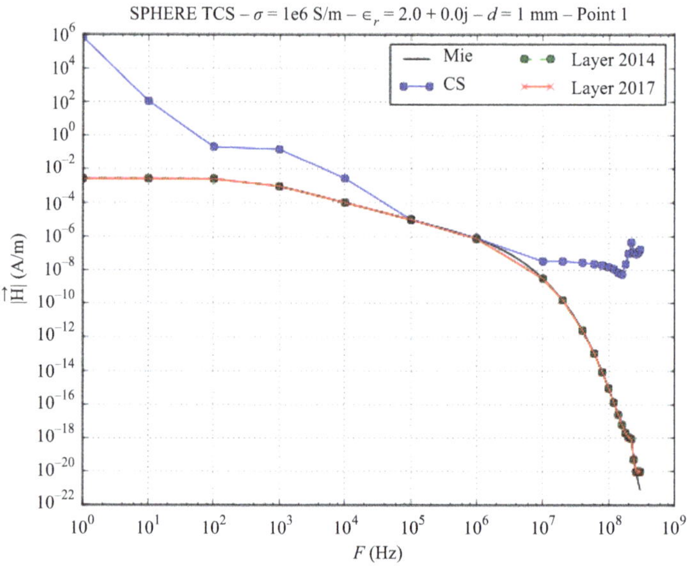

Magnetic field at the centre of the sphere

15.4 A posteriori error estimate and adaptive BEM

To improve the robustness of numerical simulation in industrial context, expert rules need to be automated. For finite element methods, a posteriori error estimate and adaptive mesh refinement have been studied for a long time and are already available in many implementations. In contrast, only few references can be found for the application of these techniques to integral equations (also known as BEM or method of moments). A full review of such techniques can be found for instance in [3] and [4]. However, most papers deal with scalar cases and non-oscillating kernels such as $1/r$ (see [5–7]). To our knowledge, the extension of these results for BEM applied to 3D Maxwell's equation has not been much studied (cf. [8]).

15.4.1 A posteriori error estimate

The aim of error estimation is to find a quantity η, which conveys some information about the local contributions of the committed error and that approximates its norm in a certain sense. In order to be reliable, this quantity must fulfil some requirements exposed in [4] and that we shall recall briefly here. In the sequel, we consider a triangular surface mesh $\mathcal{T}_h = \cup_{i=1}^{N} T_i$ of Γ. The estimator is assumed to be additive in the sense that the global estimation is related to the local contributions on each triangle $\eta(T)$ as follows:

$$\eta^2 = \sum_{i=1}^{N} \eta^2(T_i)$$

An estimator should satisfy four properties: stability, error reduction, quasi-orthogonality, and reliability properties.

From a practical point of view, the computation cost of an estimator must be smaller than the computation cost of the solution itself.

The so-called ZZ-estimator (Zienkiewicz–Zhu) is the simplest to implement and yet efficient error estimator (cf. [3]). It is based on the norm of difference between the solution $\vec{j}_h$ and some regularisation $r_h(\vec{j}_h)$:

$$\eta - \|\vec{j}_h - r_h(\vec{j}_h)\|_{V_h}$$

where V_h is the finite element space equipped with $TH^{-1/2}(div, \Gamma)$ norm. $H^{-1/2}$ norm is very complex to compute: it is a non-local norm that would require an integral operator to express. Approximate surrogate, simple to compute formulas exist which are local and additive (cf. [3] for example).

15.4.2 Adaptive mesh refinement

The error estimate colour map indicates the regions of the surface to be refined. In this example of a box with square apertures illuminated by an incident plane wave, the a posteriori estimator indicates:

singularities at free edges and dihedral angles,
zones where the mesh is not sufficiently refined by comparison with the wavelength.

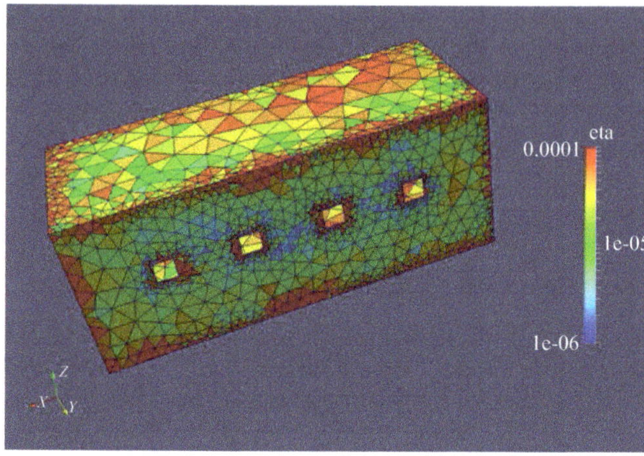

η map (red values correspond to big error estimate)

Once computed on each triangle, error estimates are sorted:

$$\eta(T_{i_1}) \geq \eta(T_{i_2}) \geq \cdots \eta(T_{i_N})$$

Dörfler marking (cf. [9]) consists of a given $\theta \in (0, 1)$ in finding N_θ

$$N_\theta = \text{argmax}_n \sum_{l=1}^{n} \eta^2(T_{i_l}) \leq \theta \eta^2$$

$(T_{i_1}, \ldots, T_{i_{N_\theta}})$ are the marked elements, candidates to be refined.

Several options exist on how to refine the mesh near these elements:

- refine these elements, propagate the refinement to adjacent elements, and then regularise the resulting mesh;
- use the marked elements to define a metric, then recreate the mesh using it. This type of procedure allows a better approximation of the geometry.

This procedure can then be repeated until a stopping criterion is satisfied.

Taking $\theta = 1$ marks all elements and the algorithm degenerates to uniform refinement. Taking $\theta = 0$ marks no elements and no refinement is performed. As it can be seen in the numerical examples, the procedure is rather robust with respect to a reasonable choice of θ around 0.5.

In this example, the adaptive procedure is particularly interesting with the uniform refinement strategy where the error estimator has been divided by 2.5, and the adaptive strategy (=0.5) has divided this error by about 12.5. The observable computed with the initial mesh (about 600 DoF) is 10 dB below the converged value. This error is less than 0.5 dB at iteration 9 (5,000 DoF) and 0.1 dB at iteration 12 (18,000 DoF). With the uniform refinement strategy, the error is still above 2.5 dB with more than 150,000 DoF. A similar precision is achieved at iteration 4 with about 1,000 DoF.

15.4.3 Stopping criterion

The stopping criterion is a practical issue that has to be driven by the user's requirements on the accuracy on some observables of interest (e.g. near field, local current, and far field). That takes into account modelling error and limitations on the computing budget (affordable computing time on available resources, etc.).

Decay of error estimator

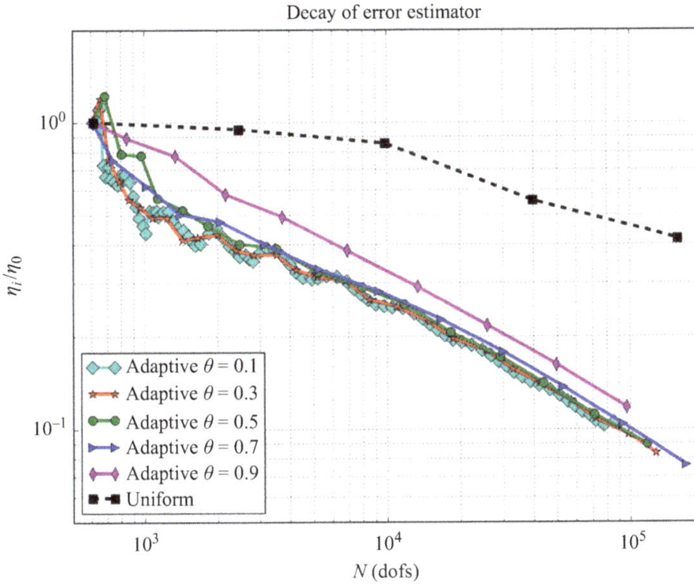

Convergence of the error estimator

Near field convergence – E

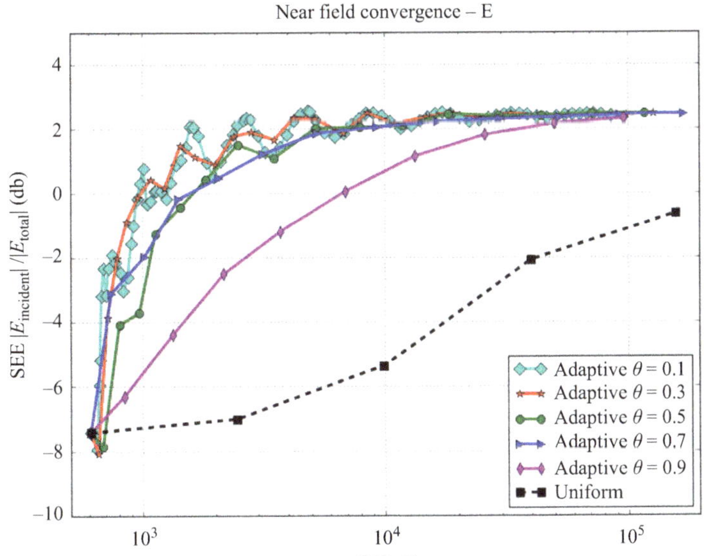

Convergence of the observable

Acknowledgements

Hi-BOX and MOSCEM projects have been supported by a grant from DGA, the French Government Defence procurement and technology agency, in the framework of RAPID program (dual innovation support scheme). RAFFINE project has been supported by a grant from the French National Research Agency (ANR).

References

[1] Terrasse, I.: "Div-Curl decomposition adapted to industrial problems in term of complexity and size", Questions of algebraic topology in numerical analysis: algorithmics and applications symposium AMAM Conference, Nice, France, Feb. 2003.

[2] Abboud, T., Barbier, D., Béreux, F. and Peres, G.: "Integral equation method for lightning indirect effects and HIRF applications", ICOLSE Proceedings 2015, pp. 416–423.

[3] Feischl, M., Fu¨hrer, T., Heuer, N., Karkulik, M., and Praetorius, D: "Adaptive boundary element methods", Archives of Computational Methods in Engineering, 2015, 22, (3), pp. 309–389.

[4] Carstensen, C., Feischl, M., Page, M., and Praetorius, D.: "Axioms of adaptivity", Computers and Mathematics with Applications, 2014, 67, pp. 1195–1253.

[5] Aurada, M., Ferraz-Leite, S. and Praetorius, D.: "Estimator reduction and convergence of adaptive BEM", Applied Numerical Mathematics, 2012, 62, pp. 787–802.

[6] Feischl, M., Karkulik, M., Melenk, J. M., and Praetorius, D.: "Quasi-optimal convergence rate for an adaptive boundary elements method", SIAM Journal of Numerical Analysis, 2013, 51, (2), pp. 1327–1348.

[7] Ferraz-Leite, S., and Praetorius, D.: "Simple a posteriori error estimators for the h-version of the boundary elements method", Computing, 2008, 83, pp. 135–162.

[8] Abboud, T., Barbier, D., Béreux, F. and Chaigne, B.: "Robust electromagnetic simulation with adaptive mesh refinement in BEM", ICOLSE Proceedings 2015, pp. 424–431.

[9] Dörfler, W.: "A convergent adaptive algorithm for Poisson's equation", SIAM Journal of Numerical Analysis, 1996, 33, (3), pp. 1106–1124.

Chapter 16
General conclusion

Olivier Maurice[1], Blaise Ravelo[2], and Zhifei Xu[2]

Abstract

The contents of these chapters illustrate the investigation on the unfamiliar tensorial analysis of networks approach proposed in this book. This final chapter summarizes the main potential contributions of all chapters (Chapters 2–15) of this book. In the fields of the electromagnetic compatibility (EMC), signal integrity and power integrity engineering, the book provides new ways to analyse the printed circuit board (PCB) problems. One of the main modellings is developed in the first 11 chapters (Chapters 2–13). Chapter 14 presents a technique of EMC conducted susceptibility analysis of digital components used regularly in the PCBs. Then, investigations on the EM NF radiations from the planar PCBs based on the scanning technique are developed in Chapters 15 and 16. Then, the concluding technical chapter proposes an overview of PCB numerical modelling.

Keywords: Methodology, analytical method, numerical method, PCB analysis, SI analysis, PI analysis, EMC analysis, experimental test techniques

16.1 Final words on the developed EMC, SI and PI analyses of PCBs based on the TAN formalism

The present book presents an advertised overview on the electromagnetic compatibility (EMC), signal integrity (SI) and power integrity (PI) modelling of printed circuit boards (PCBs) by using the tensorial analysis of networks (TAN) formalism. The main knowledge necessary to be familiar with the TAN approach for the PCB analysis is introduced. The different steps of the TAN approach are elaborated over different cases of study. The implementations of the TAN modelling in both frequency and time domains are clarified with different examples. Discussion on the strengths and also the weaknesses of the TAN approach is presented.

[1]ArianeGroup, Paris, France
[2]IRSEEM/ESIGELEC, Rouen, France

The following subsections point out the significant contributions of synthetic parts corresponding to the contents of previous chapters.

16.2 Summary on the fundamental elements to practice Kron's method

Compared to the classical circuit analysis methods as nodal and many full wave approaches, the TAN concept was not sufficiently exploited. Since Gabriel Kron has initiated the TAN formalism [1–3], until now (early 2020s), few PCB designs and analysis engineers and researches are familiar to this uncommon analytical method. However, Kron's method is expected to solve universal large-scale and complex problems [4–7].

Against this scientific cultural limitation of today's PCB design and modelling, we expect that the TAN approach will enable to provide key solutions. Chapters 2 and 3 of the present book offer the basic elements necessary to start TAN modelling of PCBs. These elements are constituted by primitive elements of PCBs as R, L and C components for lumped circuits. Further tensor objects are also proposed for the TLs, vias and pads for certain distributed circuits operating at very high frequencies.

16.3 Summary on PCB interconnect modelling in the frequency domain with TAN approach

One of the most classical ways to analyse PCB EMCs is the frequency domain approach. To solve EMC problems for complex electrical and electronic systems, Maurice *et al.* introduced the TAN concept [8–10]. Chapter 4 proposes the fundamental steps to employ the TAN formalism for the modelling of PCB interconnects in the frequency domain.

The mathematical approach, including the branch space, mesh space and metric equations, is developed. Different examples are examined to illustrate the effectiveness of the TAN formalism.

16.4 Summary on the PCB modelling in the time domain

Chapters 5 and 6 develop the time domain modelling of PCB interconnects with TAN approach. The initial step of this modelling is also based on the equivalent graph representation of the problem. Then, the methodology of time domain is based on the branch and mesh space analyses of the structure. The difference with the time domain approach is the consideration of the excitation signal discretization. Some analytical conditions must be satisfied to implement the time domain TAN depending on the geometry of the PCB structures. Validation results are discussed to demonstrate the feasibility of the concept.

16.5 Summary on the radiated EMC modelling of PCB with TAN approach

Chapters 8 and 11 describe the radiated EMC modelling of PCBs by using the TAN concept. The equivalent scenario of the PCB structure must be defined before the graph elaboration. In difference to the classical electrical modelling, these chapters are exploiting the EM field propagation in the moment space. Validation results of the radiated EMC problems of PCBs are discussed. Different POCs constituted by microstrip and multilayer PCBs are investigated. The results confirm some advantages of the PCB EMC analyses.

16.6 Summary on the conducted EMC modelling of PCBs with TAN approach

Chapters 9 and 10 focus on the TAN modelling of the conducted EMC PCBs. The proposed methods can be applied to complex PCBs. The methods offer a fast and flexible approach that can be implemented in the frequency and time domain. Different scenarios of PCB structures with conducted EMC perturbations are investigated.

16.7 Summary on the TAN modelling of PCB metallic shielding cuboid

Chapter 13 provides a theoretical study of TAN modelling applied to typical 3D structure. This case of study is focused on the magnetic field at very low frequency. The interaction between the metallic cuboid structure and the external magnetic field is formulated. The final result of the chapter can be exploited in the future to formulate innovatively the SE for any arbitrary direction quasi-static field aggression on to PCB enclosures.

16.8 Summary on TAN modelling of coaxial cable under EM NF radiation from electronic loop probe

Chapter 14 of this present book presents an outstanding analytical study of coaxial cable illuminated by EM NF radiation. This case of study illustrates the feasibility of the TAN approach for the analysis of PCB coupled with other structures as cable system. The TAN modelling of cable versus NF coupling is particularly interesting due to its flexibility to different scenarios and also its fast implementation. The model proposed in Chapter 14 needs further investigation for the improvement of SE formulation in a particularly wideband frequency.

16.9 Summary on the analysis of NL EMC effect for mixed PCBs

Chapter 15 provides a key empirical study on the EMC susceptibility of mixed PCB under RFI perturbation. The testing methodology developed in the chapter enables to realize an advance analysis of PCB for EMC engineers in the future. The technique can be extended to different families of LF, RF, microwave and digital circuits. This study opens a huge prospect as the analytical modelling allows expressing a further prediction of the circuit behaviours in the function of any *RFI* parameter.

16.10 Summary on the overview of PCB numerical modelling

Chapter 16 of the book presents a synthetic bibliography on the different popular numerical modelling of PCBs. The potential advantages and drawbacks of each numerical approach are underlined.

16.11 Concluding remark

This final conclusion provides a guidance word for students, engineers, teachers, researchers and even amateurs who would like to practice the TAN approach. In contrast to the classical circuit theory, the co-authors of the chapters would like to advice the readers to practice the different examples and follow the indicated steps to find out more about the potential of the method.

Based on the practical experiences of the co-authors, many potential scientific treasures for solving universal engineering problems [4–7] remain to be discovered behind the TAN approach.

References

[1] G. Kron, “Generalized theory of electrical machinery”, AIEE Transactions, Vol. 49, No. 2, pp. 666–683, 1930.

[2] G. Kron, “A short course in tensor analysis for electrical engineers”, Wiley, New York; Chapman & Hall, London, 1942.

[3] G. Kron, “Electrical engineering problems and topology”, Matrix and Tensor Quarterly, Vol. 2, pp. 2–4, 1951.

[4] G. Kron, “Solving extremely large and complicated physical systems in easy stages”, Matrix and Tensor Quarterly, Vol. 3, pp. 2–4, 1953.

[5] G. Kron, “A set of principles to interconnect the solutions of physical systems”, Journal of Applied Physics, Vol. 24, pp. 965–980, 1953.

[6] G. Kron, “Diakoptics – a gateway into universal engineering”, Electrical Journal, London, UK, 1956.

[7] O. Maurice, "CEM des systèmes complexes (in French)", Hermès-Lavoisier, Paris, France, 2007.

[8] O. Maurice, A. Reineix, O. Maurice, A. Reineix, "Link between the free field and the elements of frontiers in a complex structure", HAL Id: hal-00265343, 2008.

[9] O. Maurice, A. Reineix, P. Durand and F. Dubois, "Kron's method and cell complexes for magnetomotive and electromotive forces", International Journal of Applied Mathematics, Vol. 44, No. 4, pp. 183–191, 2014.

[10] B. Ravelo and O. Maurice, "Kron–Branin Modeling of Y-Y-Tree Interconnects for the PCB Signal Integrity Analysis", IEEE Transactions on Electromagnetic Compatibility, Vol. 59, No. 2, pp. 411–419, 2017.

Index

www.ingramcontent.com/pod-product-compliance
Lightning Source LLC
LaVergne TN
LVHW011834200926
848484LV00002B/356